THEORY OF ORBIT DETERMINATION

Determining orbits of natural and artificial celestial bodies is an essential step in the exploration and understanding of the Solar System. However, recent progress in the quality and quantity of data from astronomical observations and spacecraft-tracking has generated orbit determination problems which cannot be handled by classical algorithms. This book presents new algorithms capable of handling the millions of bodies which could be observed by next-generation surveys, and which can fully exploit tracking data with state-of-the-art levels of accuracy.

After a general mathematical background and summary of classical algorithms, the new algorithms are introduced using the latest mathematical tools and results, to which the authors have personally contributed. Case studies based on actual astronomical surveys and space missions are provided, with applications of these new methods. Intended for graduate students and researchers in applied mathematics, physics, astronomy, and aerospace engineering, this book is also of interest to non-professional astronomers.

ANDREA MILANI is Full Professor of Mathematical Physics in the Department of Mathematics, University of Pisa. His areas of research include the N-body problem, the stability of the Solar System, asteroid dynamics and families, satellite geodesy, planetary exploration, orbit determination, and asteroid impact risk.

GIOVANNI F. GRONCHI is a Researcher of Mathematical Physics in the Department of Mathematics, University of Pisa. His research is on Solar System body dynamics, perturbation theory, orbit determination, singularities, and periodic orbits of the N-body problem.

COVER ILLUSTRATION: The orbits of eight potentially hazardous asteroids (PHA); they have a minimum intersection distance with the orbit of the Earth of less than 0.05 astronomical units. Together with many more smaller objects, they form a swarm surrounding the orbit of our planet (represented, not to scale, in green, orbit in yellow), are observable with either telescopes or radar, and provide a good example of an orbit determination problem. The objects in this figure are the brightest PHA, with diameters larger than 2 km; thus an impact with the Earth would result in a global catastrophe. There has been interesting recent progress in the theory of orbit determination, to which the authors of this book have contributed. New algorithms have been developed to exclude the possibility that any of these objects have the possibility of impacting the Earth, at least in the next 100 years. The same result also applies to somewhat smaller PHA, but the impact of either a much smaller known asteroid or an asteroid still to be discovered is still possible; thus the orbit determination work must go on. The orbit diagram is superimposed on an actual image of the sky (courtesy of G. Rhemann, Astrostudio, Vienna) which includes a Solar System body: a comet discovered in 2008 by A. Boattini, showing its coma.

THEORY OF
ORBIT DETERMINATION

ANDREA MILANI AND
GIOVANNI F. GRONCHI
Department of Mathematics, University of Pisa

CAMBRIDGE
UNIVERSITY PRESS

Shaftesbury Road, Cambridge CB2 8EA, United Kingdom

One Liberty Plaza, 20th Floor, New York, NY 10006, USA

477 Williamstown Road, Port Melbourne, VIC 3207, Australia

314–321, 3rd Floor, Plot 3, Splendor Forum, Jasola District Centre, New Delhi – 110025, India

103 Penang Road, #05–06/07, Visioncrest Commercial, Singapore 238467

Cambridge University Press is part of Cambridge University Press & Assessment, a department of the University of Cambridge.

We share the University's mission to contribute to society through the pursuit of education, learning and research at the highest international levels of excellence.

www.cambridge.org
Information on this title: www.cambridge.org/9780521873895

© A. Milani and G. Gronchi 2010

First published 2010

A catalogue record for this publication is available from the British Library

Library of Congress Cataloging-in-Publication data
Milani, Andrea.
Theory of orbit determination / Andrea Milani, Giovanni Gronchi.
p.   cm.
ISBN 978-0-521-87389-5 (hardback)
1. Orbit determination.   2. Celestial mechanics.   I. Gronchi, Giovanni (Giovanni Federico)   II. Title.
QB355.M55   2009
521′.3 – dc22   2009034270

ISBN   978-0-521-87389-5   Hardback

Contents

PREFACE

This book is a tool for our own teaching and an opportunity to rethink and reorganize the results of our own research. However, I think such a book can be useful to others, for two main reasons. First, spaceflight is no longer the privilege of the few superpowers, but is becoming available to many nations and agencies. Orbit determination is an essential knowhow, both in the planning phase of mission analysis and in the operations of space missions. Thus its mathematical tools need to become widely available.

Second, the knowledge and skill used in orbit determination, for both natural and artificial celestial bodies, was available only among a restricted group of specialists. The prevailing attitude was a proprietary one: the knowledge and the software were protected by formal copyright and/or by secrecy, although protecting in this way the pure mathematical theory is, in the long run, impossible. This attitude might have been justified under the conditions of the world of 30–40 years ago, in the critical phases of the competition to achieve *space firsts*. Now it is time to teach and disseminate this knowledge, allowing the formation of a wider group of specialists.

I know that many of the *rules of thumb* and practical advice contained in this book will be rated as well known, even obvious, by the few experts, but this is not the point. Even well-known results may need to be presented in a rational, rigorous, and didactically effective new way, together with the outcome of recent innovative research. On the other hand, this book does not have the intent of providing a comprehensive review of all that has been done in this field, because the size would become impractical. This book is about making widely available the outcome of the research done by my group over many years, and includes methods for which there are rigorous mathematical arguments and which have been fully tested by us first hand, and found to be effective. In the last $\simeq 15$ years there has been enormous progress in this field, and several other research groups have given important contributions: we are in no way claiming that their methods would not work, we are just giving a list of methods which we know to work.

The above arguments may not be enough for the approval of all the people in this field, but I do think that to state the mathematical foundations and rules of orbit determination, thus removing a vague flavor of craftmanship, can also benefit the already existing specialists. The orbit determination expert, in the very competitive environment in which space missions and large astronomical projects are selected today, is too often under pressure to endorse claims of wonderful results to be achieved with very limited means. By ignoring the rules of good practice it is possible to claim illusory precision and/or completeness for the solution, including the orbits and other parameters which can be operationally, technologically, and scientifically relevant. Maybe being able to cite a textbook stating clearly what is appropriate and what is illusory can help in relieving this improper pressure.

This book is based on the experience accumulated in $\simeq 30$ years of research with my coworkers of the former *Space Mechanics Group* (now *Celestial Mechanics Group*) at the Department of Mathematics, University of Pisa. Thus it contains, besides the formal mathematical theory and the teaching examples, a number of "case studies" based upon actual research projects. They are about space missions and about natural objects: one of the goals is to stress the common mathematics used in satellite geodesy and in dynamical astronomy, and at the same time to present clearly the main differences.

The preparation of this book has been made possible by the collaboration of my younger colleague, Dr. Giovanni F. Gronchi. Besides classical material and original results by myself and Gronchi, this book contains the output of research done by the members of our group and by either regular or occasional external coworkers. Thus I would like to include a long, but still possibly incomplete, list of coworkers whose contributions have to be acknowledged: L. Anselmo, O. Arratia, S. Baccili, A. Boattini, C. Bonanno, M. Carpino, G. Catastini, L. Cattaneo, S.R. Chesley, S. Cicalò, L. Denneau, L. Dimare, P. Farinella, D. Farnocchia, Z. Knežević, L. Iess, R. Jedicke, A. La Spina, M. de' Michieli Vitturi, A.M. Nobili, A. Rossi, M.E. Sansaturio, G. Tommei, G.B. Valsecchi, D. Villani, D. Vokrouhlický.

This book is dedicated to two good friends and valuable coworkers: Paolo Farinella and Steve Chesley. They could have been among the authors of this book, but they both left in the year 2000, when the book project was immature. Steve went back to his home country, from where he can still advise me on these subjects. Paolo went where he can give me neither his essential scientific insight nor the warmth of his friendship. Thus I would like to thank both of them for what I learned with them and from them.

Andrea Milani Comparetti, Pisa, December 2008

Part I

Problem Statement and Requirements

1

THE PROBLEM OF ORBIT DETERMINATION

In this chapter we define the problem of orbit determination, by specifying its three basic mathematical elements: the dynamics, the observations and the error model. We state the minimum principle, the least squares principle as the main case, and attempt a classification of the types of orbit determination found in astronomy and astrodynamics. The last section contains suggestions on the reading sequence, to adapt this book to different needs.

1.1 Orbits and observations

The two essential elements of an orbit determination problem are orbits and observations. Orbits are solutions of an equation of motion:

$$\frac{d\mathbf{y}}{dt} = \mathbf{f}(\mathbf{y}, t, \boldsymbol{\mu})$$

which is an ordinary differential equation; $\mathbf{y} \in \mathbb{R}^p$ is the **state vector**, $\boldsymbol{\mu} \in \mathbb{R}^{p'}$ are the **dynamical parameters**, such as the geopotential coefficients, $t \in \mathbb{R}$ is the time. In the asteroid case the equation of motion is the N-body problem, the asteroid orbit being perturbed by the gravitational attraction of the planets; for many comets and some exceptionally accurate orbits of asteroids the non-gravitational effects are also relevant. For an artificial satellite the equation of motion is the satellite problem, the orbit being mostly perturbed by the asymmetric part of the geopotential, but also by non-gravitational perturbations.

The **initial conditions** are the value of the state vector at an epoch t_0:

$$\mathbf{y}(t_0) = \mathbf{y_0} \in \mathbb{R}^p.$$

In the two simple cases cited above we have $p = 6$, i.e., the vector of the initial condition is just formed by the position and velocity of the small body in some inertial reference system. The orbits are specific solutions, for a given value of $\mathbf{y_0}$ and $\boldsymbol{\mu}$, of the equation of motion (initial condition problem). All the orbits together form the general solution

$$\mathbf{y} = \mathbf{y}(t, \mathbf{y_0}, \boldsymbol{\mu}),$$

also known as **integral flow** when considered as a mapping from the initial conditions (and dynamical parameters) to the current state at time t:

$$\mathbf{y}(t) = \Phi_{t_0}^t(\mathbf{y_0}, \boldsymbol{\mu}).$$

For the second element we introduce an **observation function**

$$R(\mathbf{y}, t, \boldsymbol{\nu})$$

depending on the current state, directly upon time, and also upon a number of **kinematical parameters** $\boldsymbol{\nu} \in \mathbb{R}^{p''}$. The function R is assumed to be differentiable. The composition of the general solution with the observation function is the **prediction function**

$$r(t) = R(\mathbf{y}(t), t, \boldsymbol{\nu})$$

which is used to predict the outcome of a specific observation at some time t_i, with $i = 1, \ldots, m$. However, the observation result r_i is generically not equal to the prediction, the difference being the **residual**

$$\xi_i = r_i - R(\mathbf{y}(t_i), t_i, \boldsymbol{\nu}), \quad i = 1, \ldots, m.$$

The observation function can depend also upon the index i, the most common case being the use of a two-dimensional observation function like (right ascension, declination) or (range, range-rate), in which case R has two different analytical expressions, one for i even, the other for i odd. All the residuals can be assembled forming a vector in $\mathbb{R}^m$

$$\boldsymbol{\xi} = (\xi_i)_{i=1,\ldots,m}$$

which is in principle a function of all the $p + p' + p''$ variables $(\mathbf{y_0}, \boldsymbol{\mu}, \boldsymbol{\nu})$.

The above equations define a fully *deterministic* model: each residual is a single valued function of the $p+p'+p''$ parameters. This function is obtained from the observation function, for which we assume an explicit analytical expression, by using the general solution, which is not known as an analytical expression but is uniquely defined by the differential equations; both functions are assumed to be differentiable, see Chapter 2. These assumptions may not be the whole truth, as we shall see in Chapters 14 and 17, but we shall work with them for now.

The random element is introduced by the assumption that every observation contains an error. Even assuming we know with perfect accuracy all the true values $(\mathbf{y_0}^*, \boldsymbol{\mu}^*, \boldsymbol{\nu}^*)$ of the parameters, that our model is perfectly complete (both for the equation of motion and for the observations), and that our explicit computations are perfectly accurate (they are computed in "exact arithmetic", not with a realistic computer), nevertheless the residuals

$$\xi_i^* = r_i - R(\mathbf{y}(\mathbf{y_0}^*, t_i, \boldsymbol{\mu}^*), t_i, \boldsymbol{\nu}^*, i) = \epsilon_i$$

would not be zero but random variables. The joint distribution of $\epsilon = (\epsilon_i)_{i=1,\dots,m}$ needs to be modeled, that is we need some assumptions, either in the form of a probability density function or as a set of inequalities, describing the observation errors we rate as acceptable. The probabilistic approach in most cases uses Gaussian distributions, discussed in Chapter 3.

1.2 The minimum principle

The basic tool of the classical theory of orbit determination (Gauss 1809) is the definition of a **target function** $Q(\boldsymbol{\xi})$ depending on the vector of residuals $\boldsymbol{\xi}$. The target function cannot be chosen arbitrarily, but needs to satisfy suitable conditions of regularity and convexity. We shall focus on the simplest case, in which Q is proportional to the sum of squares of all the residuals:

$$Q(\boldsymbol{\xi}) = \frac{1}{m}\,\boldsymbol{\xi}^T\,\boldsymbol{\xi} = \frac{1}{m}\sum_{i=1}^{m}\xi_i^2.$$

A quadratic form of general type, provided it is non-negative, can be handled with exactly the same formalism (see Chapter 5) and often needs to be used in practical applications. Since each residual is a function of all the parameters,

$$\xi_i = \xi_i(\mathbf{y_0}, \boldsymbol{\mu}, \boldsymbol{\nu}),$$

the target function is also a function of $(\mathbf{y_0}, \boldsymbol{\mu}, \boldsymbol{\nu})$. The next step is to select the parameters to be fit to the data: let $\mathbf{x} \in \mathbb{R}^N$ be a subvector of $(\mathbf{y_0}, \boldsymbol{\mu}, \boldsymbol{\nu}) \in \mathbb{R}^{p+p'+p''}$, that is $\mathbf{x} = (x_i), i = 1, N$, with each x_i either a component of the initial conditions, or a dynamical parameter, or a kinematical parameter. Then we consider the target function

$$Q(\mathbf{x}) = Q(\boldsymbol{\xi}(\mathbf{x}))$$

as a function of $\mathbf{x}$ only, leaving the vector of the **consider parameters** $\mathbf{k} \in \mathbb{R}^{p+p'+p''-N}$ (all the parameters not included in $\mathbf{x}$) fixed at the assumed value.

The **minimum principle** selects as **nominal solution** the point $\mathbf{x}^* \in \mathbb{R}^N$ where the target function $Q(\mathbf{x})$ has its minimum value Q^*. The principle of **least squares** is the minimum principle with as target function the sum of squares $\mathcal{Q}(\boldsymbol{\xi}) = \boldsymbol{\xi}^T \boldsymbol{\xi}/m$, or some other quadratic form.

1.3 Two interpretations

The minimum principle should not be understood as if the "real" solution needs to be the point of minimum $\mathbf{x}^*$. Two interpretations can be used.

According to the **optimization interpretation**, $\mathbf{x}^*$ is the optimum point but values of the target function immediately above the minimum are also acceptable. The set of acceptable solutions can be described as the **confidence region**

$$Z(\sigma) = \left\{ \mathbf{x} \in \mathbb{R}^N \left| Q(\mathbf{x}) \leq Q^* + \frac{\sigma^2}{m} \right. \right\}$$

depending upon the **confidence parameter** $\sigma > 0$. For least squares

$$Z(\sigma) = \left\{ \mathbf{x} \in \mathbb{R}^N \left| \sum_{i=1}^{N} \xi_i^2 \leq m\,Q^* + \sigma^2 \right. \right\}.$$

The intuitive meaning of the confidence region is clear: the solutions $\mathbf{x}$ in $Z(\sigma)$ correspond to observation errors larger than those for $\mathbf{x}^*$, but still compatible with the available information on the observation procedure. The choice of the value of σ bounding the acceptable errors is not easy.

The alternative **probabilistic interpretation** describes the observation errors ϵ_i as random variables with an assumed probability density, which should be the result of an error model, justified by a priori knowledge of the observation process and/or a posteriori statistical tests. The vector $\boldsymbol{\epsilon} = (\epsilon_i), i = 1, m$, is then a set of jointly distributed random variables (see Section 3.1), and also the joint probability density function needs to be known; in particular, independence of the errors for observations at different times cannot be assumed, but needs to be justified by statistical tests.

Then the probabilistic model of the observation errors can be mapped in a probabilistic model of the result of orbit determination, with a probability density for the random variables $\mathbf{x}$ which in principle exists and can be, at least under some hypotheses, explicitly computed. The probability that the true orbit coincides exactly with the nominal solution $\mathbf{x}^*$ is zero, although under reasonable hypotheses $\mathbf{x}$ could be both the mode (point of maximum of the probability density) and the expected value.

In other words, the optimization interpretation describes the possible solutions as a subset of the **x** space where the target function has an acceptable value, surrounding the nominal solution which is the minimum point. The probabilistic interpretation regards the solutions as a probability density cloud, surrounding the point of highest probability density. Both interpretations can be useful, having different advantages and limitations.

1.4 Classification of the problem

Orbit determination appears as a number of different problems, with different dynamical systems and observation techniques. One way to classify the dynamical systems is to decompose the right-hand side of the equation of motion into three parts:

$$\frac{d\mathbf{y}}{dt} = \mathbf{f_0}(\mathbf{y}, t, \boldsymbol{\mu}) + \mathbf{f_1}(\mathbf{y}, t, \boldsymbol{\mu}) + \mathbf{f_2}(\mathbf{y}, t, \boldsymbol{\mu});$$

the *unperturbed* equation of motion has only the main term $\mathbf{f_0}$, with $|\mathbf{f_0}| \gg |\mathbf{f_1}|$. The main term may not contain unknown parameters, or very few. The perturbations are subdivided into the most relevant ones $\mathbf{f_1}$ and the negligible ones $\mathbf{f_2}$. Negligible means not only that $|\mathbf{f_1}| \gg |\mathbf{f_2}|$ but also that the effects of the $\mathbf{f_2}$ terms on the general solution are small (with respect to the observational accuracy), thus the equation of motion actually solved to compute the predictions contains only $\mathbf{f_0} + \mathbf{f_1}$. The choice of the terms to be neglected in each specific case is therefore a delicate issue, discussed in Sections 4.6, 15.3, and 17.3.

Let us focus on the main term $\mathbf{f_0}$. For a satellite of the Earth it is the monopole gravitational attraction of the Earth; for an object in heliocentric orbit it is the monopole attraction from the Sun, and so on. In most cases the unperturbed equation of motion is a two-body problem. Only in a few exceptional examples is there no dominant two-body term.

Thus we can classify orbit determination problems by the central body:

- Earth satellite orbits, for the Moon, artificial satellites, and space debris;
- heliocentric orbits, including the planets, the smaller asteroids, comets, meteoroids, trans-neptunian objects, and artificial interplanetary probes;
- satellite orbits of other planets, for the natural satellites, planetary orbiters, binary asteroids, and asteroid/comet orbiter missions;
- the orbits around another star, for binary stars and extrasolar planets;
- the cases without a dominant central body, such as orbits near the Lagrangian equilibrium points, temporary satellite captures, very small interplanetary dust with motion dominated by radiation pressure.

The orbit determination problems may differ also in the observation method, in the number and timing of the data, and in their accuracy. The main difference is between the *collaborative* and the *population* orbit determination problems.

Tracking

In **collaborative orbit determination** the object whose orbit has to be determined has a man-built device specifically intended to assist the observer. In this case the observation procedure is usually called **tracking**.

The most common case is tracking by radio waves: artificial satellites are normally equipped with a device called a **transponder**, which receives, amplifies, and retransmits the radio signal received from a ground station in a given frequency band.[1] Then the **range-rate**, the time derivative of the distance between the spacecraft and the ground station, can be measured by the Doppler shift between the signal emitted from the ground station and the one received back. If the signal also contains, beside the carrier, an encoded signal and the transponder is *regenerative*, that is it can send back this encoded signal on top of the return carrier, then also the **range**, or distance from the ground station, can be measured. This is possible also at interplanetary distance, thus the spacecraft could be in heliocentric orbit but also orbiting around another planet, or around an asteroid/comet.

In the above example the spacecraft needs to consume energy in the transponder, thus it has to be active, with a power system and possibly with attitude control to suitably point some antenna. There are examples in which the spacecraft is totally passive, such as the Earth satellites specifically launched for satellite laser ranging: they are only equipped with a special class of mirror, the *corner cubes*, to return a light ray in the same direction it came from with minimal dispersion. The ground stations are equipped with lasers capable of powerful but short-duration pulses of monochromatic light: the time interval between the emission of each pulse and the return signal detection measures the distance to the satellite.

The above examples are about artificial celestial bodies, that is man-made spacecraft. However, a tracking device can be planted on a natural body: e.g., corner cubes have been placed on the Moon by American and Soviet missions in the 1970s, thus **lunar laser ranging** has been regularly performed for more than 30 years, and the orbit of our natural satellite is known with centimeter level accuracy, actually more accurately than the

---

[1] The return signal can be shifted in frequency with respect to the received one, but this is done with *phase locking*, preserving very accurately the timing information.

orbit of any artificial satellite, affected by non-gravitational perturbations. The Viking landers have been on the surface of Mars for more than five years with operational transponders, and this has allowed the computation of the orbit of Mars with an accuracy of a few tens of meters. The interplanetary space probes like Voyager can be used to constrain the orbit of the planets they encounter. Planetary orbiters like Cassini (now around Saturn) and the future BepiColombo (around Mercury) will provide very accurate orbit determination for these planets and for the natural satellites of Saturn, thanks to the very accurate transponders on these spacecraft. Thus the main difference is not between *natural* and *artificial* orbits.

The specific properties of the collaborative cases are three.

First, the body has some built in capability to respond to tracking; thus the number of observations, their distribution in time and their accuracy are planned in the design phase of the mission. A simulation of orbit determination is a compulsory phase of **mission analysis**, the study showing that some proposed space mission is feasible from the astrodynamics point of view. If the simulated orbit determination gives poor results, the required frequency and accuracy of the observations has to be improved. Thus the most difficult cases of divergent orbit determination should not occur in the collaborative case; even strong nonlinearity and chaos should not happen. However, if there is some failure, either hardware like an antenna failing to deploy, or software like a faulty on-board computer program, or planning like an orbit determination simulation providing illusory results, then a tracking case may show some problems of the non-collaborative case, including divergence, excessive nonlinearity, and chaos.

Second, the observation data contain information on which object is being tracked. In the simplest case, there is only one spacecraft answering in a given frequency band in a given direction (within a given solid angle). Frequency bands and orbit slots (e.g., in the geosynchronous belt) are allocated by international authorities to avoid confusion and interference between signals to and from the satellites. In other cases (e.g., satellite constellations, such as navigation satellites) the satellite encodes its identity in the signal sent back to the ground. Thus we can assume we always know to which spacecraft each batch of tracking data belongs.[2] In most cases it is possible to treat each spacecraft as a separate problem of orbit determination; the exceptions are the cases of satellite-to-satellite tracking, where the radio/laser beam travels between two (or more) satellites, in which case the orbits of the two (or more) have to be solved simultaneously.

---

[2] Of course also this can occasionally fail, making orbit determination quite messy.

Third, if the amount of observational data and their accuracy exceeds what is required for the determination of the orbit in the strictest sense, that is for fitting the initial conditions $\mathbf{y_0}$, the additional information can be used to fit other parameters, either dynamical or kinematical, and in fact this is often the case. This is the key idea of **satellite geodesy**, where the gravity field of the central planet (the Earth, the Moon, another planet, an asteroid) is determined from the tracking data, rather than from the inhomogeneous ground-based gravimetry. In satellite geodesy around the Earth also the position of the ground stations can be determined with an accuracy far superior to that possible with ground-based measurements.

Catalogs

In the case of **population orbit determination** the observations are a scarce resource because the objects do not assist the observer. The total number of observations may not be small; actually it can be comparable to that of the tracking data points for a scientific space mission, e.g., tens of millions. The problem is that they refer to objects of a large population, and the average number of observations per object is small: e.g., 10^7 observations of a population of 10^6 objects (down to the minimum size observable).

The example most extensively discussed in this book is the orbit determination of the small bodies of the Solar System, including asteroids, comets, meteoroids and trans-neptunian objects. The number of objects needs to be qualified by a class of orbits and a minimum size: e.g., there are of the order of 10^6 main belt asteroids of size ≥ 1 km in diameter (this is just an estimate, extrapolated from the orbits already determined). A **survey** consists of a number of telescopes scanning the sky and looking for objects with stellar appearance which move with respect to the approximately fixed stars; this is the origin of the name asteroid, as proposed by Herschel. When such a **moving object** is detected the amount of information is minimal, typically only *astrometry*, that is angular positions, and *photometry*, that is apparent magnitude. There is a signature neither to identify the object with the ones already discovered, nor to decide it is new.

As we will see in Chapter 8, orbit determination is typically not possible with the discovery data alone. Thus the orbit determination problem cannot be disentangled from the **identification** problem, that is to find the independent discoveries referring to the same physical object: only by joining the information, contained in such separate discoveries, we can gather enough data for a solution. The output of the identification/orbit determination procedure is a *catalog* containing the list of distinct objects discovered, their best fit orbits, an estimate of their uncertainty and the little physical information available, in most cases just the absolute magnitude, a measure of the intrinsic capability of the object to reflect sunlight.[3]

The above example refers to *passive* observations detecting photons of reflected sunlight. Active observations are used in planetary radar observations, where a powerful beam of microwaves is directed towards a celestial body such as a major planet, a natural planetary satellite, an asteroid, a comet. At the present state-of-the-art, given that the signal-to-noise ratio at distance r is proportional to $1/r^4$, only the major inner planets, some very large satellites (e.g., Titan), and large asteroids can be observed by radar at interplanetary distances. Most of the targets therefore are **near-Earth asteroids**, which have the possibility of comparatively close approaches to the Earth.[4] Radar observations are a complex subject, because the radar return signal contains photons reflected from different parts of the asteroid surface, each with a different range and range-rate with respect to the radar antenna. In fact, the radar astrometry data are normal points obtained from a large fit providing also information on the size, shape, radar reflectivity, and rotation state of the object. The information constraining the orbit can be synthesized into an equivalent observation of range and range-rate. The accuracy of radar astrometry is between two and three orders of magnitude better than conventional astrometry.

The above examples are about natural bodies, but a very similar problem is obtained by considering spacecraft whose operational life is over. They can be observed in a non-collaborative way, with exactly the same techniques as asteroids, that is by astrometry and by radar. In most cases, however, these observations do not allow us to discriminate one dead spacecraft from another (actually, some care needs to be used to identify among the observations the ones belonging to operational spacecraft). As the search for this **space debris** progresses towards smaller and smaller Earth-orbiting objects, the list of bodies increases by adding spent rocket stages, pieces of exploded satellites and rocket motors, screws, bolts, and small pieces released during stage separation and antenna deployment, as well as particles of fuel, of frozen cooling liquid, all kinds of trash. A current estimate places at about 350 000 the number of orbiting debris above 1 cm of diameter. Thus the space debris problem is a population orbit determination problem, and surveys have to be set up to compile catalogs of all the particles above

---

[3] The absolute magnitude gives an indication of the diameter and mass, but the correspondence between these quantities contains unknown parameters such as the albedo and the density.

[4] With the current technology, radar astrometry for small asteroids (diameter < 1 km) is possible up to a distance of 0.2–0.3 AU.

a given size. The analogy is striking because there is an impact monitoring problem: the objects larger than a few mm could seriously damage the International Space Station by colliding at a relative speed of several km/s.

Thus the specific properties of the population cases are three, and they are opposite to those of the collaborative case.

First, the number of observations is not under our control. A survey can be designed to obtain a very large number of observations, but unavoidably the larger the data set, the larger the set of distinct objects for which the orbit has to be determined. Thus the average number of observations per object is small, typically of the order of 10.

Second, the batches of observations which can be immediately assigned to a single object are not enough to compute an orbit, thus the identification problem needs to be solved before orbit determination is possible. On the other hand, an identification can be considered reliable only if an orbit can be consistently fit to all the data believed to be of the same physical object. Thus orbit determination and identification are just a single algorithm, necessarily complex.

Third, the dynamical and kinematical parameters are normally not determined. After the reliable identifications have been established, each orbit can be solved individually, fitting just $N = p = 6$ parameters. Additionally, a separate fit of the photometric data can provide the absolute magnitude. However, this has to be performed for millions of bodies.

Planetary systems

There are a few examples of orbit determination which do not fit well into the binary classification collaborative/population. Interesting examples are the **planetary systems**. There are two main cases.

Our Solar System contains a small number N_P of planets.[5] The equation of motion for the planets needs to take into account the perturbations from the other planets, relativistic corrections, the perturbations from the larger satellites (especially the Moon), and the larger asteroids. The masses of the major planets appear as dynamical parameters $\boldsymbol{\mu}$, together with the post-Newtonian parameters describing general relativity effects.

Thus the orbits of the planets have to be determined all at once, including

---

[5] The exact definition of planet has been controversial, e.g., Pluto has size and mass comparable to those of other trans-neptunian bodies previously classified as minor planets, and it is significantly smaller than some satellites such as the Moon, Ganymede, and Titan. What matters in our discussion is the number of bodies whose masses are large enough to produce observable perturbations in the orbits of other planets, as discussed in Section 4.6; for the current accuracy in astrometric observations Pluto does not need to be included.

that of the Earth. The list of parameters may include $p = 6N_P$ initial conditions, $p' \geq N_P$ dynamical, and a number of kinematical parameters. The observations include astrometry, planetary radar, occultations, planetary landers, and spacecraft data. This example will be discussed mostly in Chapter 6 because of the rank deficiency.

Planets (and very small companion stars) orbiting around another star can be detected by measuring the star's **radial velocity**, that is the range-rate. Although the motion of the planet is much wider, the dim luminosity of the planet in most cases cannot be discriminated from the much brighter signal from the star. If we assume there is a single planet (or companion star), the dynamical system is a two-body problem and the orbit determination problem is simple, once the symmetries have been properly accounted for (see Chapter 6). These extrasolar planets can be considered as a population: indeed there is now a large data set of high-resolution radial velocity data. However, each planetary system can be solved on its own, without possibility of confusion among the data from different stars. If there are more planets around the same star the problem becomes more complex, but still not as much as the satellite geodesy and population orbit determination problems. Thus this problem is comparatively simple, but this does not mean it can be solved without care.

There is a separate theory for the orbit determination of binary stars, when both are observable from the Earth. We are not discussing this case in this book, but there are specialized textbooks such as (Aitken 1964).

There is also an intermediate case of orbit determination for artificial satellites, the **constellation orbit determination**, when a set of dozens of satellites on similar orbits is used in such a way that measurements have to be simultaneously taken from different constellation members, to exploit the higher accuracy of differences with respect to absolute measurements.[6] Then the orbits of the navigation satellites have to be determined all at once.

1.5 How to read this book

This textbook is intended for people interested in the general mathematical framework of orbit determination and in at least one of the main classes of applications (tracking, populations, and planetary systems). We expect that a significant fraction of the readers will not wish to read all the details about

---

[6] Another advantage of this method is that by measuring phase differences between the signals, from satellites also sending timing information, it is possible to solve for the time at the station, thus dispensing with accurate clocks.

applications very far from the ones they are working on. Thus we suggest three ways to select your customized path through this book.

If you are interested in satellite geodesy, mission analysis and operations of space missions, planetary exploration, and similar topics, you should check that you know the required material of Chapters 2 and 3, then read the basic theory of Chapters 5 and 6. You need the specific background on satellite orbits of Chapters 13 and 14. At this point you are ready for the satellite geodesy and planetary exploration case studies of Chapters 14, 15, 16 and 17.

If you are interested in asteroid/comet orbit determination, you should begin with Chapters 2 and 3 as in the other case, but you cannot skip Chapter 4 unless you already have a specific background in celestial mechanics. Then you should read the basic theory of Chapters 5 and 6. At this point you should be ready for the theory of simultaneous identification and orbit determination of Chapters 7–10. Chapter 11 contains a case study of an asteroid survey. Chapter 8 contains useful suggestions if you need comparable work for space debris. If you are interested in one of the most impressive applications of orbit determination, namely impact monitoring, which is a necessary tool to protect the Earth from collisions with asteroids, you may also read Chapter 12.

If you are only interested in planetary systems, the relevant examples are discussed in Chapter 6, thus you can read the first six chapters; some additional information is contained in Chapter 17.

If you are interested in all the possible applications of orbit determination, you can choose what you like, possibly using the subject index to identify where the definitions are given (on the pages with numbers in bold). We hope you will appreciate the effort done to present this subject in a unified way, based on our long and varied experience.

The appendices contain auxiliary material, which may be of significant help if you are undertaking the design and implementation of your own orbit determination software. Both for reasons of space in the book and to make the updates easier, the appendixes are not contained in this book but will be available online at the URL `http://adams.dm.unipi.it/orbdetbook`.

2

DYNAMICAL SYSTEMS

This chapter contains some basic material on dynamical systems which is required for the following. We are giving no formal proofs, but just recall the statements of the main results to be used. Several textbooks on the subject are available, including (Hartmann 1964, Milani 2002a).

2.1 The equation of motion

We shall describe the motion by an ordinary differential equation of the form

$$\frac{d\mathbf{y}}{dt} = \mathbf{f}(\mathbf{y}, t, \boldsymbol{\mu})$$

where $\mathbf{f} : \mathbb{R}^{p+p'+1} \to \mathbb{R}^p$ is a function obeying some regularity requirements, the state vector $\mathbf{y} \in \mathbb{R}^p$ is the unknown, and $\boldsymbol{\mu} \in \mathbb{R}^{p'}$ are the dynamical parameters, that we may assume as constant ($\boldsymbol{\mu} = \boldsymbol{\mu}_0$). We are interested in the solutions of the initial value problem (the *Cauchy problem*)

$$\frac{d\mathbf{y}}{dt}(t) = \mathbf{f}(\mathbf{y}(t), t, \boldsymbol{\mu}), \quad \mathbf{y}(t_0) = \mathbf{y}_0; \tag{2.1}$$

the general solution of (2.1) is a function of the time, the initial conditions, and the parameters

$$\mathbf{y}(t) = \mathbf{y}(t, t_0, \mathbf{y}_0, \boldsymbol{\mu}).$$

We can study the problem as an autonomous dynamical system, by introducing the variable $\mathbf{z}$, the initial conditions $\mathbf{z}_0$, and the function $\mathbf{g}$

$$\mathbf{z} = \begin{pmatrix} \mathbf{y} \\ t - t_0 \\ \boldsymbol{\mu} \end{pmatrix}; \qquad \mathbf{z}_0 = \mathbf{z}(t_0) = \begin{pmatrix} \mathbf{y}_0 \\ 0 \\ \boldsymbol{\mu}_0 \end{pmatrix}; \qquad \mathbf{g} = \begin{pmatrix} \mathbf{f} \\ 1 \\ 0 \end{pmatrix}$$

15

($\mathbf{0}$ is the zero vector in $\mathbb{R}^{p'}$): with these notations the problem becomes

$$\frac{d\mathbf{z}}{dt}(t) = \mathbf{g}(\mathbf{z}(t)), \qquad \mathbf{z}(t_0) = \mathbf{z}_0. \tag{2.2}$$

The general solution of (2.2) is usually denoted with either $\Phi_{t_0}^{t}(\mathbf{z}_0)$ or $\mathbf{z}(t, t_0, \mathbf{z}_0)$, and it is called the integral flow. The map $\Phi_{t_0}^{t}$ depends on the two parameters t_0, t and, for each value of t, it sends the initial conditions $\mathbf{z}_0$ into $\mathbf{z}(t)$, the value of the solution at time t. The integral flow has the following **semigroup property**: for each $t_0, t_1, t_2 \in \mathbb{R}$

$$\Phi_{t_1}^{t_2} \circ \Phi_{t_0}^{t_1} = \Phi_{t_0}^{t_2}.$$

As $\Phi_{t_0}^{t_0}$ is the identity application, the integral flow $\Phi_{t_0}^{t}$ is invertible and its inverse is $\Phi_{t}^{t_0}$. For autonomous differential equations like (2.2), we have **time-shift invariance** of the solutions:

$$\Phi_{t_0}^{t}(\mathbf{z}_0) = \Phi_{0}^{t-t_0}(\mathbf{z}_0) = \mathbf{z}(t - t_0, 0, \mathbf{z}_0);$$

thus, given the initial time t_0, we can either consider $t - t_0$ as time variable or we can assume that $t_0 = 0$, and we can use the simplified notation $\Phi^t = \Phi_0^t$.

Second-order equations

Sometimes the equation of motion is a second-order differential equation

$$\frac{d^2\mathbf{x}}{dt^2} = \mathbf{h}(\mathbf{x}, \mathbf{v}, \boldsymbol{\mu}, t), \qquad \mathbf{x}(t_0) = \mathbf{x}_0, \qquad \mathbf{v}(t_0) = \mathbf{v}_0 \tag{2.3}$$

with $\mathbf{v} = d\mathbf{x}/dt$, as is the case when the orbits are computed in Cartesian coordinates. Then the problem can be reduced to (2.1) simply by setting $\mathbf{y} = (\mathbf{x}, \mathbf{v})$, $\mathbf{y}_0 = (\mathbf{x}_0, \mathbf{v}_0)$, and $\mathbf{f}(\mathbf{y}) = (\mathbf{v}, \mathbf{h})$.

2.2 Solutions of the equation

We recall some basic results about existence, uniqueness, and regularity of the solutions of (2.2). A proof of these results can be found in several books on dynamical systems, e.g., (Hartmann 1964).

Existence and uniqueness of the solutions

Let us consider an open set $\Omega \subseteq \mathbb{R}^n$. A function $\mathbf{g} : \Omega \to \mathbb{R}^n$ is **uniformly Lipschitz-continuous** on Ω if there exists $L > 0$ such that

$$|\mathbf{g}(\mathbf{z}_1) - \mathbf{g}(\mathbf{z}_2)| < L\,|\mathbf{z}_1 - \mathbf{z}_2| \qquad \forall \mathbf{z}_1, \mathbf{z}_2 \in \Omega.$$

If $\mathbf{g}$ is uniformly Lipschitz-continuous, then for each $\mathbf{z}_0 \in \Omega$ there exists a unique solution $\mathbf{z}(t)$ of (2.2) defined in an interval $(-\epsilon, \epsilon)$ with $\epsilon > 0$ depending on $\mathbf{z}_0$. If $\mathbf{g}$ is locally Lipschitz-continuous in Ω there exists locally a unique integral flow $\mathbf{z}(t, \mathbf{z}_0)$ defined on an open set in $\mathbb{R}^{n+1}$.

Note that if $\mathbf{g}$ is differentiable of class $\mathcal{C}^1$ (with continuous partial derivatives) in a larger open set Ω_1 containing the compact K and $\Omega \subset K$, then it is also uniformly Lipschitz-continuous on Ω. In celestial mechanics the regularity of the equation of motion is guaranteed because the gravitational potential is a harmonic function (see Chapter 13). Only some non-gravitational perturbations may introduce regularity problems, see Section 14.3.

Maximal solutions

A solution of (2.2) is *maximal* if it is defined in a maximal time interval, i.e., no solution with that initial data can be defined in a larger interval.

Let $\mathbf{z}_0 \in \Omega$ and $\mathbf{z} = \mathbf{z}(t)$ be a solution of (2.2) on an open interval $I \in \mathbb{R}$ containing 0. If the solution $\mathbf{z}(t)$ defined for $t \geq 0$ on a limited interval $[0, t_1)$ is contained in a compact set $K \subset \Omega$ then $\mathbf{z}(t)$ is *not* a maximal solution; similarly for $t \leq 0$. The maximal solutions defined only in a limited interval must get out of any compact set in Ω.

Lipschitz-continuity of the flow

The integral flow $\mathbf{z}(t, \mathbf{z}_0)$ is Lipschitz-continuous as a function of the initial conditions $\mathbf{z}_0$; this can be shown using the **Gronwall lemma**: let $y : I \to \mathbb{R}$ be a non-negative function defined in an interval $I \subseteq \mathbb{R}$; if there are $\alpha, \beta > 0$ such that

$$0 \leq y(t) \leq \beta + \alpha \int_0^t y(s)\, ds, \quad \text{then} \ \ 0 \leq y(t) \leq \beta \exp(\alpha\, t).$$

An immediate consequence of the Gronwall lemma: if $\mathbf{z}_1(t), \mathbf{z}_2(t)$ are solutions of (2.2), with different initial conditions $\mathbf{z}_1(0), \mathbf{z}_2(0)$, then assuming that $\mathbf{g}$ is uniformly Lipschitz-continuous we have

$$|\mathbf{z}_1(t) - \mathbf{z}_2(t)| \leq |\mathbf{z}_1(0) - \mathbf{z}_2(0)| + L \int_0^t |\mathbf{z}_1(s) - \mathbf{z}_2(s)|\, ds.$$

Then by the Gronwall lemma

$$|\mathbf{z}_1(t) - \mathbf{z}_2(t)| \leq |\mathbf{z}_1(0) - \mathbf{z}_2(0)| \exp(L\, t), \tag{2.4}$$

i.e., the integral flow is Lipschitz-continuous with respect to the initial data.

2.3 The variational equation

Let us define the **state transition matrix** as the Jacobian matrix

$$A(t, \mathbf{z}_0) = \frac{\partial \mathbf{z}}{\partial \mathbf{z}_0}(t, \mathbf{z}_0).$$

If the integral flow $\mathbf{z}(t, \mathbf{z}_0)$ is regular enough we can differentiate it twice, with respect to the time t and the initial condition $\mathbf{z}_0$, and exchanging the order of the derivatives we obtain the same result:[1]

$$\frac{\partial}{\partial t}\left[\frac{\partial \mathbf{z}}{\partial \mathbf{z}_0}(t, \mathbf{z}_0)\right] = \frac{\partial}{\partial \mathbf{z}_0}\left[\frac{\partial \mathbf{z}}{\partial t}(t, \mathbf{z}_0)\right]. \qquad (2.5)$$

Using (2.5) and the fact that $\mathbf{z}(t, \mathbf{z}_0)$ is the solution of (2.2), we obtain the differential equation

$$\frac{\partial}{\partial t}\left[\frac{\partial \mathbf{z}}{\partial \mathbf{z}_0}(t, \mathbf{z}_0)\right] = \frac{\partial \mathbf{g}}{\partial \mathbf{z}}(\mathbf{z}(t, \mathbf{z}_0))\frac{\partial \mathbf{z}}{\partial \mathbf{z}_0}(t, \mathbf{z}_0). \qquad (2.6)$$

Equation (2.6) together with the initial condition $\partial \mathbf{z}/\partial \mathbf{z}_0 = I$ at $(t, \mathbf{z}) = (0, \mathbf{z}_0)$, where I is the identity matrix, give the Cauchy problem

$$\begin{cases} \dfrac{\partial A}{\partial t}(t, \mathbf{z}_0) &= \dfrac{\partial \mathbf{g}}{\partial \mathbf{z}}(\mathbf{z}(t, \mathbf{z}_0)) \, A(t, \mathbf{z}_0) \\[2mm] A(0, \mathbf{z}_0) &= I. \end{cases} \qquad (2.7)$$

The linear differential equation in (2.7) is the **variational equation**. It can be interpreted as the linearized equation for the relative motion. Let $\mathbf{z}^{(0)}(t, \mathbf{z}_0) = \mathbf{z}(t, \mathbf{z}_0)$ be the general solution of (2.2) and let $\mathbf{z}^{(\epsilon)}(0, \mathbf{z}_0) = \mathbf{z}_0 + \epsilon \, \mathbf{v}_0$ be the initial condition with the small perturbation $\epsilon \, \mathbf{v}_0$, where

$$\mathbf{z}^{(\epsilon)}(t, \mathbf{z}_0) = \mathbf{z}(t, \mathbf{z}_0 + \epsilon \, \mathbf{v}_0).$$

The Taylor expansion of $\mathbf{z}^{(\epsilon)}$ with respect to the small parameter ϵ gives

$$\mathbf{z}^{(\epsilon)}(t, \mathbf{z}_0) = \mathbf{z}^{(0)}(t, \mathbf{z}_0) + \epsilon \, \mathbf{z}^{(1)}(t, \mathbf{z}_0) + O(\epsilon^2)$$

where

$$\mathbf{z}^{(1)}(t, \mathbf{z}_0) = \frac{\partial \mathbf{z}}{\partial \mathbf{z}_0}(t, \mathbf{z}_0) \, \mathbf{v}_0.$$

The Taylor expansion of $\mathbf{g}\left(\mathbf{z}^{(\epsilon)}\right)$ gives

$$\mathbf{g}(\mathbf{z}^{(\epsilon)}) = \mathbf{g}(\mathbf{z}^{(0)} + \epsilon \mathbf{z}^{(1)} + O(\epsilon^2)) = \mathbf{g}(\mathbf{z}^{(0)}) + \epsilon \frac{\partial \mathbf{g}}{\partial \mathbf{z}}(\mathbf{z}^{(0)})\mathbf{z}^{(1)} + O(\epsilon^2).$$

---

[1] We use the symbol $\frac{\partial}{\partial t}$ instead of $\frac{d}{dt}$ when we consider the integral flow $\mathbf{z}(t, \mathbf{z}_0)$ of (2.2), where also $\mathbf{z}_0$ may vary, in place of the solution $\mathbf{z}(t)$ of (2.2) for $\mathbf{z}_0$ fixed.

By equating the terms of order zero and one in $\partial \mathbf{z}^{(\epsilon)}/\partial t = \mathbf{g}\left(\mathbf{z}^{(\epsilon)}\right)$ and neglecting the higher order terms in ϵ we obtain

$$\frac{\partial}{\partial t}(\mathbf{z}^{(\epsilon)} - \mathbf{z}^{(0)}) = \frac{\partial \mathbf{g}}{\partial \mathbf{z}}(\mathbf{z}^{(0)})(\mathbf{z}^{(\epsilon)} - \mathbf{z}^{(0)})$$

that is, the relative motion $\mathbf{v}(t, \mathbf{z}_0) = \mathbf{z}^{(\epsilon)}(t, \mathbf{z}_0) - \mathbf{z}^{(0)}(t, \mathbf{z}_0)$ is the solution of the system

$$\begin{cases} \dfrac{\partial \mathbf{v}}{\partial t}(t, \mathbf{z}_0) &= \dfrac{\partial \mathbf{g}}{\partial \mathbf{z}}(\mathbf{z}^{(0)}(t, \mathbf{z}_0)) \, \mathbf{v}(t, \mathbf{z}_0) \\ \mathbf{v}(0, \mathbf{z}_0) &= \mathbf{v}_0 \end{cases} \tag{2.8}$$

whose general solution is given by the variational equation (2.7).

Variational equation with dynamical parameters

Let us write explicitly the variational equation for eq. (2.1). The state transition matrix is

$$A(t) = \frac{\partial \mathbf{z}}{\partial \mathbf{z}_0} = \begin{bmatrix} \dfrac{\partial \mathbf{y}}{\partial \mathbf{y}_0} & \dfrac{\partial \mathbf{y}}{\partial \boldsymbol{\mu}} \\ \dfrac{\partial \boldsymbol{\mu}}{\partial \mathbf{y}_0} & \dfrac{\partial \boldsymbol{\mu}}{\partial \boldsymbol{\mu}} \end{bmatrix} = \begin{bmatrix} \dfrac{\partial \mathbf{y}}{\partial \mathbf{y}_0} & \dfrac{\partial \mathbf{y}}{\partial \boldsymbol{\mu}} \\ \mathbf{0} & I \end{bmatrix}$$

where $\mathbf{0}$, I are respectively the zero and the identity matrix, with dimensions suitable to their place in the matrix $A(t)$. Furthermore we have

$$\frac{\partial \mathbf{g}}{\partial \mathbf{z}} = \begin{bmatrix} \dfrac{\partial \mathbf{f}}{\partial \mathbf{y}} & \dfrac{\partial \mathbf{f}}{\partial \boldsymbol{\mu}} \\ \mathbf{0} & \mathbf{0} \end{bmatrix}$$

so that the variational equation (2.7) gives the system

$$\frac{\partial}{\partial t}\left(\frac{\partial \mathbf{y}}{\partial \mathbf{y}_0}\right) = \frac{\partial \mathbf{f}}{\partial \mathbf{y}} \frac{\partial \mathbf{y}}{\partial \mathbf{y}_0}, \quad \frac{\partial}{\partial t}\left(\frac{\partial \mathbf{y}}{\partial \boldsymbol{\mu}}\right) = \frac{\partial \mathbf{f}}{\partial \mathbf{y}} \frac{\partial \mathbf{y}}{\partial \boldsymbol{\mu}} + \frac{\partial \mathbf{f}}{\partial \boldsymbol{\mu}}$$

with initial data

$$\frac{\partial \mathbf{y}}{\partial \mathbf{y}_0}(0) = I, \quad \frac{\partial \mathbf{y}}{\partial \boldsymbol{\mu}}(0) = \mathbf{0}.$$

Variational equation for second-order equations

If the equation of motion is of second order, like eq. (2.3), then we can decompose the state transition matrix as follows

$$B = \frac{\partial \mathbf{x}}{\partial \mathbf{x}_0}, \quad C = \frac{\partial \mathbf{x}}{\partial \mathbf{v}_0}, \quad D = \frac{\partial \mathbf{x}}{\partial \boldsymbol{\mu}}.$$

We use here, and widely in this book, the dot notation $\dot{\mathbf{x}} = d\mathbf{x}/dt$. Note that

$$\dot{B} = \frac{\partial \mathbf{v}}{\partial \mathbf{x}_0}, \quad \dot{C} = \frac{\partial \mathbf{v}}{\partial \mathbf{v}_0}, \quad \dot{D} = \frac{\partial \mathbf{v}}{\partial \boldsymbol{\mu}}.$$

Then the variational equation is obtained by exchanging the derivatives: for B and C, with the double dot indicating second time derivatives,

$$\begin{cases} \ddot{B} & = \dfrac{\partial \mathbf{h}}{\partial \mathbf{v}} \dot{B} + \dfrac{\partial \mathbf{h}}{\partial \mathbf{x}} B \\ B(t_0) & = I, \dot{B}(t_0) = \mathbf{0}, \end{cases} \tag{2.9}$$

$$\begin{cases} \ddot{C} & = \dfrac{\partial \mathbf{h}}{\partial \mathbf{v}} \dot{C} + \dfrac{\partial \mathbf{h}}{\partial \mathbf{x}} C \\ C(t_0) & = \mathbf{0}, \dot{C}(t_0) = I, \end{cases} \tag{2.10}$$

while for D the linear equations are non-homogeneous

$$\begin{cases} \ddot{D} & = \dfrac{\partial \mathbf{h}}{\partial \mathbf{v}} \dot{D} + \dfrac{\partial \mathbf{h}}{\partial \mathbf{x}} D + \dfrac{\partial \mathbf{h}}{\partial \boldsymbol{\mu}} \\ D(t_0) & = \mathbf{0}, \dot{D}(t_0) = \mathbf{0}. \end{cases} \tag{2.11}$$

Differentiability of the solutions

We state another regularity property of the integral flow. Let $\Omega \subseteq \mathbb{R}^n$ be an open set and $\mathbf{g} : \Omega \to \mathbb{R}^n$ a C^1 function. Then for each $\bar{\mathbf{z}}_0 \in \Omega$ there is an open set $W \subset \mathbb{R} \times \Omega$ containing $(0, \bar{\mathbf{z}}_0)$ such that the integral flow $\Phi^t(\mathbf{z}_0)$ restricted to W is a C^1 function in the variable $(t, \mathbf{z}_0)$. Furthermore the mixed derivatives $\partial^2 \Phi^t(\mathbf{z}_0)/\partial t \, \partial \mathbf{z}_0$ exist and are continuous; in particular the procedure followed to obtain the variational equation (2.7) makes sense. The proof can be given using the variational equation itself and the Gronwall lemma. This result can also be extended to functions $\mathbf{g}$ of class C^k, getting solutions of class C^k in all the variables.

2.4 Lyapounov exponents

Given the solution $\mathbf{z}(t, \mathbf{z}_0)$ of the Cauchy problem (2.2) let $\mathbf{v}^t(\mathbf{z}_0)$ be the solution of the variational equation (2.7). If the limit

$$\lim_{t \to +\infty} \frac{1}{t} \log \frac{|\mathbf{v}^t(\mathbf{z}_0)|}{|\mathbf{v}_0|}$$

exists, we denote it by $\chi = \chi(\mathbf{z}_0, \mathbf{v}_0)$ and we call it a **Lyapounov exponent** of the dynamical system. The inverse of the maximum positive Lyapounov exponent, if it exists, is the **Lyapounov time** over which two nearby orbits diverge, on average, by a factor $\exp(1)$.

Let us consider two different solutions $\mathbf{z}_1(t), \mathbf{z}_2(t)$ of (2.2) obtained from different initial conditions $\mathbf{z}_1(0), \mathbf{z}_2(0)$. By the Gronwall lemma we have eq. (2.4):

$$\Delta(t) = |\mathbf{z}_1(t) - \mathbf{z}_2(t)| \leq \Delta(0) \cdot \exp(L\,t), \quad \text{so that} \quad \frac{1}{t}\log\frac{\Delta(t)}{\Delta(0)} \leq L.$$

Passing to the limit as $\Delta(0) \to 0$ and $t \to +\infty$ we obtain the inequality

$$\chi(\mathbf{z}_0, \mathbf{v}_0) \leq L, \tag{2.12}$$

where L is the Lipschitz constant of $\mathbf{g}$. This inequality is sharp, as shown by the following example: the equation $d\mathbf{z}/dt = \lambda\,\mathbf{z}$ with initial condition $\mathbf{z}(0) = \mathbf{z}_0$ has solution $\mathbf{z}(t) = \exp(\lambda\,t)\,\mathbf{z}_0$ and the difference $\Delta(t)$ increases in size according to

$$\frac{\Delta(t)}{\Delta(0)} = \frac{|\mathbf{z}_1(t) - \mathbf{z}_2(t)|}{|\mathbf{z}_1(0) - \mathbf{z}_2(0)|} = \exp(\lambda\,t), \quad \text{so that} \quad \frac{1}{t}\log\frac{\Delta(t)}{\Delta(0)} = \lambda,$$

that is, it has λ as Lyapounov exponent.

2.5 Model problem dynamics

We shall consider the simple nonlinear problem

$$\frac{da}{dt} = 0, \quad \frac{d\lambda}{dt} = n(a) = \frac{k}{a^{3/2}} \tag{2.13}$$

with initial conditions $a(0) = a_0, \lambda(0) = \lambda_0$; here $k > 0$ is the Gauss gravitational constant, that is $k^2 = G\,m_\odot$. This problem is modeled on the planar two-body problem, with the nonlinear dependence of the *mean motion* upon the *semimajor axis*, in a zero eccentricity approximation; it is extensively used in Chapters 5, 6 and 7 as a model problem. The integral flow is a shift map

$$a(t, a_0, \lambda_0) = a_0, \quad \lambda(t, a_0, \lambda_0) = \lambda_0 + n_0\,t \tag{2.14}$$

with $n_0 = n(a_0)$, that is two initially nearby orbits diverge linearly with time; the Lyapounov exponents are all zero. The corresponding variational equation is as follows: the state transition matrix

$$A(t, a_0, \lambda_0) = \frac{\partial(a, \lambda)}{\partial(a_0, \lambda_0)}$$

is the solution of the Cauchy problem

$$\frac{\partial A}{\partial t} = \begin{bmatrix} 0 & 0 \\ -\dfrac{3\,n_0}{2\,a_0} & 0 \end{bmatrix} A, \quad A(0) = I. \tag{2.15}$$

The integral flow of a linear Cauchy problem

$$\frac{d\mathbf{z}}{dt}(t) = M\,\mathbf{z}(t), \quad \mathbf{z}(0) = \mathbf{z}_0 \tag{2.16}$$

where M is an $n \times n$ matrix, is given by $\Phi^t(\mathbf{z}_0) = \exp(M\,t)\,\mathbf{z}_0$ where the **matrix exponential** is defined by

$$\exp(M\,t) = \sum_{i=0}^{\infty} \frac{M^i t^i}{i!}. \tag{2.17}$$

This series converges uniformly with respect to t in every compact interval. For the case of eq. (2.15)

$$A(t, a_0, \lambda_0) = \exp \begin{bmatrix} 0 & 0 \\ -\dfrac{3\,n_0}{2\,a_0} t & 0 \end{bmatrix} = \begin{bmatrix} 1 & 0 \\ -\dfrac{3\,n_0}{2\,a_0} t & 1 \end{bmatrix}.$$

The partial derivatives with respect to the dynamical parameter k

$$B(t, a_0, \lambda_0) = \left[\frac{\partial a(t, a_0, \lambda_0)}{\partial k} \,,\, \frac{\partial \lambda(t, a_0, \lambda_0)}{\partial k} \right]^T$$

are the solution of the Cauchy problem

$$\frac{\partial B}{\partial t} = \begin{bmatrix} 0 & 0 \\ -\dfrac{3\,n_0}{2\,a_0} & 0 \end{bmatrix} B + \begin{bmatrix} 0 \\ \dfrac{1}{a_0^{3/2}} \end{bmatrix}, \quad B(0) = \begin{bmatrix} 0 \\ 0 \end{bmatrix}$$

and can be written as

$$B(t, a_0, \lambda_0) = \left[0 \,,\, t/a_0^{3/2} \right]^T.$$

3

ERROR MODELS

We outline the basic tools of probability theory needed in the following chapters, with special emphasis on the Gaussian, or normal, distributions, which have an essential connection with the least squares principle (see Section 5.7). We give very few proofs; the others can be found in many textbooks, e.g., (Jazwinski 1970, Mood *et al.* 1974).

3.1 Continuous random variables

A **continuous random variable** X is defined by a **probability density** function, a real function $\mathsf{p}_X(x) \geq 0$ defined and continuous for all $x \in \mathbb{R}$ with the property

$$\int_{-\infty}^{+\infty} \mathsf{p}_X(x)\, dx = 1. \tag{3.1}$$

It follows that X has also a **distribution function** $\mathsf{d}_X(x)$, defined for all $x \in \mathbb{R}$ and continuously differentiable:

$$\mathsf{d}_X(x) = \int_{-\infty}^{x} \mathsf{p}_X(s)\, ds, \quad \text{so that} \quad \mathsf{p}_X(x) = \frac{d}{dx}\mathsf{d}_X(x),$$

by which we can compute a *probability measure* $\mathcal{P}_X$ on $\mathbb{R}$. The probability for X to be inside an open interval (a, b) is

$$\mathcal{P}_X(a < X < b) = \mathsf{d}_X(b) - \mathsf{d}_X(a) = \int_{a}^{b} \mathsf{p}_X(x)\, dx. \tag{3.2}$$

Note that $\mathcal{P}_X(a < X < b) = \mathcal{P}_X(a \leq X \leq b)$, hence we have also a probability measure for the closed intervals $[a, b]$. Because the probability density $\mathsf{p}_X(x)$ has a bounded integral, there is a large algebra of subsets B

of $\mathbb{R}$ such that the probability for X to be inside can be computed by

$$\mathcal{P}_X(X \in B) = \int_B \mathsf{p}_X(x)\, dx.$$

Indeed, finite and countable infinite disjoint unions of open intervals are included, and the algebra is closed with respect to the complement; this is called the Borel σ-algebra for $\mathbb{R}$.[1] The continuous random variables are not enough to describe all cases occurring in the applications, in particular in the measurement errors: there are also discrete distributions, and random variables X can be defined as combinations of continuous and discrete ones, such that, for some a, $\mathcal{P}_X(x = a) > 0$; they cannot be represented by continuous distribution functions and have no probability density function. These do occur in the errors, e.g., whenever the data are digitized by hand the copyist mistakes are discrete. However, for a large data set this kind of mistake is not very important, and the use of continuous random variables as error models is justified when the largest errors, which are exceptional events and hard to model, are removed (see Section 5.8).

Given a continuous random variable X, we use the following definitions:

$$E(X) = \int_{-\infty}^{+\infty} x\, \mathsf{p}_X(x)\, dx \qquad\qquad \textbf{mean (or expectation)},$$

$$\mathrm{Var}(X) = \int_{-\infty}^{+\infty} [x - E(X)]^2\, \mathsf{p}_X(x)\, dx \quad \textbf{variance},$$

$$\mathrm{RMS}(X) = \sqrt{\mathrm{Var}(X)} \qquad\qquad \begin{array}{l}\textbf{standard deviation}\\ \textbf{(or root mean square)},\end{array}$$

$$\mu_n(X) = \int_{-\infty}^{+\infty} [x - E(X)]^n\, \mathsf{p}_X(x)\, dx \quad \textbf{\textit{n}-th moment},$$

$$K(X) = \frac{\mu_4(X)}{\mathrm{Var}(X)^2} \qquad\qquad\qquad \textbf{kurtosis}.$$

Jointly distributed random variables

Two continuous random variables X, Y are *jointly distributed* if they are defined by a **joint probability density function** $\mathsf{p}_{X,Y}(x, y)$ which is con-

---

[1] The integral appearing in (3.2) should be the Lebesgue integral, to give a measure to all the elements of the Borel σ-algebra. However, in the applications the subsets of $\mathbb{R}$ whose probability has to be measured are simple: usually they are intervals, thus we can regard the integral appearing in (3.2) as the Riemann integral.

tinuous in the vector variable $(x, y) \in \mathbb{R}^2$ and such that

$$\mathsf{p}_{X,Y}(x, y) \geq 0, \qquad \int_{\mathbb{R}^2} \mathsf{p}_{X,Y}(x, y) \, dx \, dy = 1.$$

Following the same steps of the univariate case we can define a probability measure $\mathcal{P}_{X,Y}$: first we define

$$\mathcal{P}_{X,Y} \left(\begin{array}{c} a < X < b \\ c < Y < d \end{array} \right) = \int_a^b \int_c^d \mathsf{p}_{X,Y}(x, y) \, dx \, dy$$

for the rectangles $(a, b) \times (c, d)$, then we consider the subsets of $\mathbb{R}^2$ that are finite or countable infinite unions of rectangles, or the complement of them, i.e. the Borel σ-algebra for $\mathbb{R}^2$. If D is one of these subsets, we define[2]

$$\mathcal{P}_{X,Y}(D) = \int_D \mathsf{p}_{X,Y}(x, y) \, dx \, dy.$$

For the continuous random variables X, Y we define the mean and the variance of X:

$$E(X) = \int_{\mathbb{R}^2} x \, \mathsf{p}_{X,Y}(x, y) \, dx \, dy,$$

$$\text{Var}(X) = \int_{-\infty}^{+\infty} [x - E(X)]^2 \, \mathsf{p}_{X,Y}(x, y) \, dx \, dy,$$

and similarly for Y. Moreover we define the **covariance** of X, Y

$$\begin{aligned} \text{Cov}(X, Y) &= \int_{\mathbb{R}^2} [x - E(X)][y - E(Y)] \, \mathsf{p}_{X,Y}(x, y) \, dx \, dy \\ &= E\left([X - E(X)][Y - E(Y)]\right) \end{aligned}$$

and the **covariance matrix**

$$\Gamma = \left[\begin{array}{cc} \text{Var}(X) & \text{Cov}(X, Y) \\ \text{Cov}(X, Y) & \text{Var}(Y) \end{array} \right].$$

The **normal matrix** is defined as $C = \Gamma^{-1}$. The coefficient of **correlation** of X and Y is the ratio

$$\text{Corr}(X, Y) = \frac{\text{Cov}(X, Y)}{\sqrt{\text{Var}(X)} \sqrt{\text{Var}(Y)}}$$

and we shall say that the two variables are *uncorrelated* if this coefficient is zero, *correlated* if it is not zero.

It is possible to generalize these definitions to n jointly distributed continuous random variables $X_1, \ldots, X_n$. They are defined by a joint probability

---

[2] This is a Riemann integral if D is measurable according to Peano–Jordan.

density function $\mathsf{p}_{X_1,X_2,\ldots,X_n}(x_1, x_2, \ldots, x_n)$ which is non-negative, continuous in all the variables, and has unit integral on $\mathbb{R}^n$. The probability measure of a subset D of $\mathbb{R}^n$ can be computed as the multiple integral

$$\mathcal{P}_{X_1,\ldots,X_n}((X_1, X_2, \ldots, X_n) \in D) = \int_D \mathsf{p}_{X_1,X_2,\ldots,X_n}\, dx_1\, dx_2 \ldots dx_n$$

provided D satisfies conditions like those of the two-dimensional case.

Given n jointly distributed continuous random variables X_j, $j = 1, n$, the mean and the variance of each variable is

$$E(X_j) = \int_{\mathbb{R}^n} x_j \mathsf{p}_{X_1,\ldots,X_n}(x_1, \ldots, x_n)\, dx_1 \ldots dx_n,$$

$$\mathrm{Var}(X_j) = \int_{\mathbb{R}^n} [x_j - E(X_j)]^2\, \mathsf{p}_{X_1,\ldots,X_n}(x_1, \ldots, x_n)\, dx_1 \ldots dx_n,$$

the covariance of X_i, X_j is

$$\mathrm{Cov}(X_i, X_j) = \int_{\mathbb{R}^n} [x_i - E(X_i)][x_j - E(X_j)]\, \mathsf{p}_{X_1,\ldots,X_n}(x_1, \ldots, x_n)\, dx_1 \ldots dx_n,$$

and the normal matrix C is the inverse of the covariance matrix $\Gamma = (\gamma_{ij})_{i,j}$ whose coefficients are $\gamma_{ii} = \mathrm{Var}(X_i)$ and $\gamma_{ij} = \mathrm{Cov}(X_i, X_j)$ for $i \neq j$. The correlation coefficients can also be deduced from the covariance matrix

$$\mathrm{Corr}(X_i, X_j) = \frac{\gamma_{ij}}{\sqrt{\gamma_{ii}\gamma_{jj}}}.$$

Independence, marginal and conditional probability

For two jointly distributed random variables X, Y we define the **marginal density functions**

$$\mathsf{p}_X(x) = \int_{-\infty}^{+\infty} \mathsf{p}_{X,Y}(x, y)\, dy, \quad \mathsf{p}_Y(y) = \int_{-\infty}^{+\infty} \mathsf{p}_{X,Y}(x, y)\, dx,$$

which can be regarded as probability densities of one of the jointly distributed random variables, valid for each value of the other variable.

The jointly distributed X, Y are **independent random variables** if

$$\mathsf{p}_{X,Y}(x, y) = \mathsf{p}_X(x)\, \mathsf{p}_Y(y). \tag{3.3}$$

If X, Y are independent, $\mathrm{Cov}(X, Y) = 0$. The converse is not always true.

Given X, Y continuous random variables with probability density $\mathsf{p}_{X,Y}(x, y)$ the **conditional density functions** are

$$p_{X|Y}(x; y) = \frac{\mathsf{p}_{X,Y}(x, y)}{\mathsf{p}_Y(y)}, \quad p_{Y|X}(y; x) = \frac{\mathsf{p}_{X,Y}(x, y)}{\mathsf{p}_X(x)},$$

for $p_Y(y) > 0$ and $p_X(x) > 0$, respectively, where we use ";" to stress the different role of the two variables. The independence of X and Y can be expressed in terms of conditional density functions as either $p_{Y|X}(y; x) = p_Y(y)$ or $p_{X|Y}(x; y) = p_X(x)$.

3.2 Gaussian random variables

There are continuous random variables that play an important role in the least squares principle: those with density function of the type

$$p_X(x) = N(\mu, \sigma^2)(x) = \frac{1}{\sqrt{2\pi}\sigma} \exp\left(-\frac{(x-\mu)^2}{2\sigma^2}\right), \qquad (3.4)$$

where $\mu = E(X)$ and $\sigma = \mathrm{RMS}(X)$. Such variables are called **Gaussian** or **normally distributed**. The following relation is useful:

$$\int_{-\infty}^{+\infty} \exp\left(-\frac{x^2}{2\sigma^2}\right) dx = \sqrt{2\pi}\sigma;$$

it can be computed from the integral over the plane $\mathbb{R}^2$ of the function $\exp(x^2 + y^2)$ in polar coordinates. It can be easily checked that a Gaussian variable satisfies the property (3.1) of a probability density function.

Rotational invariance

A geometric characterization of the Gaussian densities is as follows. If two jointly distributed continuous random variables X, Y are independent, with equal marginal densities $p_X(x) = p_Y(x) = f(x)$ and the probability density function $p_{X,Y}(x, y)$ is invariant under rotations, i.e., there exists a function $g : \mathbb{R} \to \mathbb{R}$ such that $p_{X,Y}(x, y) = g(x^2 + y^2)$, then they are Gaussian with zero mean:

$$p_X(x) = N(0, \sigma^2)(x).$$

This can be shown as follows:

$$g(x^2 + y^2) = f(x) f(y) \Longrightarrow g(x^2) = f(x) f(0)$$

where $f(0) = k$ is constant. Thus $f(x) = g(x^2)/k = k\, h(x^2)$ and by substituting in the formula above

$$h(x^2 + y^2) = h(x^2)\, h(y^2) \Longrightarrow \log h(x^2) + \log h(y^2) = \log h(x^2 + y^2),$$

thus $\log h(z)$ is a linear function: $\log h(x^2) = s\, x^2$ and $f(x) = k\exp(s\, x^2)$. For f to have a limited integral, s must be negative, say $s = -1/2\sigma^2$ and

$$p_X(x) = f(x) = k \exp\left(-\frac{1}{2\,\sigma^2}\right).$$

Then the normalization property (3.1) implies $k = 1/\sigma\sqrt{2\pi}$. Hence the two-dimensional rotation invariant Gaussian is

$$p_{X,Y}(x,y) = N(0,\sigma^2)(x)\ N(0,\sigma^2)(y) = \frac{1}{2\pi\sigma^2}\exp\left(-\frac{x^2+y^2}{2\sigma^2}\right).$$

Two-dimensional Gaussian variables

We take two independent jointly distributed Gaussian variables X, Y with joint density function $p_{X,Y}(x,y) = p_X(x)\ p_Y(y)$ with zero mean and different standard deviations σ_x, σ_y. In this case the covariance matrix is

$$\Gamma = \begin{pmatrix} \sigma_x^2 & 0 \\ 0 & \sigma_y^2 \end{pmatrix},$$

the normal matrix is $C = \Gamma^{-1}$ and the joint probability density is

$$
\begin{aligned}
p_{X,Y}(x,y) = N(0,\Gamma)(x,y) &= \frac{\sqrt{\det C}}{2\pi}\exp\left[-\frac{1}{2}(x,y)\,C\begin{pmatrix} x \\ y \end{pmatrix}\right] \\
&= \frac{1}{2\pi\sigma_x\sigma_y}\exp\left[-\frac{1}{2}\left(\frac{x^2}{\sigma_x^2}+\frac{y^2}{\sigma_y^2}\right)\right].
\end{aligned}
$$

More generally, let us consider two correlated Gaussian random variables X, Y, with normal matrix C and covariance matrix Γ defined by

$$C = \frac{1}{1-\rho^2}\begin{pmatrix} 1/\sigma_x^2 & -\rho/(\sigma_x\sigma_y) \\ -\rho/(\sigma_x\sigma_y) & 1/\sigma_y^2 \end{pmatrix},\quad \Gamma = \begin{pmatrix} \sigma_x^2 & \rho\sigma_x\sigma_y \\ \rho\sigma_x\sigma_y & \sigma_y^2 \end{pmatrix}$$

where $\rho = \mathrm{Corr}(X,Y)$. The marginal probability densities are the same as in the independent case:

$$p_X(x) = N(0,\sigma_x^2)(x),\quad p_Y(y) = N(0,\sigma_y^2)(y),$$

but the joint probability density is different:

$$
\begin{aligned}
p_{X,Y}(x,y) &= N(0,\Gamma)(x,y) \\
&= \frac{1}{2\pi\,\sigma_x\sigma_y\,\sqrt{1-\rho^2}}\exp\left[-\frac{1}{2(1-\rho^2)}\left(\frac{x^2}{\sigma_x^2}-\frac{2\rho\,x\,y}{\sigma_x\,\sigma_y}+\frac{y^2}{\sigma_y^2}\right)\right].
\end{aligned}
$$

The result is valid also for jointly distributed continuous random variables with non-zero mean: in this case the density function is

$$
\begin{aligned}
p_{X,Y}(x,y) &= N(\mathbf{m},\Gamma)(x,y) \qquad\qquad\qquad\qquad\qquad (3.5) \\
&= \frac{\sqrt{\det C}}{2\pi}\exp\left[-\frac{1}{2}(x-m_x,y-m_y)\,C\begin{pmatrix} x-m_x \\ y-m_y \end{pmatrix}\right]
\end{aligned}
$$

where

$$\mathbf{m} = (m_x, m_y) = (E(X), E(Y));$$

the marginal density functions of X and Y are normal

$$\mathsf{p}_X(x) = N(m_x, \sigma_x^2), \qquad \mathsf{p}_Y(y) = N(m_y, \sigma_y^2).$$

Moreover, if $\mathrm{Corr}(X, Y) = 0$ then both the normal and the covariance matrix are diagonal, and X, Y are independent.

Regression line

Given two jointly distributed Gaussian variables X, Y, with probability density (3.6), the conditional probability density of X given Y is also Gaussian:

$$p_{X|Y}(x; y) = N\left(m_x + \rho \frac{\sigma_x}{\sigma_y}(y - m_y), \sigma_x^2(1 - \rho^2)\right)(x).$$

The above formula uses the **regression line**

$$y = m_y + \frac{\sigma_y}{\sigma_x}\rho(x - m_x),$$

giving the expected conditional value $E[Y|X](x) = \int_{\mathbb{R}} p_{Y|X}(y; x)\, dy$. A similar formula gives the conditional probability density of Y given X

$$p_{Y|X}(y; x) = N\left(m_y + \rho \frac{\sigma_y}{\sigma_x}(x - m_x), \sigma_y^2(1 - \rho^2)\right)(y),$$

which uses the other regression line

$$x = m_x + \frac{\sigma_x}{\sigma_y}\rho(y - m_y).$$

The regression lines are shown in Figure 5.1.

Multidimensional Gaussian variables

Given n jointly distributed random variables $X_1, X_2, \ldots, X_n$, we say that they are Gaussian, or normally distributed, if their joint density function is of the form

$$\mathsf{p}_{X_1, X_2, \ldots, X_n}(x_1, x_2, \ldots, x_n) = \frac{\sqrt{\det C}}{(2\pi)^{n/2}} \exp\left[-\frac{1}{2}(\mathbf{x} - \mathbf{m})^T\, C\,(\mathbf{x} - \mathbf{m})\right]$$

where $\mathbf{m} = (m_1, \ldots, m_n)^T$ is the vector of the means and C, the normal matrix, is symmetric and positive definite. The notation $N(\mathbf{m}, \Gamma)$ is used for the above probability density function, where $\Gamma = C^{-1}$. Again, if C is diagonal so is Γ, and the X_j are all independent: for Gaussian variables, being independent and uncorrelated is equivalent.

For a multidimensional Gaussian, we need to generalize the result on marginal probability densities: let us consider the two vector random variables

$$\mathbf{X} = (X_1, \ldots, X_n), \quad \mathbf{Y} = (Y_1, \ldots, Y_m)$$

jointly distributed, Gaussian with probability density

$$p_{\mathbf{X},\mathbf{Y}}(\mathbf{x}, \mathbf{y}) = N((\mathbf{m_x}; \mathbf{m_y}), \Gamma_{\mathbf{xy}}),$$

where $\mathbf{x} = (x_1, \ldots, x_n)$, $\mathbf{y} = (y_1, \ldots, y_m)$, $(\mathbf{m_x}; \mathbf{m_y})$ is the stacking of the two vectors, and the covariance matrix can be decomposed as

$$\Gamma = \begin{bmatrix} \Gamma_{\mathbf{x}} & \Gamma_{\mathbf{xy}} \\ \Gamma_{\mathbf{yx}} & \Gamma_{\mathbf{y}} \end{bmatrix}, \quad \Gamma_{\mathbf{yx}} = \Gamma_{\mathbf{xy}}^T,$$

where $\Gamma_{\mathbf{x}}$ is an $n \times n$ matrix, $\Gamma_{\mathbf{y}}$ is $m \times m$, and $\Gamma_{\mathbf{xy}}$ is $n \times m$. Then the marginal probability densities are

$$p_{\mathbf{X}} = N(\mathbf{m_x}, \Gamma_{\mathbf{x}}), \quad p_{\mathbf{Y}} = N(\mathbf{m_y}, \Gamma_{\mathbf{y}}), \tag{3.6}$$

that is the **marginal covariance matrix** is the restriction, to the corresponding linear subspace, of the covariance matrix. For the **conditional covariance matrix** the following formula applies

$$p_{X|Y}(x; y) = N(\mathbf{m_x} + \Gamma_{\mathbf{xy}}\Gamma_{\mathbf{y}}^{-1}(\mathbf{y} - \mathbf{m_y}), \Gamma_{\mathbf{x}} - \Gamma_{\mathbf{xy}}\Gamma_{\mathbf{y}}^{-1}\Gamma_{\mathbf{yx}}), \tag{3.7}$$

which can be described by the statement that the **conditional normal matrix** $C^{\mathbf{x}}$ is the restriction, to the corresponding linear subspace, of the normal matrix $C = \Gamma^{-1}$. Similarly for $p_{Y|X}(y; x)$, see Section 5.4.

3.3 Expected values and transformations

Given a continuous random variable X and a continuous real function $f(x)$, we can define the random variable $Y = F(X)$ with probability measure

$$\mathcal{P}_Y(a < Y < b) = \mathcal{P}_X(X : a < F(X) < b).$$

The question is whether Y is a continuous random variable, i.e., whether a continuous probability density function $p_Y(y)$ can be defined. Under the assumptions

(i) $y = f(x)$ is bijective from $W = \{x \in \mathbb{R} : p_X(x) > 0\}$ to a set $D \subset \mathbb{R}$;

(ii) $x = f^{-1}(y)$ has in D continuous derivative, different from 0;

$Y = F(X)$ is a continuous random variable with probability density

$$p_Y(y) = \left| \frac{df^{-1}(y)}{dy} \right| p_X(f^{-1}(y))$$

inside D and 0 outside.

This definition can be generalized to n variables as follows. Given a set $\mathbf{X} = (X_1, \ldots, X_n)$ of jointly distributed continuous random variables with probability density function $p_\mathbf{X}(\mathbf{x})$, where $\mathbf{x} = (x_1, \ldots, x_n)$, and given a continuous function $\mathbf{f}(\mathbf{x}) = \mathbf{y}$, with $\mathbf{y} = (y_1, \ldots, y_n)$, let

$$W = \{\mathbf{x} \in \mathbb{R}^n \text{ such that } p_\mathbf{X}(\mathbf{x}) > 0\}$$

and $D = \mathbf{f}(W)$, with $\mathbf{f} : W \to D$ a bijective function. If $\mathbf{f}^{-1} \in C^1(D)$ with Jacobian J different from zero, then

$$p_\mathbf{Y}(\mathbf{y}) = |\det J| \, p_\mathbf{X}(\mathbf{f}^{-1}(\mathbf{y})) \tag{3.8}$$

inside D, 0 outside, is the probability density that defines the continuous vector random variable $\mathbf{Y} = \mathbf{F}(\mathbf{X})$.

Linear transformations of Gaussians

Given a continuous random variable $\mathbf{X}$, let $\mathbf{y} = \mathbf{f}(\mathbf{x}) = A\mathbf{x} + \mathbf{b}$, with A an $n \times n$ matrix, be an affine transformation in $\mathbb{R}^n$. If $\mathbf{X}$ is Gaussian, with probability density $p_\mathbf{X}(\mathbf{x}) = N(\mathbf{m}, \Gamma)$, $\mathbf{m} \in \mathbb{R}^n$ and Γ a symmetric positive definite $n \times n$ matrix, then $\mathbf{Y} = \mathbf{F}(\mathbf{X})$ also has a normal distribution, with probability density

$$p_\mathbf{Y}(\mathbf{y}) = N\left(A\mathbf{m} + \mathbf{b}, A\,\Gamma\,A^T\right), \tag{3.9}$$

that is, with expected value $\mathbf{f}(\mathbf{m})$ and covariance matrix $A\,\Gamma\,A^T$. This is called the **covariance propagation** rule.

In this case the transformation is invertible: $\mathbf{x} = A^{-1}(\mathbf{y} - \mathbf{b})$ with $\det \partial \mathbf{x}/\partial \mathbf{y} = \det^{-1}(A)$. Then eq. (3.8), by using $\mathbf{x} - \mathbf{m} = A^{-1}[\mathbf{y} - \mathbf{f}(\mathbf{m})]$ and $(A^T)^{-1}\,C\,A^{-1} = (A\,\Gamma\,A^T)^{-1}$, gives

$$\begin{aligned}
p_\mathbf{Y}(\mathbf{y}) &= \frac{\sqrt{\det C\,(\det A)^{-2}}}{(2\pi)^{n/2}} \exp\left[-\frac{1}{2}[\mathbf{y} - \mathbf{f}(\mathbf{m})]^T \, (A^T)^{-1}\,C\,A^{-1}\,[\mathbf{y} - \mathbf{f}(\mathbf{m})]\right] \\
&= \frac{\sqrt{\det(A\,\Gamma\,A^T)^{-1}}}{(2\pi)^{n/2}} \exp\left[-\frac{1}{2}[\mathbf{y} - \mathbf{f}(\mathbf{m})]^T \, (A\,\Gamma\,A^T)^{-1}\,[\mathbf{y} - \mathbf{f}(\mathbf{m})]\right] \\
&= N\left(\mathbf{f}(\mathbf{m}), A\,\Gamma\,A^T\right).
\end{aligned}$$

A generalization to transformations $\mathbf{y} = \mathbf{f}(\mathbf{x}) = B\,(\mathbf{x} + \mathbf{b})$ of the Gaussian variable $\mathbf{X}$ with density $N(\mathbf{m}, \Gamma)$, where B is an $m \times n$ matrix $(m < n)$ with maximal rank m, can be obtained as follows. Let $\Pi = [I|0]$ be the matrix of the projection onto the subspace of the first m coordinates and let A be an invertible $n \times n$ matrix with $B = \Pi A$: we use relation (3.9) to compute the probability density $N(A(\mathbf{m}+\mathbf{b}), A \Gamma A^T)$ of the random variable defined by the invertible transformation $\mathbf{z} = A(\mathbf{x} + \mathbf{b})$, and then relation (3.6), about marginal densities, to obtain

$$\mathsf{p_Y}(\mathbf{y}) = N\left(\Pi A(\mathbf{m}+\mathbf{b}), \Pi A \Gamma A^T \Pi^T\right) = N\left(\mathbf{f}(\mathbf{m}), B\Gamma B^T\right). \qquad (3.10)$$

Conditional probability density on a linear subspace

We need to generalize the formula for the conditional probability density of a Gaussian to an ambient space of arbitrary dimension m and to an affine subset of arbitrary dimension $N < m$; this is used in Section 5.7.

Let W be an N-plane, image of $\mathbb{R}^N$ by a linear (non-homogeneous) map defined by the $m \times N$ matrix B and the reference point $\boldsymbol{\xi}^*$

$$W = \{\boldsymbol{\xi} \in \mathbb{R}^m : \boldsymbol{\xi} = B\mathbf{x} + \boldsymbol{\xi}^*, \mathbf{x} \in \mathbb{R}^N\};$$

moreover, we may assume that $\boldsymbol{\xi}^*$ is a vector orthogonal to W (otherwise the component parallel to W can be subtracted). Let $\mathsf{p}_\Xi(\boldsymbol{\xi}) = N(0, I)$ be a rotation-invariant Gaussian probability density; we need to compute the conditional probability density of the random variable Ξ on W.

We use a rotation matrix R such that

$$R(\boldsymbol{\xi} - \boldsymbol{\xi}^*) = \begin{bmatrix} \boldsymbol{\xi}' \\ \boldsymbol{\xi}'' \end{bmatrix} \Longrightarrow R^T \begin{bmatrix} 0 \\ \boldsymbol{\xi}'' \end{bmatrix} + \boldsymbol{\xi}^* \in W, \qquad (3.11)$$

that is, $\boldsymbol{\xi}'' \in \mathbb{R}^N$ parameterizes W. The probability density of Ξ'' is the conditional probability density of $R(\Xi)$ given $\boldsymbol{\xi}' = R\boldsymbol{\xi}^*$, but the distribution $N(\mathbf{0}, I)$ is rotation invariant, thus the probability density of Ξ'' can be computed from eq. (3.7) and Ξ'' is Gaussian with as normal matrix the restriction of the normal matrix of Ξ

$$\mathsf{p}_{\Xi''} = N(\mathbf{0}, I)$$

with I the $N \times N$ identity matrix. Geometrically, the intersection of $(m-1)$-spheres with N-planes can only be $(N-1)$-spheres, and these are the level surfaces of the probability density of Ξ''.

4

THE *N*–BODY PROBLEM

This chapter presents the basic theory of the gravitational N-body problem, the coordinate systems used for both theoretical investigations and practical applications, and how to select the dynamical model for a Solar System orbit.

4.1 Equation of motion and integrals

By $(N+1)$-body problem we mean the ordinary differential equation defining the motion of $N+1$ point masses with positions $\mathbf{r}_j$, velocities $\dot{\mathbf{r}}_j$, and masses m_j, interacting only through the mutual gravitational attraction

$$m_j \ddot{\mathbf{r}}_j = \sum_{i \neq j} \frac{G m_i m_j}{|\mathbf{r}_i - \mathbf{r}_j|^3} (\mathbf{r}_i - \mathbf{r}_j), \qquad j = 0, \ldots, N \qquad (4.1)$$

where the dots indicate time derivatives and G is the universal gravitational constant; this is the equation of motion in Newtonian form. We need to express it in another form, more suitable both to discuss symmetries and integrals and to perform coordinate changes. The mutual gravitational forces admit a potential, thus we can define the potential energy

$$V = - \sum_{0 \leq i < j \leq N} \frac{G m_i m_j}{|\mathbf{r}_i - \mathbf{r}_j|};$$

we introduce the kinetic energy T and the **Lagrange function** (or **Lagrangian**) L:

$$T = \frac{1}{2} \sum_{i=0}^{N} m_i |\dot{\mathbf{r}}_i|^2, \qquad L = T - V. \qquad (4.2)$$

The Newton equation of motion is equivalent to the **Lagrange equation**

$$\frac{d}{dt} \left(\frac{\partial L}{\partial \dot{\mathbf{r}}_j} \right) - \frac{\partial L}{\partial \mathbf{r}_j} = \mathbf{0} \qquad (4.3)$$

33

with two important properties. The first one has to do with integrals of motion, the second is discussed in Section 4.2. A **first integral** of the Lagrange equation (4.3) is a function of all the positions and velocities

$$I = I(\mathbf{R}, \dot{\mathbf{R}}), \quad \mathbf{R} = (\mathbf{r}_0, \mathbf{r}_1, \ldots, \mathbf{r}_N), \quad \dot{\mathbf{R}} = (\dot{\mathbf{r}}_0, \dot{\mathbf{r}}_1, \ldots, \dot{\mathbf{r}}_N)$$

such that the total time derivative along the solutions is identically zero:

$$\frac{dI}{dt} = \frac{\partial I}{\partial \mathbf{R}} \dot{\mathbf{R}} + \frac{\partial I}{\partial \dot{\mathbf{R}}} \ddot{\mathbf{R}} = 0;$$

thus the value of I is constant along the orbits.

Symmetries and integrals

A one-parameter **group of symmetries** of the Lagrange function L is a diffeomorphism F^s of the positions $\mathbf{R}$ depending (in a differentiable way) upon a parameter $s \in \mathbb{R}$ so that $F^s \circ F^z = F^{s+z}$ and the Lagrange function is invariant:

$$L\left(F^s(\mathbf{R}), \frac{d}{dt}F^s(\mathbf{R})\right) = L\left(F^s(\mathbf{R}), \frac{\partial F^s}{\partial \mathbf{R}}\dot{\mathbf{R}}\right) = L(\mathbf{R}, \dot{\mathbf{R}}).$$

F^0 is the identity transformation; we also assume the mixed derivatives $\partial^2 F^s/\partial \mathbf{R}\partial s$ are continuous. A *local one-parameter group of symmetries* of the Lagrange function is defined by the same properties for s in a neighborhood of 0. The main result we need is the **Noether theorem**, stating that if the Lagrange function L admits a local one-parameter group of symmetries F^s then

$$I(\mathbf{R}, \dot{\mathbf{R}}) = \frac{\partial L}{\partial \dot{\mathbf{R}}} \cdot \frac{\partial F^s(\mathbf{R})}{\partial s}\bigg|_{s=0} \tag{4.4}$$

is a first integral of the Lagrange equation (4.3).

To apply this theorem to the $(N+1)$-body problem we look for symmetries of the Lagrange function in (4.2), a function of the mutual distances $|\mathbf{r}_i - \mathbf{r}_j|$ and of the velocities $|\dot{\mathbf{r}}_j|$. Thus every **isometry** of the space of positions, preserving distances and independent of time, preserves the Lagrange function. The isometries of the Euclidean space $\mathbb{R}^3$ are the functions

$$G(\mathbf{x}) = R\,\mathbf{x} + \mathbf{q}, \quad \frac{dG}{dt}(\mathbf{x}) = R\,\dot{\mathbf{x}},$$

where R is an orthogonal matrix ($R^T R = I$) and $\mathbf{q}$ a constant vector, both independent of time. The symmetry group of three-dimensional space has dimension 6 and is generated by six one-parameter subgroups.[1] There are three one-parameter symmetry groups of translations ($R = I$):

$$F^s(\mathbf{x}) = \mathbf{x} + s\,\hat{\mathbf{v}}_h, \qquad \frac{\partial F^s(\mathbf{x})}{\partial s} = \hat{\mathbf{v}}_h$$

where $\hat{\mathbf{v}}_h$ is the unit vector along one coordinate axis, for $h = 1, 2, 3$. If equal translations are applied to all bodies, then the integral of (4.4) is

$$p_h = \hat{\mathbf{v}}_h \cdot \sum_{j=0}^{N} m_j\,\dot{\mathbf{r}}_j,$$

the component along the axis $\hat{\mathbf{v}}_h$ of the total **linear momentum p**. The latter is a vector integral, and the **center of mass** $\mathbf{b}_0$ moves with constant velocity:

$$\mathbf{b}_0 = \frac{1}{M_0} \sum_{j=0}^{N} m_j\,\mathbf{r}_j; \quad M_0 = \sum_{j=0}^{N} m_j \text{ (total mass)}; \quad \mathbf{b}_0(t) = \frac{t}{M_0}\,\mathbf{p} + \mathbf{b}_0(0).$$

$$(4.5)$$

In the above formula, $\mathbf{b}_0(0)$ is a constant vector which can be obtained as a combination of positions and velocities, but with coefficients depending upon time: each of its components is a **time-dependent first integral**.

The other three one-parameter symmetry groups are groups of rotations ($\mathbf{q} = \mathbf{0}$). A three-dimensional $\mathbf{x}$ rotates by an angle of s radians around an axis $\hat{\mathbf{v}}_h$; the rotation is counterclockwise for $s > 0$, as seen from the tip of $\hat{\mathbf{v}}_h$,

$$F^s(\mathbf{x}) = R_{s\hat{\mathbf{v}}_h}\,\mathbf{x}, \qquad \left.\frac{\partial F^s(\mathbf{x})}{\partial s}\right|_{s=0} = \hat{\mathbf{v}}_h \times \mathbf{x}$$

and the integral of the Noether theorem

$$c_h = \sum_{j=0}^{N} (\hat{\mathbf{v}}_h \times \mathbf{r}_j) \cdot m_j\,\dot{\mathbf{r}}_j = \hat{\mathbf{v}}_h \cdot \sum_{j=0}^{N} m_j\,(\mathbf{r}_j \times \dot{\mathbf{r}}_j)$$

is the component along $\hat{\mathbf{v}}_h$ of the total angular momentum

$$\mathbf{c} = \sum_{j=0}^{N} m_j\,(\mathbf{r}_j \times \dot{\mathbf{r}}_j), \qquad (4.6)$$

thus the motion preserves the angular momentum vector integral.

There is one additional integral, the total energy integral, which is not deduced from the Noether theorem.[2] By computing the total time deriva-

---

[1] The tangent space to the unit element, the *Lie Algebra*, is generated by the tangents to these subgroups. Only orientation preserving isometries are included in the one-parameter subgroups.

[2] It could be interpreted, with the Hamiltonian formalism, as a consequence of the invariance with respect to time, thus it corresponds to the symmetry $t \mapsto t + s$.

tives

$$\frac{dT}{dt} = \sum_{j=0}^{N} m_j \, \ddot{\mathbf{r}}_j \cdot \dot{\mathbf{r}}_j, \qquad \frac{dV}{dt} = \sum_{j=0}^{N} \frac{\partial V}{\partial \mathbf{r}_j} \cdot \dot{\mathbf{r}}_j$$

and by eq. (4.1) they are opposite, thus $E = T + V$ is a first integral.

There is one additional symmetry in the $(N + 1)$-body problem, which involves not only the coordinates but also the time and possibly the masses: the **change of scale**. It is also associated with a first integral, which is not independent of the previous ones. If the lengths are changed by a factor λ, the times by a factor τ, the masses by a factor μ, then

$$m_j \ddot{\mathbf{r}}_j \mapsto \frac{\mu \, \lambda}{\tau^2} \, m_j \ddot{\mathbf{r}}_j, \qquad \frac{\partial V}{\partial \mathbf{r}_j} \mapsto \frac{\mu^2}{\lambda^2} \frac{\partial V}{\partial \mathbf{r}_j},$$

and the equation of motion is satisfied by the scaled orbits if and only if

$$\lambda^3 = \mu \, \tau^2, \tag{4.7}$$

the dimensional version of Kepler's third law. If $\tau = 1$ it is possible to scale the lengths compensating with a scaling of the masses $\lambda^3 = \mu$; this may imply the impossibility of determining masses and lengths (see Section 6.2).

When a scaling with $\lambda^3 = \mu \, \tau^2$ is applied, the energy integral is scaled

$$T \mapsto \frac{\mu \lambda^2}{\tau^2} \, T, \qquad V \mapsto \frac{\mu^2}{\lambda} \, V \Longrightarrow E \mapsto \frac{\mu \lambda^2}{\tau^2} \, E$$

and the angular momentum vector integral scales as $\mathbf{c} \mapsto \mu \lambda^2 / \tau \, \mathbf{c}$, thus the combination $E \, c^2$, where $c = |\mathbf{c}|$, scales as

$$E \, c^2 \mapsto \frac{\mu \lambda^2}{\tau^2} \frac{\mu^2 \lambda^4}{\tau^2} E \, c^2 = \mu^5 \, E \, c^2;$$

thus $E \, c^2$ is invariant if $\mu = 1$, that is, if masses are not scaled.

A deep result obtained by the celestial mechanicians of the late nineteenth century states that for $N \geq 3$ there are no first integrals in the $(N + 1)$-body problem independent of the 10 classical ones of the linear and angular momentum and total energy (seven time independent and three time dependent).

4.2 Coordinate changes

The first integrals have to be exploited to reduce the dimensionality of the equation of motion, and this is for two reasons. First, the dimensions $3N + 3$ of the configuration space, and $6N + 6$ of the phase space (of the initial conditions), are too large to understand the properties of the solutions. Second, the symmetries associated with the integrals may result in degeneracy of the orbit determination problem, as discussed in Chapter 6; one of the possible remedies is to decrease the number of variables. Also for the above purpose, we need to know how the equation of motion transforms under a coordinate change: this is easier for the Lagrange equation.

Let $\mathbf{B} = (\mathbf{b}_0, \mathbf{b}_1, \dots, \mathbf{b}_n)$ be another set of coordinates for the positions of the $N + 1$ bodies, and $\mathbf{R} = \mathbf{R}(\mathbf{B})$ a coordinate change which is a diffeomorphism (with continuous second derivatives) of the $(3N + 3)$-dimensional space; we are thus assuming that the Jacobian matrix $A(\mathbf{B}) = \partial \mathbf{R}/\partial \mathbf{B}$ is invertible at each point $\mathbf{B}$. The corresponding change in the velocities is

$$\dot{\mathbf{R}} = \frac{\partial \mathbf{R}}{\partial \mathbf{B}}(\mathbf{B}) \, \dot{\mathbf{B}} = A(\mathbf{B}) \, \dot{\mathbf{B}}.$$

Let $L(\mathbf{R}, \dot{\mathbf{R}})$, $\mathcal{L}(\mathbf{B}, \dot{\mathbf{B}})$ be Lagrange functions corresponding by value:

$$\mathcal{L}(\mathbf{B}, \dot{\mathbf{B}}) = L\left(\mathbf{R}(\mathbf{B}), \dot{\mathbf{R}}(\mathbf{B}, \dot{\mathbf{B}})\right) = L\left(\mathbf{R}(\mathbf{B}), A(\mathbf{B}) \, \dot{\mathbf{B}}\right);$$

then the left-hand side of the Lagrange equation is transformed as follows:

$$\frac{d}{dt}\left(\frac{\partial \mathcal{L}}{\partial \dot{\mathbf{B}}}\right) - \frac{\partial \mathcal{L}}{\partial \mathbf{B}} = \left[\frac{d}{dt}\left(\frac{\partial L}{\partial \dot{\mathbf{R}}}\right) - \frac{\partial L}{\partial \mathbf{R}}\right] A(\mathbf{B}). \tag{4.8}$$

The Lagrange equations in the two coordinate systems are equivalent

$$\frac{d}{dt}\left(\frac{\partial \mathcal{L}}{\partial \dot{\mathbf{B}}}\right) - \frac{\partial \mathcal{L}}{\partial \mathbf{B}} = 0 \iff \frac{d}{dt}\left(\frac{\partial L}{\partial \dot{\mathbf{R}}}\right) - \frac{\partial L}{\partial \mathbf{R}} = 0;$$

solutions of one are transformed by $\mathbf{R} = \mathbf{R}(\mathbf{B})$ into solutions of the other.

Reduction of the two-body problem

We shall start from the simplest case, the two-body problem, to get some ideas to be exploited in the general case. The Lagrange function is

$$L = \frac{1}{2}m_0 \, |\dot{\mathbf{r}}_0|^2 + \frac{1}{2}m_1 \, |\dot{\mathbf{r}}_1|^2 + \frac{G \, m_0 \, m_1}{|\mathbf{r}_0 - \mathbf{r}_1|}.$$

We can change coordinates by using, in place of $\mathbf{r}_0, \mathbf{r}_1$, the coordinates of the center of mass and the relative position of $\mathbf{r}_1$ with respect to $\mathbf{r}_0$

$$\mathbf{b}_0 = \mu_1 \, \mathbf{r}_1 + (1 - \mu_1)\mathbf{r}_0, \quad \mu_1 = \frac{m_1}{m_0 + m_1}, \quad \mathbf{b}_1 = \mathbf{r}_1 - \mathbf{r}_0. \tag{4.9}$$

Then $V = \mathcal{V}(b_1) = -Gm_0m_1/b_1$, with $b_1 = |\mathbf{b}_1|$; to write T as a function of $\mathbf{b}_0, \mathbf{b}_1$ we express $\dot{\mathbf{r}}_0$ and $\dot{\mathbf{r}}_1$ as a function of $\dot{\mathbf{b}}_0, \dot{\mathbf{b}}_1$ and substitute in T:

$$2T = m_0 \, \dot{\mathbf{r}}_0^2 + m_1 \, \dot{\mathbf{r}}_1^2 = (m_0 + m_1) \, \dot{\mathbf{b}}_0^2 + \frac{m_0 m_1}{m_0 + m_1} \, \dot{\mathbf{b}}_1^2$$

the mixed terms canceling. The Lagrange function as a function of $\mathbf{b}_0, \mathbf{b}_1$ is

$$L = \frac{1}{2} M_0 \, \dot{\mathbf{b}}_0^2 + \frac{1}{2} M_1 \, \dot{\mathbf{b}}_1^2 + \frac{G \, M_0 \, M_1}{b_1}$$

with $M_0 = m_0 + m_1$ the total mass and M_1 the **reduced mass**:

$$M_1 = \frac{m_0 m_1}{m_0 + m_1}. \tag{4.10}$$

Then the Lagrange function L can be decomposed as the sum of two Lagrange functions $L = M_0 \, L_0(\dot{\mathbf{b}}_0) + M_1 \, L_1(\mathbf{b}_1, \dot{\mathbf{b}}_1)$, one containing only $\mathbf{b}_0$, the other containing only $\mathbf{b}_1$, and the Lagrange equation decouples:

$$M_0 \, \ddot{\mathbf{b}}_0 = 0, \quad M_1 \, \ddot{\mathbf{b}}_1 = -\frac{\partial \mathcal{V}(\mathbf{b}_1)}{\partial \mathbf{b}_1}.$$

The first equation states that the center of mass moves with constant velocity along a straight line, the second equation is the **Kepler problem**, with a particle of mass M_1 attracted by a fixed center of mass M_0.

By repeating the same computations done for T, we find that also the angular momentum has a simple expression in the **B** coordinates:

$$\mathbf{c} = m_0 \, \mathbf{r}_0 \times \dot{\mathbf{r}}_0 + m_1 \, \mathbf{r}_1 \times \dot{\mathbf{r}}_1 = M_0 \, \mathbf{b}_0 \times \dot{\mathbf{b}}_0 + M_1 \, \mathbf{b}_1 \times \dot{\mathbf{b}}_1.$$

When $\mathbf{b}_0(t)$ from eq. (4.5) is substituted, the $\mathbf{b}_0$ contribution is constant

$$\mathbf{c}_0 = \mathbf{b}_0 \times \dot{\mathbf{b}}_0 = \frac{1}{M_0} \, \mathbf{b}_0(0) \times \mathbf{p}, \quad \mathbf{c} = M_0 \, \mathbf{c}_0 + M_1 \, \mathbf{c}_1$$

and the contribution from $\mathbf{b}_1$ is $\mathbf{c}_1 = \mathbf{b}_1 \times \dot{\mathbf{b}}_1$, the angular momentum per unit (reduced) mass of $\mathbf{r}_1$ with respect to the center $\mathbf{r}_0$, which is also a vector first integral. Thus $\mathbf{b}_1, \dot{\mathbf{b}}_1$ will lie for each t in the orbital plane normal to $\mathbf{c}_1$.

Solution of the two-body problem

The two-body problem has another vector integral, not occurring in the $N \geq 3$-body problem: the **Laplace–Lenz vector**

$$\mathbf{e} = \frac{1}{G \, M_0} \, \dot{\mathbf{b}}_1 \times \mathbf{c}_1 - \frac{1}{b_1} \, \mathbf{b}_1. \tag{4.11}$$

This can be shown by using a reference frame formed by three mutually orthogonal unit vectors, $\mathbf{v}_z = \mathbf{c}_1/c_1$ ($c_1 = |\mathbf{c}_1|$), $\mathbf{v}_r = \mathbf{b}_1/b_1$, and $\mathbf{v}_\theta$ such that $\dot{\mathbf{b}}_1 \cdot \mathbf{v}_\theta > 0$. If θ is the angle between the vector $\mathbf{v}_r$ and a fixed direction in the orbital plane, and $r = b_1$, we have

$$\mathbf{c}_1 = r \, \mathbf{v}_r \times \frac{d}{dt}(r \, \mathbf{v}_r) = r \, \mathbf{v}_r \times (\dot{r} \, \mathbf{v}_r + r\dot{\theta} \, \mathbf{v}_\theta) = r^2 \, \dot{\theta} \, \mathbf{v}_r \times \mathbf{v}_\theta = r^2 \, \dot{\theta} \, \mathbf{v}_z,$$

$$G\,M_0\,\mathbf{e} = -r^2\,\dot{r}\,\dot{\theta}\,\mathbf{v}_\theta + (r^3\,\dot{\theta}^2 - G\,M_0)\,\mathbf{v}_r. \tag{4.12}$$

Along the solutions we have

$$\dot{c}_1 = 0, \quad 2\dot{r}\dot{\theta} + r\ddot{\theta} = 0, \quad \ddot{r} = -\frac{GM_0}{r^2} + \frac{c_1^2}{r^3},$$

so that

$$G\,M_0\,\dot{\mathbf{e}} = \ddot{\mathbf{b}}_1 \times \mathbf{c}_1 - G\,M_0\,\dot{\theta}\,\mathbf{v}_\theta = -G\,M_0\,\dot{\theta}\,(\mathbf{v}_r \times \mathbf{v}_z + \mathbf{v}_\theta) = \mathbf{0}.$$

Thus $\mathbf{e}$ contains two integrals independent of $\mathbf{c}_1$ (not three because $\mathbf{e}\cdot\mathbf{c}_1 = 0$). We define the **true anomaly** v as the angle between $\mathbf{e}$ and $\mathbf{v}_r$ on the orbital plane, that is

$$e\cos v = \mathbf{e} \cdot \mathbf{v}_r = \frac{r^3\,\dot{\theta}^2}{G\,M_0} - 1 = \frac{c_1^2}{G\,M_0\,r} - 1$$

where $r^2\dot{\theta} = c_1$ is the (scalar) angular momentum of $\mathbf{b}_1$ and is constant. From this we find the familiar formula of a conic section

$$r = \frac{c_1^2/G\,M_0}{1 + e\cos v}$$

and the interpretation of the two additional **two-body integrals** as **eccentricity** $e = |\mathbf{e}|$ and **argument of pericenter** ω, that is the angle of $\mathbf{e}$ with a fixed direction in the orbital plane, in such a way that $\theta = v + \omega$. The eccentricity e is an integral depending upon angular momentum and energy. The energy integral of the two-body problem in $(\mathbf{b}_0, \mathbf{b}_1)$ coordinates is

$$E(\mathbf{B}, \dot{\mathbf{B}}) = M_0\,E_0 + M_1\,E_1, \quad E_0 = \frac{1}{2}\,|\dot{\mathbf{b}}_0|^2, \quad E_1 = \frac{1}{2}\,|\dot{\mathbf{b}}_1|^2 - \frac{G\,M_0}{|\mathbf{b}_1|}$$

and the eccentricity squared, computed from eq. (4.12), is

$$e^2 = \mathbf{e} \cdot \mathbf{e} = \frac{r^4\,\dot{\theta}^2\,\dot{r}^2 + \left(r^3\,\dot{\theta}^2 - G\,M_0\right)^2}{G^2\,M_0^2} = 1 + \frac{2\,E_1\,c_1^2}{G^2\,M_0^2}.$$

If the energy of the relative motion E_1 is negative, then $e < 1$ and the trajectory of $\mathbf{b}_1$ is an ellipse with semimajor axis

$$a = \frac{q + Q}{2} = \frac{1}{2}\left[\frac{c_1^2/G\,M_0}{1 + e} + \frac{c_1^2/G\,M_0}{1 - e}\right] = \frac{G\,M_0}{-2\,E_1},$$

where q, Q are the pericenter and apocenter distances, and the scalar angular momentum of the relative motion is $c_1 = \sqrt{G\,M_0\,a\,(1 - e^2)}$. Formulae to express explicitly the solutions of the two-body problem are available in Appendix A.

4.3 Barycentric and heliocentric coordinates

The set of positions of the $N + 1$ bodies can be represented in different coordinates; we are interested in the linear coordinate changes of the form

$$\mathbf{b}_j = \sum_{i=0}^{N} a_{ji}\,\mathbf{r}_i, \qquad A = (a_{ji}),\; i, j = 0, N \tag{4.13}$$

where the matrix A is a function of the masses only. The purpose is to exploit the integrals of the center of mass to reduce the number of equations, generalizing the results of the two-body case. A natural choice is to use the center of mass as $\mathbf{b}_0$, thus by (4.5) the first row of the matrix A is

$$a_{0i} = \frac{m_i}{M_0}, \qquad i = 0, N. \tag{4.14}$$

The choice of the other $\mathbf{b}_i,\, i = 1, N$, is not as simple as in the two-body case. Different choices have different advantages, and can be used for different purposes. We shall review in this and in the next section the most common coordinate systems used for the $(N + 1)$-body problem.

Barycentric coordinates

The **barycentric coordinate** system uses the fact that a reference system with a constant velocity translation with respect to an inertial system is also inertial. Thus a reference system with $\mathbf{b}_0 = \mathbf{0}$ as origin and barycentric positions $\mathbf{b}_i = \mathbf{r}_i - \mathbf{b}_0$ for $i = 1, N$ is inertial, and the equation of motion is the same as eq. (4.1). The change to barycentric is not just a change of coordinates, but also a reduction of the dimension of the problem: we write three differential equations less. The barycentric coordinates of body 0 (e.g., the Sun) are not dynamical variables, but are deduced from the coordinates of the other bodies and $\mathbf{b}_0$, by eq. (4.5):

$$\mathbf{s} = \mathbf{s}(\mathbf{B}) = \mathbf{r}_0 - \mathbf{b}_0 = -\sum_{i=1}^{N} \frac{m_i}{m_0}\,\mathbf{b}_i, \tag{4.15}$$

where the first term is assumed to be zero. The equation of motion is

$$m_j\,\ddot{\mathbf{b}}_j = \sum_{i\neq j, i=1}^{N} \frac{G\,m_i\,m_j}{|\mathbf{b}_i - \mathbf{b}_j|^3}\,(\mathbf{b}_i - \mathbf{b}_j) + \frac{G\,m_0\,m_j}{|\mathbf{b}_j - \mathbf{s}|^3}\,(\mathbf{s} - \mathbf{b}_j) \qquad j = 1, \ldots, N \tag{4.16}$$

and can be written in conservative form

$$m_j\,\ddot{\mathbf{b}}_j = -\frac{\partial \mathcal{V}(\mathbf{s}, \mathbf{b}_1, \mathbf{b}_2, \ldots, \mathbf{b}_n)}{\partial \mathbf{b}_j}, \qquad j = 1, N,$$

with the potential energy $\mathcal{V}(\mathbf{B}) = V(\mathbf{R}(\mathbf{B}))$, where the partial derivatives of $\mathcal{V}$ have to be computed before substituting $\mathbf{s} - \mathbf{s}(\mathbf{B})$. The integrals of energy and angular momentum have a less simple expression, including the contributions from $\dot{\mathbf{s}}$.

Barycentric coordinates are efficient to be used for numerical integrations: only the $3N$ equations (4.16) have to be integrated, and the only additional computation to be performed at each step is $\mathbf{s}$ according to (4.15). The computed orbit does not need to be used in barycentric coordinates: to change the output back to heliocentric coordinates is the normal procedure.

Barycentric coordinates need to be used when the inertial velocities are directly observable: this is the case when the radial velocity of some star is measured (either by radio astronomy, for pulsars, or by spectroscopy, for normal stars). This is used to detect the small velocity of the star as a result of the presence of a small companion, such as a planet, see Section 6.5. The measured radial velocity is the difference between $\dot{\mathbf{s}}$ of the star and $\dot{\mathbf{b}}_3$ of the Earth; to use heliocentric coordinates for the Earth would result in a serious mistake.[3] The barycentric coordinates also play a role in the general relativistic corrections to the Newton equation, see Section 6.6.

On the other hand, barycentric coordinates are seldom used in analytical developments and in theoretical discussions, because of the lack of symmetry of the equation and of the less simple expressions for the classical integrals.

Heliocentric coordinates

A possible choice to represent the motion of planets and asteroids is the use of **heliocentric coordinates**. These follow the same idea used in the two-body case, eq. (4.9), namely use the motion of bodies $j = 1, N$ relative to the one with index 0, usually the Sun. Since $m_0 \gg m_j$, $j = 1, N$, the Sun moves little, but this motion cannot be neglected in the differential equations. The positions are thus represented by the vectors $\mathbf{b}_i = \mathbf{r}_i - \mathbf{r}_0$ and the equation of motion can be simply derived from eq. (4.1), taking into account the non-inertial frame, that is adding the apparent force exactly opposite to the acceleration of the Sun times the mass of the body:

$$m_j \ddot{\mathbf{b}}_j = \sum_{i \neq j, i=0}^{N} \frac{G m_i m_j}{|\mathbf{r}_i - \mathbf{r}_j|^3} (\mathbf{r}_i - \mathbf{r}_j) - m_j \ddot{\mathbf{r}}_0.$$

[3] It would lead to a pretended discovery of a companion with a period of one year!

The equation can be written in terms of the heliocentric vectors, since they contain only the differences $\mathbf{b}_i - \mathbf{b}_j = \mathbf{r}_i - \mathbf{r}_j$ and $\mathbf{b}_i = \mathbf{r}_i - \mathbf{r}_0$

$$m_j \ddot{\mathbf{b}}_j = -\frac{G m_0 m_j}{|\mathbf{b}_j|^3} \mathbf{b}_j + \sum_{i \neq j, i=1}^{N} \frac{G m_i m_j}{|\mathbf{b}_i - \mathbf{b}_j|^3} (\mathbf{b}_i - \mathbf{b}_j) - m_j \ddot{\mathbf{r}}_0.$$

The value of the acceleration of the Sun, resulting from the gravitational attraction of all the planets, is obtained from eq. (4.1) for $j = 0$; by substituting into the equation and removing the common factor m_j

$$\ddot{\mathbf{b}}_j = -\frac{G m_0}{|\mathbf{b}_j|^3} \mathbf{b}_j + \sum_{i \neq j, i=1}^{N} \frac{G m_i}{|\mathbf{b}_i - \mathbf{b}_j|^3} (\mathbf{b}_i - \mathbf{b}_j) - \sum_{i=1}^{N} \frac{G m_i}{|\mathbf{b}_i|^3} \mathbf{b}_i. \qquad (4.17)$$

The equations above allow us to compute a solution for each heliocentric vector $\mathbf{b}_i$, $i = 1, n$, without the need to compute the position of the Sun in an inertial frame. Taking into account that in the acceleration of the Sun there is also a component due to the same planet

$$\ddot{\mathbf{b}}_j = -\frac{G (m_0 + m_j)}{|\mathbf{b}_j|^3} \mathbf{b}_j + \sum_{i \neq j, i=1}^{N} \frac{G m_i}{|\mathbf{b}_i - \mathbf{b}_j|^3} (\mathbf{b}_i - \mathbf{b}_j) - \sum_{i \neq j, i=1}^{N} \frac{G m_i}{|\mathbf{b}_i|^3} \mathbf{b}_i. \qquad (4.18)$$

In this way the equation of motion is split into the two-body part, with the planet orbiting around a fixed center with mass $m_0 + m_j$ (as in the reduction of a two-body problem with the Sun and the planet j only), the **direct perturbations** by the attraction of the other planets, and the **indirect perturbations**, resulting from the other planets accelerating the Sun.

The heliocentric coordinates are a natural choice for Solar System orbits. The relative positions $\mathbf{r}_j - \mathbf{r}_k = \mathbf{b}_j - \mathbf{b}_k$ generate the only quantities observable inside our Solar System, e.g., the direction angles in optical astrometry and the range and range-rate in radar observations. The center of mass $\mathbf{b}_0$ and the barycentric position $\mathbf{s}$ of the Sun are derived quantities containing the mass ratios m_j/m_0. Thus, a catalog of asteroid orbital elements, computed from Cartesian coordinates in a barycentric system, would contain values dependent upon the planetary masses: every time the masses are corrected, the catalog should be revised. If the orbital elements are computed from heliocentric coordinates, there is no need for revision when the estimated values of the planetary masses change, with the exception of the asteroids having close approaches to a planet whose mass has been revised.

4.4 Jacobian coordinates

The **Jacobian coordinates** are obtained by selecting, among the linear coordinate changes of the form (4.13), the ones with the center of mass as first vector, thus fulfilling eq. (4.14), with the simplest equation of motion. This requires a matrix A, thus a set of *Jacobian vectors* $\mathbf{b}_0, \mathbf{b}_1, \mathbf{b}_2, \ldots, \mathbf{b}_N$, and a set of reduced masses $M_0, M_1, M_2, \ldots, M_N$ with the properties

[1] the first vector $\mathbf{b}_0$ is the center of mass, M_0 is the total mass;

[2] the Lagrange equation in the $\mathbf{R}$ coordinates is transformed into the Lagrange equation in the Jacobian coordinates of the same form:

$$m_i\, \ddot{\mathbf{r}}_i = -\frac{\partial V}{\partial \mathbf{r}_i} \iff M_i\, \ddot{\mathbf{b}}_i = -\frac{\partial \mathcal{V}}{\partial \mathbf{b}_i}$$

where $\mathcal{V}(\mathbf{B}) = V(\mathbf{R})$ is the potential energy in the Jacobian coordinates.

The conditions on A resulting from [1] are given in (4.14), the ones resulting from [2] require that the kinetic energy remains in diagonal form:

$$2T = \sum_{i=0}^{N} m_i\, |\dot{\mathbf{r}}_i|^2 = \sum_{j=0}^{N} M_j\, |\dot{\mathbf{b}}_j|^2;$$

then the Jacobian momentum is $M_j\, \dot{\mathbf{b}}_j$ and the equation is in the simple form required by [2]. By substituting eq. (4.13) in the above formula

$$2T = \sum_{i,k=0}^{N} \dot{\mathbf{r}}_i \cdot \dot{\mathbf{r}}_k \sum_{j=0}^{N} a_{ji} M_j a_{jk} = \sum_{i,k=0}^{N} \dot{\mathbf{r}}_i \cdot \dot{\mathbf{r}}_k\, m_i \delta_{ik}$$

where $\delta_{ik} = 1$ for $i = k$, and $\delta_{ik} = 0$ for $i \neq k$. Thus the equations for A are

$$m_i \delta_{ik} = \sum_{j=0}^{N} a_{ji} M_j a_{jk} \qquad i, k = 0, N. \tag{4.19}$$

In matrix form, if m, M are the diagonal matrices with the masses and the reduced masses, respectively, as coefficients

$$m = \mathrm{diag}[m_0, m_1, \ldots, m_N], \quad M = \mathrm{diag}[M_0, M_1, \ldots, M_N]$$

then eq. (4.19) can be written with A^T, the transposed matrix

$$m = A^T M A. \tag{4.20}$$

The Jacobian coordinates have another property, which is a consequence of [2]: the total angular momentum (4.6) has also a simple expression

$$\mathbf{c} = \sum_{i=0}^{N} \mathbf{r}_i \times m_i \, \dot{\mathbf{r}}_i = \sum_{j=0}^{N} \mathbf{b}_j \times M_j \, \dot{\mathbf{b}}_j,$$

i.e., the total angular momentum of the $(N+1)$-body system is the angular momentum of the free motion of the center of mass $\mathbf{b}_0 \times M_0 \, \dot{\mathbf{b}}_0$ plus the sum of the angular momentum of the two-body subsystems $\mathbf{b}_j \times M_j \, \dot{\mathbf{b}}_j$, $j = 1, \ldots, N$.

Equation (4.20) implies $\det(m) = \det(M) \det(A)^2$, where the determinants of m, M are the product of all masses and the product of all reduced masses, respectively. Thus [1] and [2] allow rescaling of the masses; a change of orientation is also possible. To avoid this, two additional properties have to be added to the definition of Jacobian coordinates:

[3] the product of the masses is equal to the product of the reduced masses

$$\prod_{i=0}^{N} m_i = \prod_{j=0}^{N} M_j; \tag{4.21}$$

[4] the linear transformation defined by A preserves orientation: $\det(A) > 0$.

Properties [2], [3], and [4] imply $\det(A) = +1$.

Existence and conditional uniqueness of Jacobian coordinates

If the transformation (4.13) fulfills [1], [2], [3], and [4], it defines a system of Jacobian coordinates. Matrices A with all these properties exist but they are not unique for a given N and for the given set of masses m_i. To obtain a unique selection we proceed as follows.

Let $\mathbf{b}_0^N, \ldots, \mathbf{b}_N^N$ be a set of Jacobian vectors satisfying [1]–[4], with reduced masses $M_0^N, \ldots, M_N^N$. Let $m_{N+1}, \mathbf{r}_{N+1}$ be the mass and position of an additional body. Then there are unique Jacobian coordinates, satisfying [1]–[4], with N unchanged Jacobian vectors and N unchanged reduced masses

$$\mathbf{b}_j^{N+1} = \mathbf{b}_j^N, \quad M_j^{N+1} = M_j^N \quad j = 1, N.$$

The new reduced masses are

$$M_{N+1} = \frac{m_{N+1} M_0^N}{M_0^{N+1}}, \quad M_0^{N+1} = M_0^N + m_{N+1} \tag{4.22}$$

and the new Jacobian vectors are

$$\mathbf{b}_{N+1} = \mathbf{r}_{N+1} - \mathbf{b}_0^N, \qquad \mathbf{b}_0^{N+1} = \frac{1}{M_0^{N+1}} \sum_{j=0}^{N+1} m_j \, \mathbf{r}_j. \qquad (4.23)$$

This can be shown by comparing eqs. (4.19) and (4.21) for $N+1$ and $N+2$ bodies (Milani and Nobili 1983).

The solutions (4.23) and (4.22) can be described as follows. A Jacobian coordinate system is a way to decompose an $(N+1)$-body system into free motion of the center of mass and N two-body subsystems. To add a new body, the new Jacobian vector is the position of the new body $\mathbf{r}_{N+1}$ relative to the center of mass $\mathbf{b}_0^N$ of the previous system, and the new reduced mass is the harmonic mean of the new mass m_{N+1} and of the previous total mass M_0^N. This generalizes the reduction of the two-body problem (4.9), (4.10).

As for uniqueness, the reduction of the two-body problem to the central force problem gives the Jacobian coordinates for $N+1 = 2$ bodies. However, if the list of bodies was $\{\mathbf{r}_1, \mathbf{r}_0\}$ the Jacobian vector would be $\mathbf{b}_1 = \mathbf{r}_0 - \mathbf{r}_1$. For $N+1 = 3$ the standard solution is to first couple (m_0, m_1), that is

$$\mathbf{b}_1 = \mathbf{r}_1 - \mathbf{r}_0, \qquad M_1 = \frac{m_0 \, m_1}{m_0 + m_1}$$

then use the vector $\mathbf{b}_2$ relative to the center of mass of (m_0, m_1), that is

$$\mathbf{b}_2 = \mathbf{r}_2 - \frac{m_0}{m_0 + m_1} \mathbf{r}_0 - \frac{m_1}{m_0 + m_1} \mathbf{r}_1, \qquad M_2 = \frac{m_2 \, (m_0 + m_1)}{m_0 + m_1 + m_2}.$$

This solution is not unique: it is possible to form first the binary (m_2, m_0), that is $\mathbf{b}_1 = \mathbf{r}_0 - \mathbf{r}_2$ and then join $\mathbf{r}_1$ to the center of mass of (m_2, m_0). A third solution corresponds to the sequence of couplings $((m_1, m_2), m_0)$; there are three more solutions violating [4].

The choice of a solution depends upon the sequence of coupling operations, which can be represented by a symbol like $((m_0, m_1), m_2)$ for the standard three-body solution. At a purely formal level, each of the $(N+1)!$ ways to order the $N+1$ bodies results, by applying recursively the procedure above, in a set of Jacobian coordinates. When the relative size of the perturbation is computed, as in the next section, the solutions are found to be by no means equivalent. As an example, if m_0 corresponds to the Sun, m_1 to the Earth, m_2 to the Moon, the best Jacobian system is the one with $((m_1, m_2), m_0)$, that is the center of mass of the Earth–Moon system is orbiting around the Sun, while the Moon is orbiting around the Earth–Moon center of mass.

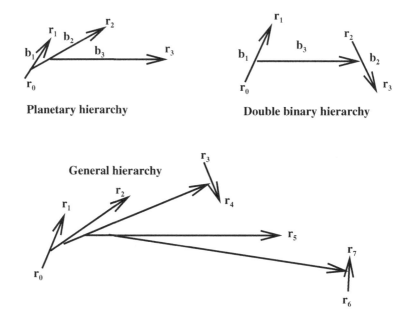

Fig. 4.1. Three examples of hierarchies and of the corresponding Jacobian vectors. The planetary hierarchy and the double binary hierarchy are described in the text. The hierarchy in the lower part of the figure could be used to describe a planetary system around the star r_0, with planets r_2, r_3, r_5 and r_6; planet r_3 has one satellite, planet r_6 has two satellites.

Planetary and binary type hierarchies

The non-uniqueness of the Jacobian coordinates becomes more significant for $N + 1 = 4$ bodies. Let us assume some Jacobian coordinates have been selected for the first three bodies, e.g., according to the coupling order $((m_0, m_1), m_2)$. When a body $m_4, \mathbf{r}_4$ is added, there are two options. One is the recursive procedure of the previous subsection, that is $\mathbf{b}_3 = \mathbf{r}_3 - \mathbf{b}_0^3$. The other is to set $\mathbf{b}_2 = \mathbf{r}_3 - \mathbf{r}_2$ and to replace $\mathbf{r}_2$ with the center of mass of the binary (m_2, m_3), that is to use as $\mathbf{b}_3$ the vector joining the two centers of mass of the binary subsystems (m_0, m_1) and (m_2, m_3)

$$\mathbf{b}_1 = \mathbf{r}_1 - \mathbf{r}_0, \quad \mathbf{b}_2 = \mathbf{r}_3 - \mathbf{r}_2, \quad \mathbf{b}_3 = [(1 - \mu_2)\,\mathbf{r}_2 + \mu_2\,\mathbf{r}_3] - [(1 - \mu_1)\,\mathbf{r}_0 + \mu_1\,\mathbf{r}_1]$$

where $\mu_2 = m_3/(m_2 + m_3)$. Then the reduced mass M_2 is the harmonic mean of the masses m_2 and m_3, M_3 is the harmonic mean of the masses $(m_0 + m_1)$ and $(m_2 + m_3)$:

$$M_1 = \frac{m_0\,m_1}{m_0 + m_1}, \quad M_2 = \frac{m_2\,m_3}{m_2 + m_3}, \quad M_3 = \frac{(m_0 + m_1)\,(m_2 + m_3)}{m_0 + m_1 + m_2 + m_3}.$$

The first option is called a **planetary hierarchy** and is represented by the coupling symbol $(((m_0, m_1), m_2), m_3)$; the second is a **double binary hierarchy** and is represented by $((m_0, m_1), (m_2, m_3))$. Formally, both choices are equivalent, in that both provide a Jacobian coordinate system satisfying [1], [2], [3], and [4]. The planetary hierarchy suggests that all the "planets" of masses m_1, m_2, m_3 orbit around the "star" of much larger mass m_0, at increasing distances $|\mathbf{r}_1 - \mathbf{r}_0|$, $|\mathbf{r}_2 - \mathbf{r}_0|$, and $|\mathbf{r}_3 - \mathbf{r}_0|$. The double binary hierarchy suggests that the "interior planet" m_1 orbits around the "star" m_0 at a smaller distance than the "exterior planet" m_2, the latter having a "satellite" m_3. To give rigorous meaning to this suggestion, we need to show that dynamical configurations, with different mass and distance ratios, are better represented in either one or the other hierarchy.

In general, given two subsystems with N' and N'' bodies, each with Jacobian coordinates, centers of mass $\mathbf{b}_0'$, $\mathbf{b}_0''$ and total masses M_0', M_0'', respectively, there is a Jacobian system for the joint system of $N'+N''$ masses with a new Jacobian vector joining the two centers of mass, and a new reduced mass equal to the harmonic mean of the two total masses

$$\mathbf{b}_{N'+N''} = \mathbf{b}_0'' - \mathbf{b}_0', \qquad M_{N'+N''} = \frac{M_0' \, M_0''}{M_0' + M_0''};$$

$\mathbf{b}_0$ is the center of mass of all bodies, and the other $(N' - 1) + (N'' - 1)$ vectors coincide with the previously defined ones. This is the only way to combine the two subsystems, preserving $N' + N'' - 2$ Jacobian vectors (not including the centers of mass of the subsystems). In this way we can build a Jacobian system for an arbitrary coupling symbol. For example, for the hierarchy shown in the lower portion of Figure 4.1 the coupling symbol is $(((((m_0, m_1), m_2), (m_3, m_4)), m_5), (m_6, m_7))$.

4.5 Small parameter perturbation

We would like to assess how relevant are the perturbations resulting from each additional body included in the dynamical model of a planetary system. The Jacobian coordinates provide a direct way to estimate the relative size of the perturbations, with the Roy–Walker parameters.

The perturbing function

We shall discuss first a three-body case, in Jacobian coordinates, with standard hierarchy $((m_0, m_1), m_2)$. The Lagrange function is

$$\mathcal{L}(\mathbf{B}, \dot{\mathbf{B}}) = \sum_{i=1}^{3} \frac{1}{2} M_i \, |\dot{\mathbf{b}}_i|^2 + \frac{G m_0 m_1}{|\mathbf{b}_1|} + \frac{G m_1 m_2}{|\mathbf{r}_2 - \mathbf{r}_1|} + \frac{G m_0 m_2}{|\mathbf{r}_2 - \mathbf{r}_0|},$$

with

$$r_2 - r_1 = b_2 - \frac{m_0}{m_0 + m_1} b_1, \quad r_2 - r_0 = b_2 + \frac{m_1}{m_0 + m_1} b_1.$$

Our goal is to express the Lagrange function as a sum of three "unperturbed" Lagrange functions and a **perturbing function**. Since the kinetic energy, in Jacobian coordinates, is already decomposed as needed, only the potential needs to be transformed. We use the sum of masses

$$N_j = \sum_{i=0}^{j} m_i, \quad N_1 M_1 = m_0 m_1, \quad N_2 M_2 = M_0 M_2 = m_2 (m_0 + m_1)$$

to form the three unperturbed Lagrange functions

$$L_0(b_0, \dot{b}_0) = \frac{1}{2} |\dot{b}_0|^2, \quad L_i(b_i, \dot{b}_i) = \frac{1}{2} |\dot{b}_i|^2 + \frac{G N_i}{|b_i|}, \quad i = 1, 2$$

corresponding to the free motion of the center of mass and to the two-body motion of b_i around an attracting center of mass N_i, for $i = 1, 2$. The perturbing function is simply what is left:

$$\mathcal{L}(B, \dot{B}) = M_0 L_0(b_0, \dot{b}_0) + M_1 L_1(b_1, \dot{b}_1) + M_2 L_2(b_2, \dot{b}_2) + R_{12}(b_1, b_2),$$

$$R_{12}(b_1, b_2) = m_2 G N_1 \left\{ \frac{\mu_1}{|r_2 - r_1|} + \frac{1 - \mu_1}{|r_2 - r_0|} - \frac{1}{|b_2|} \right\} \tag{4.24}$$

where $\mu_1 = m_1/(m_0 + m_1)$ is the mass ratio of the b_1 binary. R_{12} has three terms, corresponding to the potential (the opposite of the gravitational potential energy) at the position r_2 of the mass m_0, of the mass m_1, and to the opposite of the potential of a mass $m_0 + m_1$ placed in the center of mass of (m_0, m_1). This is because the hypothetical potential of a mass $m_0 + m_1$ in the center of mass of the first binary has been used to form the unperturbed Lagrange function of the binary $(m_0 + m_1, m_2)$. Thus the perturbing function is the gravitational potential of a mass distribution consisting of three masses, one of which is negative, with total mass zero.

Expansions in spherical harmonics

We expand the perturbing function R_{12} in spherical harmonics. Three masses m_0, m_1, and $-(m_0 + m_1)$, located at r_0, r_1 and at the center of mass $b_0^1 = \mu_1 r_1 + (1 - \mu_1) r_0$ form the mass distribution generating the perturbing function: since these three masses are aligned, the potential is axially symmetric and can be expressed by zonal spherical harmonics only (see Section 13.2). We are only interested in the first few terms in the harmonics expansion, and shall compute the harmonic coefficients directly.

Let the angle between $\mathbf{b}_1$ and $\mathbf{b}_2$ be ψ, and $\theta = \pi/2 - \psi$ the latitude (with respect to an equatorial plane through $\mathbf{b}_0^1$ and perpendicular to the $\mathbf{b}_1$ axis). We shall now compute the three distances appearing in the denominators in $\mathcal{R}_{12}(\mathbf{b}_1, \mathbf{b}_2)$ as functions of the lengths $b_1 = |\mathbf{b}_1|$ and $b_2 = |\mathbf{b}_2|$ and of the angle θ. The distance between m_2 and of m_1 is

$$|\mathbf{r}_2 - \mathbf{r}_1|^2 = |\mathbf{b}_2 - (1 - \mu_1)\mathbf{b}_1|^2 = b_2^2 + (1 - \mu_1)^2 b_1^2 - 2(1 - \mu_1) b_1 b_2 \sin\theta.$$

Then we shall express the power -1 of the distance by means of the unperturbed distance b_2 and the ratio $\alpha_1 = b_1/b_2$, with $\alpha_1 < 1$ (if not, the hierarchy should be changed):

$$\frac{1}{|\mathbf{r}_2 - \mathbf{r}_1|} = \frac{1}{b_2} \left\{ 1 - 2(1 - \mu_1)\alpha_1 \sin\theta + (1 - \mu_1)^2 \alpha_1^2 \right\}^{-1/2}$$

and by using the Taylor formula $(1 + x)^{-1/2} = 1 - 1/2\,x + 3/8\,x^2 + O(x^3)$ we obtain the expansion, with respect to the small parameter α_1,

$$\frac{1}{|\mathbf{r}_2 - \mathbf{r}_1|} = \frac{1}{b_2} \left\{ 1 + (1 - \mu_1)\alpha_1 P_1(\sin\theta) + (1 - \mu_1)^2 \alpha_1^2 P_2(\sin\theta) + O(\alpha_1^3) \right\},$$

where we have used the first and the second Legendre polynomials (for a discussion of the Legendre functions, see Section 13.2)

$$P_1(\sin\theta) = \sin\theta, \qquad P_2(\sin\theta) = \frac{3}{2} \sin^2\theta - \frac{1}{2}.$$

The formula for the distance between the position of m_2 and of m_0 is

$$|\mathbf{r}_2 - \mathbf{r}_0|^2 = |\mathbf{b}_2 + \mu_1 \mathbf{b}_1|^2 = b_2^2 + \mu_1^2 b_1^2 + 2\mu_1 b_1 b_2 \sin\theta,$$

so that

$$\frac{1}{|\mathbf{r}_2 - \mathbf{r}_0|} = \frac{1}{b_2} \left\{ 1 - \mu_1 \alpha_1 P_1(\sin\theta) + \mu_1^2 \alpha_1^2 P_2(\sin\theta) + O(\alpha_1^3) \right\}.$$

The perturbing function R_{12} is a linear combination of the previous expressions minus $1/b_2$, in which both the monopole term $1/b_2$ and the dipole term containing P_1 cancel out, as it has to be expected when the expansion in spherical harmonics is centered at the center of mass

$$\frac{1}{m_2} R_{12}(\mathbf{b}_1, \mathbf{b}_2) = \frac{G\,N_1}{b_2} \mu_1 (1 - \mu_1) \left[\alpha_1^2 P_2(\sin\theta) + O(\alpha_1^3) \right] \qquad (4.25)$$

where the remainder has been indicated taking into account that the $O(\alpha_1^3)$ term also contains the coefficient $\mu_1 (1 - \mu_1)$; this can be checked by computing of the degree 3 zonal harmonics.

Perturbations in Jacobian coordinates

The effect of the perturbing function on each of the binaries can be measured
as a change in the related two-body energy. The integral of energy

$$E(\mathbf{B}, \dot{\mathbf{B}}) = M_0\, E_0 + M_1\, E_1 + M_2\, E_2 - R_{12}$$

contains a linear combination of the two-body energies (per unit mass) of
the subsystems

$$E_0 = \frac{1}{2}\,|\dot{\mathbf{b}}_0|^2, \quad E_i = T_i + V_i = \frac{1}{2}\,|\dot{\mathbf{b}}_i|^2 - \frac{G\,N_i}{|\mathbf{b}_i|}, \quad i = 1, 2. \tag{4.26}$$

The perturbing potential R_{12} has the relative effect $R_{12}/(M_2\,E_2)$ on the $\mathbf{b}_2$
subsystem: this ratio can be approximated, for order of magnitude com-
putations, assuming $V_2 = -G\,N_2/b_2 \simeq 2\,E_2$, which is exact for a circular
orbit

$$\frac{R_{12}}{M_2\,E_2} \simeq \frac{2\,R_{12}}{M_2\,V_2} = -2\mu_1\,(1-\mu_1)\left[\alpha_1^2\,P_2(\sin\theta) + O(\alpha_1^3)\right]$$

and for $\alpha_1 \ll 1$ this leads to the approximate upper bound

$$\left|\frac{R_{12}}{M_2\,E_2}\right| \le 2\,\epsilon_{12}, \quad \epsilon_{12} = \mu_1\,(1-\mu_1)\,\alpha_1^2.$$

The same argument applied to the $\mathbf{b}_1$ subsystem is $V_1 = -G\,N_1/b_1 \simeq 2\,E_1$
and gives

$$\left|\frac{R_{12}}{M_1\,E_1}\right| \le 2\,\frac{M_2\,N_2\,\mu_1\,(1-\mu_1)}{M_1\,N_1}\,\alpha_1^3 = 2\,\epsilon_{21}, \quad \epsilon_{21} = \frac{\mu_2}{1-\mu_2}\,\alpha_1^3.$$

Thus the size of the perturbing function, relative to the size of the unperturbed potential energy, is estimated by the **Roy–Walker parameters**
$\epsilon_{12}, \epsilon_{21}$. Note they both contain mass ratios and the ratio α_1 of the Jacobian vector lengths: the exterior perturbation decreases like the cube of
α_1, the interior one like the square. The effect of such a perturbation on
the semimajor axes a_j of the two orbits of $\mathbf{b}_j$, $j = 1, 2$, assuming ϵ_{ij} small,
could be estimated by the simple rule $\Delta a_j/a_j = -\Delta E_j/E_j$.

The four-body case

A **hierarchy** can be understood just as a combinatorial structure, represented by either a symbol like $((m_0, m_1), (m_2, m_3))$ or a graph like the one
of Figure 4.1, top right. Not all graphs are suitable to represent a hierarchy:
each vector $\mathbf{b}_j$ must have one and only one "superior" vector $\mathbf{b}_{s(j)}$, with the
exception of the "top" vector with no superior: in the example above of the

double binary hierarchy the vector $\mathbf{b}_3$ is at the top and is the superior for both $\mathbf{b}_2$ and $\mathbf{b}_1$. Then for each Jacobian vector (not at the top) $\mathbf{b}_j$ we can define a length ratio $\alpha_j = b_j/b_{s(j)}$. A hierarchy becomes more than a combinatorial device if the superior vectors are also longer, that is, if the length ratios α_j are small. Then we can estimate the relative size of the perturbing functions, describing the interaction of each two binaries, by using powers of the α_j and mass ratios to form generalized Roy–Walker parameters.

For a double binary hierarchy, the potential energy is the sum of three two-body terms and a perturbing function with three terms

$$V = -\sum_{0 \le i < j \le 4} \frac{G\,m_i\,m_j}{|\mathbf{r}_i - \mathbf{r}_j|} = M_1\,V_1 + M_2\,V_2 + M_3\,V_3 - R_{13} - R_{23} - R_{12}$$

where the perturbing terms are (Milani and Nobili 1983)

$$R_{13} = N_2\,\mathcal{R}_1(\mathbf{b}_{23}), \quad \mathcal{R}_1(\mathbf{x}) = G\,N_1 \left[\frac{\mu_1}{|\mathbf{x} - \mathbf{r}_1|} + \frac{1 - \mu_1}{|\mathbf{x} - \mathbf{r}_0|} - \frac{1}{|\mathbf{x} - \mathbf{b}_{01}|} \right]$$

$$R_{23} = N_1\,\mathcal{R}_2(\mathbf{b}_{01}), \quad \mathcal{R}_2(\mathbf{x}) = G\,N_2 \left[\frac{\mu_2}{|\mathbf{x} - \mathbf{r}_3|} + \frac{1 - \mu_3}{|\mathbf{x} - \mathbf{r}_2|} - \frac{1}{|\mathbf{x} - \mathbf{b}_{23}|} \right]$$

for the perturbations between each of the two binaries with state vector $\mathbf{b}_1, \mathbf{b}_2$ and the "handle" with state vector $\mathbf{b}_3$, and

$$\begin{aligned} R_{12} &= N_1 \left\{ \mu_1\,\mathcal{R}_2(\mathbf{r}_1) + (1 - \mu_1)\,\mathcal{R}_2(\mathbf{r}_0) - \mathcal{R}_2(\mathbf{b}_{01}) \right\} \\ &= N_2 \left\{ \mu_2\,\mathcal{R}_1(\mathbf{r}_3) + (1 - \mu_2)\,\mathcal{R}_1(\mathbf{r}_2) - \mathcal{R}_1(\mathbf{b}_{23}) \right\} \end{aligned}$$

for the perturbations between binaries, with $\mathbf{b}_{ik} = (m_i\,\mathbf{b}_i + m_k\,\mathbf{b}_k)/(m_i + m_k)$ the centers of mass of the binaries for $(i, k) = (0, 1), (2, 3)$. By using essentially the same formalism as in the three-body case, it is possible to estimate the ratio of the perturbing functions to the two-body potential energies: e.g., for the perturbations between $\mathbf{b}_3$ and $\mathbf{b}_1$

$$\begin{aligned} \left| \frac{R_{13}}{M_1\,V_1} \right| &= \frac{G\,M_1\,N_1}{b_1} \left[\mu_1\,(1 - \mu_1)\,\alpha_1^2\,P_2(\sin\theta) + O(\alpha_1^3) \right] \frac{b_1}{G\,N_1\,N_2} \\ &= \frac{N_2\,\mu_1\,(1 - \mu_1)}{N_1}\,\alpha_1^3 + O(\alpha_1^4) = \epsilon_{31} + O(\alpha_1^4), \end{aligned}$$

$$\begin{aligned} \left| \frac{R_{31}}{M_3\,V_3} \right| &= \frac{G\,N_1\,N_2}{b_3} \left[\mu_1\,(1 - \mu_1)\,\alpha_1^2\,P_2(\sin\theta) + O(\alpha_1^3) \right] \frac{b_3}{G\,M_3\,N_3} \\ &= \mu_1\,(1 - \mu_1)\,\alpha_1^2 + O(\alpha_1^3) = \epsilon_{13} + O(\alpha_1^3) \end{aligned}$$

and the Roy–Walker parameters have the same expression as in the three-body case; the same occurs for $\epsilon_{32}, \epsilon_{23}$.

The case of ϵ_{12} estimating the perturbations of the binary $\mathbf{b}_1$ onto $\mathbf{b}_2$, is more complicated. It can be shown (Milani and Nobili 1983) that the lowest order term in the expansion (in powers of α_1, α_2) of the perturbing function R_{12} contains $\mu_1 (1 - \mu_1) \mu_2 (1 - \mu_2) \alpha_1^2 \alpha_2^2$. That is, ϵ_{12} is of the same order in the small parameters as the product $\epsilon_{13} \epsilon_{23}$. Thus the mutual perturbation of the two binaries in a double binary hierarchy is negligible in many practical cases, such as the perturbations of the satellites of Jupiter on the orbits of the inner planets.

Perturbations in heliocentric coordinates

To estimate the size of the perturbations to the two-body orbital elements in heliocentric coordinates we need to take into account separately the indirect perturbations resulting from the non-inertial origin in the Sun.

Let us consider the simplest case $N = 2$ and use the analogs of eq. (4.26) for the two-body energies of $\mathbf{b}_j = \mathbf{r}_j - \mathbf{r}_0$ orbiting around the Sun

$$E_j = \frac{1}{2} |\dot{\mathbf{b}}_j|^2 - \frac{G(m_0 + m_j)}{|\mathbf{b}_j|} = -\frac{G(m_0 + m_j)}{2 a_j},$$

with a_j the semimajor axis of the osculating heliocentric orbit. In the equation of motion (4.17) ($j = 1, 2$) the first term is two-body like, not affecting E_j; the time derivative of the two-body energies is the power of the perturbing forces

$$\dot{E}_j = \dot{E}_j^{dir} + \dot{E}_j^{ind}, \quad \dot{E}_j^{dir} = G m_i \frac{\mathbf{b}_i - \mathbf{b}_j}{|\mathbf{b}_i - \mathbf{b}_j|^3} \cdot \dot{\mathbf{b}}_j, \quad \dot{E}_j^{ind} = -G m_i \frac{\mathbf{b}_i}{|\mathbf{b}_i|^3} \cdot \dot{\mathbf{b}}_j.$$

The indirect part can be estimated with a circular orbit approximation

$$|\mathbf{b}_j| \simeq a_j, \quad |\dot{\mathbf{b}}_j| = n_j a_j, \quad n_j = \sqrt{G m_0 m_j / a_j^3}$$

taking also into account that the main terms of this perturbation have the frequency $n_j - n_i$, resulting in an approximate amplitude of oscillation in E_j with this frequency, due to the indirect part only

$$|\Delta^{ind} E_j| \lesssim \frac{1}{|n_j - n_i|} \frac{G m_i a_j n_j}{a_i^2}.$$

The oscillation amplitude of the heliocentric semimajor axis is estimated by

$$\frac{|\Delta^{ind} a_j|}{a_j} \simeq \frac{|\Delta^{ind} E_j|}{E_j} = 2 \frac{m_i}{(m_0 + m_j)} \frac{n_j}{|n_j - n_i|} \frac{a_j^2}{a_i^2}.$$

For the indirect perturbation by an exterior planet $j = 1, i = 2, n_1 \gg n_2$

and, assuming $m_0 \gg m_1, m_2$,

$$\frac{|\Delta_2^{ind} a_1|}{a_1} \lesssim 2 \frac{m_2}{m_0} \frac{a_1^2}{a_2^2}.$$

For the indirect perturbation of an interior planet $j = 2, i = 1$

$$\frac{|\Delta_1^{ind} a_2|}{a_2} \lesssim 2 \frac{m_1}{m_0} \frac{a_2^2 \, n_2}{a_1^2 \, n_1} \simeq 2 \frac{m_1}{m_0} \frac{\sqrt{a_2}}{\sqrt{a_1}}.$$

For an upper bound to the direct perturbations we use the triangular inequality and the circular orbit approximation $|\mathbf{b}_i - \mathbf{b}_j| \geq |a_j - a_i|$

$$|\Delta^{dir} E_j| \lesssim \frac{1}{|n_j - n_i|} \frac{G \, m_i}{|a_j - a_i|^2} n_j \, a_j$$

and for the semimajor axis amplitude

$$\frac{|\Delta^{dir} a_j|}{a_j} \lesssim \frac{|\Delta^{dir} E_j|}{E_j} = 2 \frac{m_i}{(m_0 + m_j)} \frac{n_j}{|n_j - n_i|} \frac{a_j^2}{|a_i - a_j|^2}.$$

For the direct perturbation by an exterior planet $j = 1, i = 2$

$$\frac{|\Delta_2^{dir} a_1|}{a_1} \lesssim 2 \frac{m_2}{m_0} \frac{a_1^2}{a_2^2},$$

that is the same estimate obtained for the indirect part. For the direct perturbation of an interior planet $j = 2, i = 1$

$$\frac{|\Delta_1^{dir} a_2|}{a_2} \lesssim 2 \frac{m_1}{m_0} \frac{a_2^2 \, n_2}{a_2^2 \, n_1} \simeq 2 \frac{m_1}{m_0} \frac{a_1^{3/2}}{a_2^{3/2}},$$

which is qualitatively different from the estimate for the indirect part: for $a_2/a_1 \to +\infty$ the direct perturbation of the interior planet $\to 0$, the indirect perturbation $\to +\infty$. The indirect perturbation of an interior planet may perturb the semimajor axis to arbitrarily large values, even to an apparent hyperbolic orbit. For growing a_2 the attraction of the other bodies $\to 0$ while the acceleration of the Sun due to $m_1, \mathbf{b}_1$ remains constant, until the indirect perturbation is larger than the attraction from the Sun.

For the perturbations from an exterior planet the estimates for the perturbations on the heliocentric semimajor axis contain the ratio a_1^2/a_2^2, while the corresponding estimate computed in Jacobian coordinates would contain the Roy–Walker parameter ϵ_{21}, proportional to a_1^3/a_2^3. That is, for large a_2/a_1 the heliocentric perturbations are larger, but still $\to 0$ for $a_2/a_1 \to +\infty$.

The conclusion from the discussion in this section and in Section 4.3 is that it may be necessary to use heliocentric coordinates to express planetary orbits with elements independent of the values of the mass ratios m_j/m_0,

but these elements could be sharply changing with time as a result of perturbations from interior planets. Jacobian coordinates could provide orbital
elements more stable in time, but dependent upon the masses. Barycentric
coordinates have an intermediate behavior, with perturbations larger than
the Jacobian ones but without the divergence for $a_2/a_1 \to +\infty$ of the heliocentric coordinates. Thus there is no choice optimal for all purposes: we
need to use coordinate changes to exploit the best properties of each system.

4.6 Solar System dynamical models

The equation of motion for an $(N+1)$-body system needs to be used as a
dynamic model for the orbit determination of objects belonging to a planetary system, especially our own Solar System. Which terms have to be
included in the equation of motion for a given orbit determination problem?
This depends upon the orbits and upon the accuracy of the observations.

This section discusses the number of bodies to be included in the gravitational perturbation model and the non-gravitational perturbations. The
effect of a non-spherical shape of the bodies is discussed in Chapter 13. The
general relativistic perturbations are discussed in Section 6.6.

How many bodies?

The first question about an $(N+1)$-body model for the orbit of a given
object is: how to choose N? Our Solar System contains the Sun, a number
of major planets,[4] the natural satellites of the planets, and a large population
of minor bodies (asteroids, comets, trans-neptunian objects, Centaurs, even
meteoroids). Of course we need to cut off at some level, and to use a
consistent approximation we need to select some order of magnitude of the
perturbations and to neglect the bodies resulting in lesser effects on the
target body (or bodies) of the orbit determination.

The most efficient method to do this is to use the Jacobian coordinates
and the Roy–Walker small parameters to estimate the perturbative effects.
This does not imply that we have to use Jacobian coordinates as variables
to be determined. The values of the ϵ_{ij} parameters for perturbations by
superior and inferior planets are given in Table II of (Walker *et al.* 1980);
here we either reproduce or recompute some values.[5]

The conclusions to be drawn from Table 4.1, and from similar computations, of course depend upon the application. The accuracy required in the

[4] The number of objects to be considered major planets requires some discussion, see below.
[5] The data of the 1980 paper are not up to date for the perturbations by Pluto, whose mass has
been reassessed to a value 200 times smaller with the observations of the Pluto–Charon binary.

Table 4.1. Roy–Walker parameters for three-body subsystems of the Solar System.

Subsystem	ϵ_{12}	ϵ_{21}
Sun–Mercury–Earth	2.5×10^{-8}	1.7×10^{-7}
Sun–Venus–Earth	1.3×10^{-6}	1.1×10^{-6}
Sun–Earth–Mars	1.3×10^{-6}	9.2×10^{-8}
Sun–Earth–Ceres	4.0×10^{-7}	2.2×10^{-11}
Sun–Earth–Jupiter	1.1×10^{-7}	6.8×10^{-6}
Sun–Earth–Saturn	3.3×10^{-8}	3.3×10^{-7}
Sun–Earth–Uranus	8.2×10^{-9}	6.2×10^{-9}
Sun–Earth–Neptune	3.3×10^{-9}	1.9×10^{-9}
Sun–Earth–Pluto	4.8×10^{-10}	7.5×10^{-13}
Earth–Moon–Sun	7.9×10^{-8}	5.7×10^{-3}
Jupiter–Ganymede–Sun	1.5×10^{-10}	2.7×10^{-6}

orbit computation has to be adequate for the accuracy of the observations. As an example, this table contains the parameters to be used to discuss the dynamical model for the orbit of the Earth, allowing the following conclusions: the perturbations from Ceres are more important than the ones from Pluto, the perturbations from Ganymede are negligible, and the problem of the orbit of the Moon is strongly coupled to the orbit of the Earth–Moon center of mass, thus it cannot be solved independently.

Non-gravitational perturbations

Gravitation is the most penetrating interaction, in that it is coupled to the entire mass of a celestial body, without distinction between the near surface and the central portions. The other perturbations act essentially on the surface only. For example, electrostatic forces cannot be too important for macroscopic bodies even if they become highly charged, because charges tend to migrate towards the surface. Electromagnetic radiation interacts only with a comparatively thin layer near the surface, with thickness comparable to the wavelength. Drag is an interaction, due to electromagnetic forces, of external particles with the surface. Thus a small parameter appearing in all non-gravitational perturbations is the **area-to-mass ratio**

$$\frac{A}{m} \simeq \frac{\pi R^2}{\frac{4\pi}{3} \rho R^3} = \frac{3}{4 \rho R}$$

where A is the cross-sectional area, ρ the average density, R the radius of the perturbed body, and the approximate formula becomes exact for a spherical body. The simplest example is the direct radiation pressure force due to sunlight $F = (\Phi/c)\,A$, where Φ is the radiation energy flow (per unit cross-section) at the given distance from the Sun $r_\odot$ and c is the speed of light. The ratio of radiation pressure force to gravitational attraction is

$$\beta = \frac{\Phi\,A\,r_\odot^2}{G\,m_\odot\,m\,c}$$

where $m_\odot = m_0$ is the Sun mass. The energy flow from the Sun, because of $E = m\,c^2$, carries away mass from the Sun at a rate $\dot{m}_\odot = 4\pi\,r_\odot^2\,\Phi/c^2 \simeq 7 \times 10^{-14}\ m_\odot/\mathrm{y}$, that is, the Sun decreases its mass due to shedding of photons with a characteristic time[6] $t_\odot = m_\odot/\dot{m}_\odot \simeq 1.5 \times 10^{13}$ y. Thus

$$\beta \simeq \frac{A}{m}\frac{\dot{m}_\odot}{m_\odot}\frac{c}{4\pi\,G}.$$

In CGS units, $c/4\pi\,G \simeq 3 \times 10^{16}$ and 3×10^{16} s $\simeq 10^9$ y, thus

$$\beta = \frac{A}{m}\frac{1}{t_\odot} \simeq \frac{A}{m}\frac{1}{15\,000} \qquad \text{(unit of } t_\odot \text{ in billion years).}$$

Let us use this estimate to assess when radiation pressure can be a significant perturbation of a heliocentric orbit.

- For a planet, e.g., Mercury, $\rho \simeq 5$ and $R \simeq 2\,400$ km, $A/m \simeq 6 \times 10^{-10}$ and $\beta \simeq 4 \times 10^{-14}$: radiation pressure is almost at the rounding off level.
- For a small asteroid with $\rho \simeq 1.5$ and $R \simeq 500$ m, $A/m \simeq 2 \times 10^{-5}$ and $\beta \simeq 1.3 \times 10^{-9}$: radiation pressure is small, negligible for astrometric observations, but it cannot be neglected if very accurate observations (e.g., radar, tracking of an orbiter) are available.
- For a spacecraft, $A \simeq 5$ m^2 and $m \simeq 500$ kg, $A/m \simeq 0.1$ and $\beta \simeq 7 \times 10^{-6}$ is not negligible at all. Radiation pressure and other non-gravitational forces acting on spacecraft are discussed in Chapter 14.
- For a dust particle of given density, e.g., $\rho = 2$, there is a critical radius at which $\beta = 1$: for spherical shape $A/m = 3/8\,R$ and $\beta = 2.5 \times 10^{-5}/R = 1$ implies $R = 0.25$ micron. For a particle of this size this is a simplistic model of the particle/wave interaction, but the order of magnitude is right, particles in the sub-micron range released from a Solar System orbit at low relative velocity are not bound to the Solar System: they are called β-particles.

---

[6] This time is related to the time span over which the Sun will remain a main sequence star, which corresponds to the conversion in radiation of about $1/1000$ of the mass. The Sun also sheds mass as charged particles in the *solar wind*.

Part II
Basic Theory

5

LEAST SQUARES

In this chapter we give the basic formulation of orbit determination as a
nonlinear least squares problem. First we introduce the linear least squares
problem and the classical iterative methods: Newton's method and differential corrections. The uncertainty of the result is described by confidence
ellipsoids, with the optimization interpretation. We show that the probabilistic interpretation gives strictly analogous results, if the observation error
is Gaussian; this assumption is also discussed. This chapter contains mostly
classical material: the main reference is (Gauss 1809). Only Section 5.8
contains recent results, based on (Carpino *et al.* 2003).

5.1 Linear least squares

The basic idea of the least squares problem is to fit some model of an unknown function $f(t)$ of time, given a finite number of observations. The
problem is linear if the model can be expressed as a linear combination

$$f(t) = \sum_{k=1}^{N} x_k \, f_k(t)$$

of a set of N base functions f_k; then the coefficients x_k of the linear combinations are the **fit parameters**. The observational data are the $m \geq N$ pairs
$(t_i, \lambda_i) < i = 1, m$: let us introduce the vectors $\mathbf{x} = (x_k) \in \mathbb{R}^N, \mathbf{t} = (t_i), \boldsymbol{\lambda} =
(\lambda_i) \in \mathbb{R}^m$. Given the observations, we compute the vector of residuals[1]
$\boldsymbol{\xi} = (\xi_i) \in \mathbb{R}^m$,

$$\xi_i = \lambda_i - f(t_i) = \lambda_i - \sum_{k=1}^{N} x_k \, f_k(t_i) = \xi_i(\mathbf{x}).$$

---

[1] The minus sign in front of the prediction is an old convention: residual= observed − computed.

The problem is converted to an optimization one by defining a target function which is proportional to the sum of squares of the residuals

$$Q(\mathbf{x}) = \frac{1}{m} \sum_{i=1}^{m} \left[\lambda_i - \sum_{k=1}^{N} x_k f_k(t_i) \right]^2 .$$

By using the design matrix

$$B = \frac{\partial \boldsymbol{\xi}}{\partial \mathbf{x}} = (b_{ik}), \quad b_{ik} = -f_k(t_i), \quad i = 1, m; \ k = 1, N$$

we have the target function in vector/matrix notation:

$$Q(\mathbf{x}) = \frac{1}{m} (\boldsymbol{\lambda} + B\mathbf{x})^T (\boldsymbol{\lambda} + B\mathbf{x}) = \frac{1}{m} \left[\boldsymbol{\lambda}^T \boldsymbol{\lambda} + 2\boldsymbol{\lambda}^T B\mathbf{x} + \mathbf{x}^T B^T B\mathbf{x} \right].$$

The stationary points of the target function are the solutions of

$$m\, \frac{\partial Q}{\partial \mathbf{x}} = 2\left[\boldsymbol{\lambda}^T B + \mathbf{x}^T B^T B \right] = \mathbf{0}$$

that is the **normal equation** $B^T B\, \mathbf{x} = -B^T \boldsymbol{\lambda}$ where the normal matrix $C = B^T B$ is symmetric and defines a non-negative quadratic form. If this quadratic form is positive, the quadratic form $Q(\mathbf{x})$ defines level hypersurfaces $m\, Q(\mathbf{x}) = \sigma^2$ in $\mathbb{R}^N$ which are ellipsoids. The inverse $\Gamma = C^{-1}$ is the covariance matrix and provides the solution $\mathbf{x}^* = -\Gamma\, B^T \boldsymbol{\lambda}$. The center of all these ellipsoids is $\mathbf{x}^*$, that is

$$m\, Q(\mathbf{x}) = m\, Q^* + (\mathbf{x} - \mathbf{x}^*)^T C (\mathbf{x} - \mathbf{x}^*)$$

where $Q^* = Q(\mathbf{x}^*)$ is the minimum value of the target function; its value can be computed by comparing the expansions of $Q(\mathbf{x})$

$$m\, Q^* = \boldsymbol{\lambda}^T \boldsymbol{\lambda} - \boldsymbol{\lambda}^T B \Gamma B^T \boldsymbol{\lambda} \le \boldsymbol{\lambda}^T \boldsymbol{\lambda}.$$

The vector of the residuals after the fit is

$$\boldsymbol{\xi} = \boldsymbol{\lambda} + B\mathbf{x}^* = \boldsymbol{\lambda} - B \Gamma B^T \boldsymbol{\lambda} \tag{5.1}$$

and is $\ne \mathbf{0}$, unless $\boldsymbol{\lambda}$ belongs to the subspace spanned by the columns of B.

Model problem

Let us consider the model problem introduced in Section 2.5, by using the variables (n, λ) instead of (a, λ): the general solution is

$$n(t) = n_0, \quad \lambda(t) = n_0\, t + \lambda_0$$

where (n_0, λ_0) are the initial conditions. The residuals and their partial derivatives are

$$\xi_i = \lambda_i - n_0\, t_i - \lambda_0, \quad i = 1, m, \quad \frac{\partial \xi}{\partial(n_0, \lambda_0)} = B = [-\mathbf{t} \ -1]$$

where the first column is just minus the times, the second one has -1 in all the entries. The normal matrix is

$$C = B^T B = \begin{bmatrix} \mathbf{t} \cdot \mathbf{t} & \mathbf{t} \cdot 1 \\ 1 \cdot \mathbf{t} & 1 \cdot 1 \end{bmatrix} = \begin{bmatrix} \sum t_i^2 & \sum t_i \\ \sum t_i & m \end{bmatrix}.$$

Assuming as initial conditions $n_0 = 0$, $\lambda_0 = 0$, the residuals are the same as the observations and the right-hand side of the normal equation is

$$D = -B^T \boldsymbol{\xi} = -B^T \boldsymbol{\lambda} = \begin{bmatrix} \sum t_i \lambda_i \\ \sum \lambda_i \end{bmatrix}.$$

By using the definition of mean, variance, and covariance for a finite set

$$\bar{t} = \frac{1}{m} \sum_{i=1}^{m} t_i, \quad \mathrm{Var}(\mathbf{t}) = \frac{1}{m} \sum_{i=1}^{m} (t_i - \bar{t})^2, \quad \mathrm{Cov}(\mathbf{t}, \boldsymbol{\lambda}) = \frac{1}{m} \sum_{i=1}^{m} (t_i - \bar{t})(\lambda_i - \bar{\lambda}),$$

with $\bar{\lambda}$ the mean of the λ_i, and by the identities $\mathbf{t} \cdot \mathbf{t} = m\, \mathrm{Var}(\mathbf{t}) + m\bar{t}^2$, $\mathbf{t} \cdot \boldsymbol{\lambda} = m\, \mathrm{Cov}(\boldsymbol{\lambda}, \mathbf{t}) + m\bar{t}\bar{\lambda}$, we get

$$C = m \begin{bmatrix} \mathrm{Var}(\mathbf{t}) + \bar{t}^2 & \bar{t} \\ \bar{t} & 1 \end{bmatrix}, \quad D = m \begin{bmatrix} \mathrm{Cov}(\mathbf{t}, \boldsymbol{\lambda}) + \bar{t}\bar{\lambda} \\ \bar{\lambda} \end{bmatrix}. \tag{5.2}$$

If $\det C = m^2 \mathrm{Var}(\mathbf{t}) > 0$ then the covariance matrix is

$$\Gamma = \frac{1}{m\, \mathrm{Var}(\mathbf{t})} \begin{bmatrix} 1 & -\bar{t} \\ -\bar{t} & \mathrm{Var}(\mathbf{t}) + \bar{t}^2 \end{bmatrix}$$

and the solution (n^*, λ^*) is the regression line, such that

$$n^* = \mathrm{Cov}(\mathbf{t}, \boldsymbol{\lambda})/\mathrm{Var}(\mathbf{t}), \quad \lambda^* = \bar{\lambda} - n^* \bar{t}.$$

The residuals $\boldsymbol{\xi} = \boldsymbol{\lambda} - n^* \mathbf{t} - \lambda^* \mathbf{1}$ are such that the mean $\bar{\xi} = 0$ and

$$\mathrm{Var}(\boldsymbol{\xi}) = Q^* = \mathrm{Var}(\boldsymbol{\lambda}) - \frac{\mathrm{Cov}^2(\mathbf{t}, \boldsymbol{\lambda})}{\mathrm{Var}(\mathbf{t})} = \mathrm{Var}(\boldsymbol{\lambda})\left[1 - \mathrm{Corr}^2(\mathbf{t}, \boldsymbol{\lambda})\right]$$

where the correlation

$$\mathrm{Corr}(\mathbf{t}, \boldsymbol{\lambda}) = \frac{\mathrm{Cov}(\mathbf{t}, \boldsymbol{\lambda})}{\sqrt{\mathrm{Var}(\mathbf{t})\, \mathrm{Var}(\boldsymbol{\lambda})}}$$

is a parameter between -1 and 1 measuring the decrease in target function with respect to the pre-fit value $Q(0) = \mathrm{Var}(\boldsymbol{\lambda})$.

5.2 Nonlinear least squares

The target function of the nonlinear least squares problem

$$Q(\mathbf{x}) = \frac{1}{m} \, \boldsymbol{\xi}^T(\mathbf{x}) \, \boldsymbol{\xi}(\mathbf{x})$$

is a differentiable function of the fit parameters $\mathbf{x}$, although it is not just a quadratic function. The partial derivatives of the residuals with respect to the fit parameters are assembled in the arrays

$$B = \frac{\partial \boldsymbol{\xi}}{\partial \mathbf{x}}(\mathbf{x}), \qquad H = \frac{\partial^2 \boldsymbol{\xi}}{\partial \mathbf{x}^2}(\mathbf{x})$$

where the **design matrix** B is an $m \times N$ matrix, with $m \geq N$, and H is a three-index array of shape $m \times N \times N$. In the context of orbit determination, the partial derivatives of the residuals are the partials of the prediction function (with sign changed). These can be computed by using the chain rule from the partials of the observation function R and the partials of the general solution $\mathbf{y}(t) = \mathbf{y}(\mathbf{y_0}, t, \boldsymbol{\mu}, \boldsymbol{\nu})$ of the equation of motion

$$\frac{\partial \xi_i}{\partial x_k} = -\frac{\partial R}{\partial \mathbf{y}} \frac{\partial \mathbf{y}(t_i)}{\partial x_k} - \frac{\partial R}{\partial x_k}$$

where the first term is relevant if x_k is a component of the vector $(\mathbf{y_0}, \boldsymbol{\mu})$ (either an initial condition or a dynamical parameter), the second one if x_k is a component of $\boldsymbol{\nu}$ (a kinematical parameter). The formula for H is less simple, containing first and second derivatives of the general solution of the equation of motion.

To find the minimum, we look for stationary points of $Q(\mathbf{x})$:

$$\frac{\partial Q}{\partial \mathbf{x}} = \frac{2}{m} \, \boldsymbol{\xi}^T B = \mathbf{0}.$$

Two problems contribute in making this case not as simple as the linear one. First, the equation above is a system of nonlinear equations, and generally does not have an explicit solution. Second, a stationary point does not need to be the absolute minimum point: it could be a saddle, or a local minimum. The first problem can be handled by using some iterative method, such as the Newton method, or some modification of it. The second one requires us to check the Hessian matrix of second derivatives to exclude saddles; the methods to ensure that a local minimum found by some iterative method is the absolute minimum are computationally expensive.

The Newton method

The standard **Newton method** involves the computation of the second derivatives of the target function:

$$\frac{\partial^2 Q}{\partial \mathbf{x}^2} = \frac{2}{m} \left(B^T B + \boldsymbol{\xi}^T H \right) = \frac{2}{m} C_{new} \tag{5.3}$$

where C_{new} is an $N \times N$ matrix, non-negative in the neighborhood of a local minimum.[2] Given the residuals $\boldsymbol{\xi}(\mathbf{x}_k)$ obtained from the value $\mathbf{x}_k$ of the parameters at iteration k, the (non-zero) gradient is expanded around $\mathbf{x}_k$

$$\frac{\partial Q}{\partial \mathbf{x}}(\mathbf{x}) = \frac{\partial Q}{\partial \mathbf{x}}(\mathbf{x}_k) + \frac{\partial^2 Q}{\partial \mathbf{x}^2}(\mathbf{x}_k)\,(\mathbf{x} - \mathbf{x}_k) + \cdots$$

where the dots stand for terms of higher order in $(\mathbf{x} - \mathbf{x}_k)$. If this gradient has to be zero in $\mathbf{x} = \mathbf{x}^*$

$$\mathbf{0} = \frac{\partial Q}{\partial \mathbf{x}}(\mathbf{x}_k) + \frac{\partial^2 Q}{\partial \mathbf{x}^2}(\mathbf{x}_k)\,(\mathbf{x}^* - \mathbf{x}_k) + \cdots$$

that is

$$C_{new}\,(\mathbf{x}^* - \mathbf{x}_k) = -B^T \boldsymbol{\xi} + \cdots$$

Neglecting the higher order terms, if the matrix C_{new}, as computed at the point $\mathbf{x}_k$, is invertible then the iteration $k + 1$ of the Newton method provides a correction $\mathbf{x}_k \longrightarrow \mathbf{x}_{k+1}$ with

$$\mathbf{x}_{k+1} = \mathbf{x}_k + C_{new}^{-1} D, \qquad D = -B^T \boldsymbol{\xi},$$

where also $D = D(\mathbf{x}_k)$. The point $\mathbf{x}_{k+1}$ should be a better approximation to $\mathbf{x}^*$ than $\mathbf{x}_k$. In practice, the Newton method may converge or not, depending upon the choice of the **first guess** $\mathbf{x}_0$ selected to start the iterations.

Differential corrections

The most used method is a variant of the Newton method, known in this context as **differential corrections**, with each iteration making the correction

$$\mathbf{x}_{k+1} = \mathbf{x}_k - (B^T B)^{-1} B^T \boldsymbol{\xi}$$

where the normal matrix $C = B^T B$, computed at $\mathbf{x}_k$, replaces the matrix C_{new}. This amounts to neglecting, on top of the terms of order ≥ 2 in $(\mathbf{x}^* - \mathbf{x}_k)$, also the term $\boldsymbol{\xi}^T H\,(\mathbf{x}^* - \mathbf{x}_k)$. The additional neglected term is of first order in $(\mathbf{x}^* - \mathbf{x}_k)$ but contains also the residuals, thus it is

---

[2] By $\boldsymbol{\xi}^T H$ we mean the matrix with components $\sum_i \xi_i\, \partial^2 \xi_i / \partial x_j \partial x_k$.

smaller than $C\,(\mathbf{x}^* - \mathbf{x}_k)$ if the residuals are small enough. However, this qualitative argument does not always apply (see Section 10.2).

The main practical motivation for this simplification of the Newton method is that the computation of the three-index arrays of second derivatives $\partial B / \partial \mathbf{x} = \partial^2 \boldsymbol{\xi} / \partial \mathbf{x}^2$ for $P = p' + p''$ dynamical parameters (p' initial conditions and p'' parameters to be solved appearing in the equation of motion) requires us to solve $p'\,P^2$ scalar differential equations on top of the usual $p' + p'\,P$ for the equation of motion and the variational equation.

One iteration of differential corrections is just the solution of a linear least squares problem, with normal equation

$$C\,(\mathbf{x}_{k+1} - \mathbf{x}_k) = D$$

where the right-hand side $D = -B^T\,\boldsymbol{\xi}$ is the same as in the Newton method. This linear problem can be obtained by truncation of the target function

$$Q(\mathbf{x}) \simeq Q(\mathbf{x}_k) + \frac{2}{m}\,\boldsymbol{\xi}^T\,B\,(\mathbf{x} - \mathbf{x}_k) + \frac{1}{m}\,(\mathbf{x} - \mathbf{x}_k)^T\,C\,(\mathbf{x} - \mathbf{x}_k),$$

which is not the Taylor expansion to order 2, since C_{new} is replaced by C.

Convergence and comparison with the linear case

An iteration, that is a differential correction step, is possible if the covariance matrix, the inverse of the normal matrix $\Gamma = C^{-1}$, can be computed. Since $C = B^T\,B$, it is always positive semidefinite, and indeed positive definite if B has rank N (this requires $m \geq N$). All this applies in exact arithmetic: numerical problems can arise for badly conditioned matrices C and Γ. The **conditioning number** $cond(A)$, for a symmetric positive definite matrix A, is[3] the ratio of the largest to the smallest eigenvalue of A.

If C is a **badly conditioned matrix**, that is its conditioning number is very large, comparable to the inverse of the rounding off, the computation of its inverse may become numerically unstable. There are methods such as the Cholewsky algorithm and the eigenvalues algorithm allowing us to handle badly conditioned cases in a numerically stable way. For a linear least squares fit this is a solution, although also the covariance matrix is badly conditioned (the small eigenvalues of C corresponding to the large eigenvalues of Γ).

In an iterative procedure, to succeed in accurately inverting a badly conditioned normal matrix C is by no means a guarantee of success. If Γ has large eigenvalues then one step of differential corrections could apply a large correction $\mathbf{x}_{k+1} - \mathbf{x}_k = \Gamma\,D$, in particular with large components along the

---

[3] This is only one of the alternative definitions found in the literature; it is the most intuitive and is enough for our purposes. For a more detailed discussion, see Section 6.4.

weak direction corresponding to the largest eigenvalue of Γ (see Chapter 10). With large corrections the approximations done by truncating the equation $\partial Q/\partial \mathbf{x} = \mathbf{0}$ are poor and the value $Q(\mathbf{x}_{k+1})$ may fail to decrease with respect to $Q(\mathbf{x}_k)$. If the target function begins to increase, often it goes on increasing, then the size of the successive corrections also increases, until physically meaningless values are reached for some of the parameters $\mathbf{x}$.

In conclusion there are two main differences between solving a linear least squares problem and a nonlinear one. First, the linear problem always has a solution; if the normal matrix is badly conditioned, even with zero eigenvalues, it is always possible to find a solution algorithm.[4]

Second, the nonlinear problem requires a number of iterations, e.g., of the differential corrections step: experience shows that convergence may fail either catastrophically, by divergence passing from absurd values, or by undamped oscillations, or by correction steps of approximately constant size moving in the same weak direction. Thus we need criteria to terminate the iteration and to proclaim success, that is a good approximation to the convergence which would be achieved only at the limit for $k \to +\infty$. We also need criteria to decide when it is better to give up and to proclaim failure, i.e., when the iterations do not show any tendency to converge.

We can use two criteria to terminate the iterations with an acceptable approximation to $\mathbf{x}^*$. One is based on the size of the last correction $\Delta \mathbf{x} = \mathbf{x}_{k+1} - \mathbf{x}_k$. To decide $\Delta \mathbf{x}$ is small we need a metric in the N-dimensional space of fit parameters. One such metric is defined by the normal matrix

$$||\Delta \mathbf{x}||_C = \sqrt{\Delta \mathbf{x}^T\, C \Delta \mathbf{x}/N},$$

with an immediate interpretation either in terms of confidence ellipsoids (see Section 5.4) or probabilistic (see Section 5.7). If $||\Delta \mathbf{x}||_C \ll 1$ the following iterations will not provide significant improvements of the solution. The second criterion uses the target function at each iteration $Q_k = Q(\mathbf{x}_k)$: if

$$|Q_{k+1} - Q_k|/Q_{k+1} \ll 1$$

the change of the last step has not been very useful for the goal of minimizing the target function. However, one iteration with a small relative change in the value of Q is not enough to predict that the value of Q will not change significantly in the following iterations: to terminate the iterations, it is better to require either that there is no significant change in Q for a number (3–5) of iterations, or that $||\Delta \mathbf{x}||$ is small.

---

[4] For the case with $\det C = 0$ a (non-unique) solution always exists, and the pseudo-inverse algorithm can be used to compute it, see Section 6.1.

The criteria to give up orbit determination can be selected depending upon the circumstances. If many orbits have to be computed, and failure in a fraction of them is acceptable, then the iterative procedure could be terminated under weaker conditions, e.g., when either the target function has increased for a number (3–5) of iterations, or the fit parameters $\mathbf{x}$ are outside of some acceptable region. If the orbit determination failure is considered unacceptable, the differential corrections iterations should be continued until either convergence or catastrophic divergence, in which case some other initial guess and/or some other iterative method should be attempted, as discussed in Chapters 7–10.

There are additional problems specific to nonlinear least squares. First, the nominal solutions are in general local minima of the target function and they could be more than one. In some cases there is only one minimum with an acceptable value of $Q(\mathbf{x}^*)$, but it is possible that two local minima have comparable values: see in particular the cases discussed in Chapter 9 and Section 10.2; then both nominal solutions, and the points in their neighborhoods, are possible solutions and neither of them can be discarded arbitrarily.

Second, the differential corrections search for stationary points of $Q(\mathbf{x})$, thus there could be convergence to a saddle point, that is a stationary point in which the Hessian matrix $\partial^2 Q/\partial \mathbf{x}^2$ has some negative eigenvalues. From eq. (5.3) we find that negative eigenvalues are due to the H term containing the second derivatives of the residuals, not to the $B^T B = C$ part, thus the presence of saddles can occur either for comparatively large $\boldsymbol{\xi}$ or when the normal matrix C is badly conditioned. If differential corrections are used, the array H is not computed and the data to decide if the convergence point $\mathbf{x}^*$ is a saddle rather than a local minimum are not available. There are very few examples documented in the literature of a saddle point in an orbit determination problem (Sansaturio *et al.* 1996).

5.3 Weighting of the residuals

A simple generalization of the least squares problem is the *weighted* least squares problem, with a non-negative quadratic form as target function:

$$Q(\boldsymbol{\xi}) = \frac{1}{m}\,\boldsymbol{\xi}^T\,W\,\boldsymbol{\xi} = \frac{1}{m}\sum_{i=1}^{m}\sum_{k=1}^{m} w_{ik}\,\xi_i\,\xi_k$$

where $W = (w_{ik})$ is the **weight matrix**, a symmetric matrix with non-negative eigenvalues. The only change in the formulae established so far is that the normal matrix and the right-hand side of the normal equation become

$$C = B^T W B, \qquad D = -B^T W \, \boldsymbol{\xi}.$$

In the simplest case $W = 1/s^2 \, I$ (I the $m \times m$ identity matrix),

$$C = \frac{1}{s^2} \, B^T B, \qquad D = -\frac{1}{s^2} \, B^T \boldsymbol{\xi}$$

and the parameter s appears through the factor s^2 in the covariance matrix $\Gamma = C^{-1}$ and disappears in the differential correction ΓD. That is, a uniform weight does not matter in the solution, although it matters in the uncertainty. A uniform weight is implicitly used anyway, to express the residuals in some appropriate units: e.g., residuals of angular observations could be expressed in arcseconds, of distances in km.

Non-uniform weights express the assumption that different observations are rated to have different accuracy: this changes the nominal solution as well as the covariance matrix. Non-uniform weights can be formally introduced by using a **normalization of the residuals**. If the weight matrix is diagonal $W = \mathrm{diag}[s_1^{-2}, s_2^{-2}, \ldots, s_m^{-2}]$ we can change notation and use $\boldsymbol{\xi}' = (\xi'_i)$ for the true residuals, $\boldsymbol{\xi} = (\xi_i)$ for the normalized ones:

$$\boldsymbol{\xi} = \sqrt{W} \, \boldsymbol{\xi}', \qquad \xi_i = \frac{\xi'_i}{s_i}$$

and then $C = B^T B$ and $D = -B^T \boldsymbol{\xi}$ as in the simple least squares case.

A slightly more complicated case can occur if the observation errors are correlated (see Section 5.8) and the weight matrix W is not diagonal. In this case we cannot abuse the notation $\sqrt{W}$ because there are many possible "matrix square roots" which can be computed with a number of well known algorithms: we shall mention only two of them.

The **Cholewsky algorithm** is a procedure to find an upper triangular matrix P such that $P^T P = W$ (Bini *et al.* 1988, Section 4.17). The **eigenvalues algorithm** uses a rotation matrix R to diagonalize the matrix W, then computes the square root of the diagonal matrix (that is, the square root of each eigenvalue), and rotates back to the original reference system (Bini *et al.* 1988, Section 4.15). In this way a matrix $P = R\sqrt{D}R^T$ is obtained with the property $P^2 = W$ (also $P^T P = W$ because P is symmetric). Both methods are used to solve the normal equation, because they have significant advantages with respect to computing the inverse of the matrix W. With the Cholewsky decomposition the triangular P can be inverted by successive substitutions, then $[P^T P]^{-1} = P^{-1} [P^T]^{-1}$; in this way badly conditioned matrices can be inverted by an algorithm with denominators of the order of the square root of the conditioning number. With the eigenvalue method, if $R^T W R = \text{diag}[\lambda_k]$ then $W^{-1} = R \text{ diag}[\lambda_k^{-1}] R^T$. Given a numerically robust method to find the eigenvalues and eigenspaces, this allows us to handle cases with even higher conditioning numbers than Cholewsky.

Having computed such a matrix P we can again change notation, with the normalized residuals $\boldsymbol{\xi}$ obtained from the true residuals $\boldsymbol{\xi}'$ by $\boldsymbol{\xi} = P \boldsymbol{\xi}'$. The matrix of partial derivatives B of the normalized residuals is obtained from the matrix of partial derivatives B' of the true residuals

$$
\begin{aligned}
B &= \frac{\partial \boldsymbol{\xi}}{\partial \mathbf{x}} = \frac{\partial \boldsymbol{\xi}}{\partial \boldsymbol{\xi}'} \frac{\partial \boldsymbol{\xi}'}{\partial \mathbf{x}} = P \, B', \\
C &= (B')^T W B' = (B')^T P^T P B' = B^T B, \\
D &= -(B')^T W \boldsymbol{\xi}' = -(B')^T P^T P \boldsymbol{\xi}' = -B^T \boldsymbol{\xi}
\end{aligned}
$$

and the weight matrix again disappears from the normal equation. Thus we may use the formulae in which the weight matrix W does not appear but still assume that the observations have been weighted, either with a diagonal matrix expressing non-uniform individual weights, or possibly with a full matrix W expressing also correlations.

We have assumed the weight matrix W has no negative eigenvalues (otherwise some combination of residuals would give a negative target function, and the existence of a minimum would be doubtful); however, W could have a 0 eigenvalue. This can be used to handle an observation to be discarded, e.g., observation number i is given a weight $w_{ii} = 0$; if the observations are correlated, we also need to set $w_{ij} = w_{ji} = 0$ for $j \neq i$. This is either because they are known a priori to be faulty as a result of some quality control (which precedes their use for orbit determination), or because they are found a posteriori to have residuals too large to be acceptable (see Section 5.8).

5.4 Confidence ellipsoids

In a neighborhood of a nominal solution $\mathbf{x}^*$, the target function Q has a value somewhat above the minimum $Q^* = Q(\mathbf{x}^*)$, that is $Q(\mathbf{x}) = Q^* + \Delta Q(\mathbf{x})$; we call $\Delta Q(\mathbf{x})$ the **penalty**. By expanding Q around $\mathbf{x}^*$, where the gradient $-\boldsymbol{\xi}^T B$ is zero, the lowest order part of the penalty is a quadratic form in $\Delta \mathbf{x}$, with $C_{new}(\mathbf{x}^*)/m$ as coefficient matrix. As shown by eq. (5.3), if $\boldsymbol{\xi}$ is small enough, this quadratic form can be replaced by

$$
\Delta Q(\mathbf{x}) = \frac{1}{m} (\mathbf{x} - \mathbf{x}^*)^T C (\mathbf{x} - \mathbf{x}^*) + \cdots,
$$

where the dots stand for both terms of degree ≥ 3 in $\Delta \mathbf{x}$ and for those of degree 2 in $\Delta \mathbf{x}$ containing also $\boldsymbol{\xi}$.

In Chapter 1 we have defined the confidence region as the set of $\mathbf{x}$ such that the penalty does not exceed a control value. By the expansion above the confidence region can be approximated by a **confidence ellipsoid**

$$Z_L(\sigma) = \{\mathbf{x} \in \mathbb{R}^N | (\mathbf{x} - \mathbf{x}^*)^T C (\mathbf{x} - \mathbf{x}^*) \leq \sigma^2\}$$

which is indeed the inside of an $(N-1)$-dimensional ellipsoid if and only if C is positive definite. How do the confidence ellipsoids describe the uncertainty of the parameters x_k, $k = 1, N$, both one by one and by subsets? (See Section 3.2.) Let us suppose the vector of the parameters to be solved is split into two components, along orthogonal linear subspaces of the parameter space:

$$\mathbf{x} = \begin{bmatrix} \mathbf{h} \\ \mathbf{g} \end{bmatrix}, \quad \mathbf{x}^* = \begin{bmatrix} \mathbf{h}^* \\ \mathbf{g}^* \end{bmatrix}.$$

By decomposing the normal and covariance matrices as

$$C = \begin{bmatrix} C_{hh} & C_{hg} \\ C_{gh} & C_{gg} \end{bmatrix}, \quad \Gamma = \begin{bmatrix} \Gamma_{hh} & \Gamma_{hg} \\ \Gamma_{gh} & \Gamma_{gg} \end{bmatrix}$$

the quadratic approximation to the penalty is

$$\begin{aligned} m \, \Delta Q \quad &\simeq \quad (\mathbf{h} - \mathbf{h}^*) \cdot C_{hh} (\mathbf{h} - \mathbf{h}^*) \\ &+ 2(\mathbf{h} - \mathbf{h}^*) \cdot C_{hg} (\mathbf{g} - \mathbf{g}^*) + (\mathbf{g} - \mathbf{g}^*) \cdot C_{gg} (\mathbf{g} - \mathbf{g}^*). \end{aligned}$$

The uncertainty of the component $\mathbf{g}$ of the solution is expressed by three different formulae, depending upon the assumption we make on the role of the orthogonal $\mathbf{h}$ subspace. In the particular case in which $\mathbf{g}$ has dimension 1 we obtain the uncertainty of one coordinate x_k.

Conditional ellipsoids for nominal values

Case 1: uncertainty of $\mathbf{g}$ for fixed $\mathbf{h} = \mathbf{h}^*$. We have

$$m \, \Delta Q \simeq (\mathbf{g} - \mathbf{g}^*) \cdot C_{gg} (\mathbf{g} - \mathbf{g}^*)$$

and the **conditional confidence ellipsoid** in the $\mathbf{g}$ subspace has matrix C_{gg}, the submatrix of C corresponding to the subspace. C_{gg} is the normal matrix of the fit obtained by selecting $\mathbf{x} = \mathbf{g}$ and moving the $\mathbf{h}$ variables in the consider parameters, left at their nominal value. Note that the covariance matrix $\Gamma_{\mathbf{g}} = C_{gg}^{-1}$ of the variables $\mathbf{g}$ considered in isolation does not coincide with the restriction Γ_{gg} of the covariance Γ, unless $C_{hg} = \Gamma_{hg} = \mathbf{0}$.

Geometrically this corresponds to the intersection of the confidence ellipsoid with the affine subspace parallel to the linear subspace of the $\mathbf{g}$ variables and passing through $\mathbf{h}^*$, see Figure 5.1 and Section 3.3.

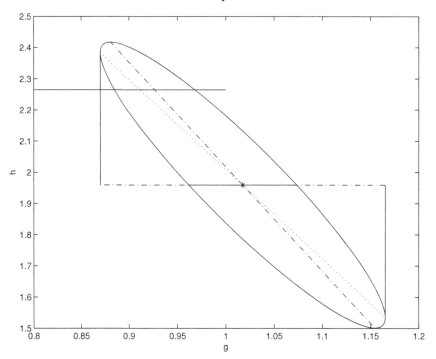

Fig. 5.1. The regression line of $\mathbf{g}$ given $\mathbf{h}$ (dash dot) contains the centers of the $\mathbf{h} = const$ sections, in this $N = 2$ case the midpoint of the horizontal intersection segments. The regression line of $\mathbf{h}$ given $\mathbf{g}$ (dotted) contains the midpoints of the vertical intersections segments, including the points of tangency of the ellipse with vertical lines.

Marginal ellipsoids

Case 2: uncertainty of $\mathbf{g}$ for arbitrary $\mathbf{h}$. Geometrically we are projecting the confidence ellipsoid onto the $\mathbf{g}$ subspace. To find the boundary of the projection we find the values of $\mathbf{g}$ corresponding to the points on the ellipsoids where the tangent linear space is parallel to the $\mathbf{h}$ subspace:

$$\frac{\partial}{\partial \mathbf{h}} [m\, \Delta Q] \simeq 2(\mathbf{h} - \mathbf{h}^*)^T C_{\mathbf{hh}} + 2(\mathbf{g} - \mathbf{g}^*)^T C_{\mathbf{gh}} = \mathbf{0}$$

that is, if $C_{\mathbf{hh}}$ is invertible, which is the case if C is positive definite,

$$\mathbf{h} - \mathbf{h}^* = -C_{\mathbf{hh}}^{-1} C_{\mathbf{hg}}\, (\mathbf{g} - \mathbf{g}^*).$$

This is the parametric equation of the **regression subspace** of $\mathbf{h}$ given $\mathbf{g}$, whose intersection with the confidence ellipsoid projects onto the **marginal confidence ellipsoid** of $\mathbf{g}$, for arbitrary $\mathbf{h}$ (see Figure 5.1 and Section 3.3). Substituting into the quadratic approximation to ΔQ

$$m\, \Delta Q \simeq (\mathbf{g} - \mathbf{g}^*) \cdot C^{\mathbf{gg}}\, (\mathbf{g} - \mathbf{g}^*), \qquad C^{\mathbf{gg}} = C_{\mathbf{gg}} - C_{\mathbf{gh}}\, C_{\mathbf{hh}}^{-1}\, C_{\mathbf{hg}}.$$

The marginal confidence ellipsoid in the $\mathbf{g}$ subspace has matrix $C^{\mathbf{gg}}$, which is not the corresponding submatrix of C. The minus sign in the above formula, in front of a non-negative quadratic form, implies that the confidence ellipsoid for arbitrary $\mathbf{h}$ contains the ellipsoid of confidence for $\mathbf{h} = \mathbf{h}^*$. The definition of $C^{\mathbf{gg}}$ can be deduced by a purely algebraic approach. We partition the normal system into two equations restricted to the subspaces:

$$\begin{cases} C_{\mathbf{hh}}\,\Delta\mathbf{h} + C_{\mathbf{hg}}\,\Delta\mathbf{g} = D_{\mathbf{h}} \\ C_{\mathbf{gh}}\,\Delta\mathbf{h} + C_{\mathbf{gg}}\,\Delta\mathbf{g} = D_{\mathbf{g}} \end{cases}$$

and then we solve it by eliminating $\Delta\mathbf{h}$. From the first equation

$$\Delta\mathbf{h} = C_{\mathbf{hh}}^{-1}\,[D_{\mathbf{h}} - C_{\mathbf{hg}}\,\Delta\mathbf{g}]\,,$$

by substituting into the other equation, we obtain another linear system

$$C^{\mathbf{gg}}\,\Delta\mathbf{g} = D_{\mathbf{g}} - C_{\mathbf{gh}}\,C_{\mathbf{hh}}^{-1}\,D_{\mathbf{h}}$$

with the matrix $C^{\mathbf{gg}}$ as defined above; if $C^{\mathbf{gg}}$ is invertible, we have $\Gamma_{\mathbf{gg}} = (C^{\mathbf{gg}})^{-1}$ and the solution, in terms of a partitioned covariance matrix, is

$$\begin{cases} \Delta\mathbf{g} = \Gamma_{\mathbf{gg}}\,D_{\mathbf{g}} - \Gamma_{\mathbf{gg}}\,C_{\mathbf{gh}}\,C_{\mathbf{hh}}^{-1}\,D_{\mathbf{h}} \\ \Delta\mathbf{h} = \Gamma_{\mathbf{hh}}\,D_{\mathbf{h}} - C_{\mathbf{hh}}^{-1}\,C_{\mathbf{hg}}\,\Gamma_{\mathbf{gg}}\,D_{\mathbf{g}} \end{cases}$$

where $\Gamma_{\mathbf{hh}} = C_{\mathbf{hh}}^{-1} + C_{\mathbf{hh}}^{-1}\,C_{\mathbf{hg}}\,\Gamma_{\mathbf{gg}}\,C_{\mathbf{gh}}\,C_{\mathbf{hh}}^{-1}$. This solution by substitution is possible only if $C_{\mathbf{hh}}$ and $C^{\mathbf{gg}}$ are invertible; it is not required that $C_{\mathbf{gg}}$ be invertible. In this hypothesis, it is possible to describe the matrix of the confidence ellipsoid in the $\mathbf{g}$ space for arbitrary $\mathbf{h}$ in terms of the covariance matrix $C^{\mathbf{gg}} = \Gamma_{\mathbf{gg}}^{-1}$, where $\Gamma_{\mathbf{gg}}$ is the restriction of Γ to the subspace. This has a probabilistic interpretation in terms of marginal probability distribution, see Section 3.2. By exchanging the role of $\mathbf{h}$ and $\mathbf{g}$ we obtain

$$C^{\mathbf{hh}} = C_{\mathbf{hh}} - C_{\mathbf{hg}}\,C_{\mathbf{gg}}^{-1}\,C_{\mathbf{gh}}$$

and if it is invertible the inverse is $\Gamma_{\mathbf{hh}}$, the restriction of the covariance matrix to the $\mathbf{h}$ subspace. This provides a complete solution computable for $C_{\mathbf{gg}}$ and $C^{\mathbf{hh}}$ invertible, but not requiring $C_{\mathbf{hh}}$ to be invertible. The choice of the matrices to be inverted can become important for numerical stability reasons when C and Γ are badly conditioned matrices.

Conditional ellipsoids for non-nominal values

Case 3: uncertainty of $\mathbf{g}$ for fixed $\mathbf{h} = \mathbf{h}_0 \neq \mathbf{h}^*$. The approximation is

$$m\,\Delta Q \simeq (\mathbf{h}_0 - \mathbf{h}^*)\cdot C_{\mathbf{hh}}\,(\mathbf{h}_0 - \mathbf{h}^*)$$
$$+ 2(\mathbf{h}_0 - \mathbf{h}^*)\cdot C_{\mathbf{hg}}\,(\mathbf{g} - \mathbf{g}^*) + (\mathbf{g} - \mathbf{g}^*)\cdot C_{\mathbf{gg}}\,(\mathbf{g} - \mathbf{g}^*),$$

and the minimum with respect to $\mathbf{g}$ is a solution of

$$\frac{\partial}{\partial \mathbf{g}}\,[m\,\Delta Q] = 2(\mathbf{h}_0 - \mathbf{h}^*)^T\,C_{\mathbf{hg}} + 2\,(\mathbf{g} - \mathbf{g}^*)^T\,C_{\mathbf{gg}} = 0.$$

Provided $C_{\mathbf{gg}}$ is invertible, the minimum point for fixed $\mathbf{h} = \mathbf{h}_0$ is

$$\mathbf{g}_0 = \mathbf{g}^* - C_{\mathbf{gg}}^{-1}\,C_{\mathbf{gh}}\,(\mathbf{h}_0 - \mathbf{h}^*)$$

which is the regression subspace of $\mathbf{g}_0$ given $\mathbf{h}_0$. $\mathbf{g}_0$ is in general different from $\mathbf{g}^*$, unless the subspaces are uncorrelated, that is unless $C_{\mathbf{gh}}$ is a zero matrix. We can now compute the penalty ΔQ as a function of the displacement $\mathbf{g} - \mathbf{g}_0$ with respect to this minimum: upon substitution of

$$\mathbf{g} - \mathbf{g}^* = (\mathbf{g} - \mathbf{g}_0) + (\mathbf{g}_0 - \mathbf{g}^*) = (\mathbf{g} - \mathbf{g}_0) - C_{\mathbf{gg}}^{-1}\,C_{\mathbf{gh}}\,(\mathbf{h}_0 - \mathbf{h}^*),$$

$$m\,\Delta Q \simeq (\mathbf{g} - \mathbf{g}_0)\cdot C_{\mathbf{gg}}\,(\mathbf{g} - \mathbf{g}_0) + (\mathbf{h}_0 - \mathbf{h}^*)\cdot C^{\mathbf{hh}}\,(\mathbf{h}_0 - \mathbf{h}^*).$$

This means that the constraint $\mathbf{h} = \mathbf{h}_0$ implies a minimum penalty

$$m\,\Delta Q_{\mathbf{h}} = (\mathbf{h}_0 - \mathbf{h}^*)\cdot C^{\mathbf{hh}}\,(\mathbf{h}_0 - \mathbf{h}^*)$$

which is a quadratic form in the difference $\mathbf{h}_0 - \mathbf{h}^*$ with matrix $C^{\mathbf{hh}} = \Gamma_{\mathbf{hh}}^{-1}$. The quadratic form expressing the supplementary penalty, for moving $\mathbf{g}$ from the constrained minimum $\mathbf{g}_0$, has matrix $C_{\mathbf{gg}}$ as in the $\mathbf{h} = \mathbf{h}^*$ case, but the conditional confidence ellipsoid in the $\mathbf{g}$ space is smaller because

$$m\,\Delta Q \simeq \sigma^2 \Leftrightarrow (\mathbf{g} - \mathbf{g}_0)\cdot C_{\mathbf{gg}}\,(\mathbf{g} - \mathbf{g}_0) = \sigma^2 - (\mathbf{h}_0 - \mathbf{h}^*)\cdot C^{\mathbf{hh}}\,(\mathbf{h}_0 - \mathbf{h}^*)$$

and the last term to be subtracted is positive (for $\mathbf{h}_0 \neq \mathbf{h}^*$) when $C^{\mathbf{hh}}$ is positive definite, that is when C and Γ are positive definite. This has a probabilistic interpretation in terms of the conditional probability distribution.

5.5 Propagation of covariance

Let $\mathbf{y}$ represent the state vector at some time t, the solution of the equation of motion. The differential of the integral flow $\Phi_{t_0}^t(\mathbf{y_0})$, where $\mathbf{y_0} = \mathbf{y}(t_0)$, is expressed by a matrix of partial derivatives, the state transition matrix

$$\frac{\partial \mathbf{y}(t)}{\partial \mathbf{y_0}} = D\Phi_{t_0}^{t_1}(\mathbf{y_0})$$

which is in turn the solution of the variational equation, a system of linear ordinary differential equations. The variational equation has a solution which can be computed numerically, simultaneously with the solution of the equation of motion. We shall use the semigroup property

$$\Phi_t^{t_0}\left(\Phi_{t_0}^t(\mathbf{y_0})\right) = \Phi_{t_0}^{t_0}(\mathbf{y_0}) = \mathbf{y_0}, \qquad \frac{\partial \mathbf{y}}{\partial \mathbf{y_0}} \frac{\partial \mathbf{y_0}}{\partial \mathbf{y}} = I.$$

Let us first assume that the vector of fit variables $\mathbf{x}$ coincides with $\mathbf{y_0}$. By the use of the state transition matrix, the normal and covariance matrices for $\mathbf{y_0}$ can be propagated from time t_0 to an arbitrary time t. We shall indicate with subscript 0 the quantities referring to the epoch t_0, with subscript t the quantities for epoch t; for the normal matrix

$$C_0 = \frac{\partial \boldsymbol{\xi}}{\partial \mathbf{y_0}}^T \frac{\partial \boldsymbol{\xi}}{\partial \mathbf{y_0}}$$

the propagation to time t is obtained by assuming the fit variables are $\mathbf{y}(t)$, then by applying to the Jacobian matrix the chain rule

$$
\begin{aligned}
C_t &= \frac{\partial \boldsymbol{\xi}}{\partial \mathbf{y}}^T \frac{\partial \boldsymbol{\xi}}{\partial \mathbf{y}} = \left(\frac{\partial \boldsymbol{\xi}}{\partial \mathbf{y_0}} \frac{\partial \mathbf{y_0}}{\partial \mathbf{y}}\right)^T \left(\frac{\partial \boldsymbol{\xi}}{\partial \mathbf{y_0}} \frac{\partial \mathbf{y_0}}{\partial \mathbf{y}}\right) \\
&= \frac{\partial \mathbf{y_0}}{\partial \mathbf{y}}^T C_0 \frac{\partial \mathbf{y_0}}{\partial \mathbf{y}} = \left(\frac{\partial \mathbf{y}}{\partial \mathbf{y_0}}^T\right)^{-1} C_0 \left(\frac{\partial \mathbf{y}}{\partial \mathbf{y_0}}\right)^{-1}.
\end{aligned}
\tag{5.4}
$$

The covariance matrices are the inverse of the normal matrices, thus

$$\Gamma_0 = C_0^{-1}, \qquad \Gamma_t = C_t^{-1} = \frac{\partial \mathbf{y}}{\partial \mathbf{y_0}} \Gamma_0 \frac{\partial \mathbf{y}}{\partial \mathbf{y_0}}^T, \tag{5.5}$$

giving the covariance propagation formula, corresponding to eq. (3.9). In conclusion to propagate the normal and covariance matrix, and to compute the confidence ellipsoid for another epoch, it is not necessary to solve again the least square problem, but only to solve the variational equation. However, as we can see already in our model problem, the assumption of linearity is often questionable for this step of the computation.

If the fit parameters $\mathbf{x}$ include, besides $\mathbf{y_0}$, some of the constants μ, ν, then the state transition matrix for $\mathbf{x}$ is of the form

$$\frac{\partial \mathbf{x}(t)}{\partial \mathbf{x}(t_0)} = \begin{bmatrix} \partial \mathbf{y}(t)/\partial \mathbf{y_0} & 0 \\ 0 & I \end{bmatrix}$$

and the propagation formulae can be used in exactly the same way.

Sources of nonlinearity

The problem of nonlinearity arises from two phenomena. First, the linear approximation may result in normal and covariance matrices with a large conditioning number. Then the confidence ellipsoids have a very elongated shape: if λ_j, $j = 1, N$, are the eigenvalues of the propagated normal matrix C_t, the lengths of the semiaxes of the confidence ellipsoid are $\sigma_j = 1/\sqrt{\lambda_j}$, and the ratio of the longest to the shortest is $\sqrt{\text{cond}(C)}$. Ratios such as $10^5 \simeq 10^6$ are often found, thus the distance of points in the confidence ellipsoid from the nominal solution can become large. Second, the quadratic approximation can fail when it is used to describe a large region in the parameter space: that is, the confidence region is significantly different from the confidence ellipsoid. The two phenomena act together: nonlinearity is important when the longest axis of the ellipsoid is too long.

Both the wild increase in the size of the longest semiaxis, and the dominance of the nonlinear effects, can occur in each of the two steps in the computation of the propagated confidence region. The first step is the computation of the confidence region for the fit parameters $\mathbf{x}$ at epoch t_0, supposedly near to the center of the observations time span. Still, when the observational data are inadequate the normal matrix C_0 (for epoch t_0) is badly conditioned.

The second step is the propagation of the uncertainty to the time t_1. The integral flow $\Phi_{t_0}^{t_1}$ is nonlinear, and its derivative, the state transition matrix, has some coefficients growing at least linearly with time. Then the propagation of the normal and covariance matrices, eq. (5.4) and (5.5), results in conditioning numbers increasing at least quadratically with time. In a chaotic dynamics, that is with positive Lyapounov exponents, some coefficients of the state transition matrix grow exponentially with time and the effects described above are enormously enhanced. This is the case for an asteroid on a planet-crossing orbit, see Chapter 12.

5.6 Model problem

Let us use as an example the model problem, discussed in Section 5.1, in the nonlinear formulation with variables (a, λ). The general solution is

$$a(t) = a_0, \qquad \lambda(t) = n(a_0)\, t + \lambda_0$$

with $n(a)$ a monotonically decreasing and convex function defined for $a > 0$

$$n(a) = \frac{k}{a^{3/2}}, \qquad \frac{dn}{da} = -\frac{3}{2}\frac{n}{a} < 0, \qquad \frac{d^2 n}{da^2} = \frac{15}{4}\frac{n}{a^2} > 0.$$

The partial derivatives and the design matrix are

$$\frac{\partial \xi_i}{\partial a_0} = \frac{3}{2}\frac{n_0}{a_0} t_i, \qquad \frac{\partial \xi_i}{\partial \lambda_0} = -1, \qquad B = \left[\frac{3}{2}\frac{n_0}{a_0}\mathbf{t} \;\; -\mathbf{1}\right],$$

where $n_0 = n(a_0)$. Let us select a suitable origin of time in such a way that $\bar{t} = 0$, then $\sum t_i^2 = m\,\mathrm{Var}(\mathbf{t})$ and $\sum t_i \lambda_i = m\,\mathrm{Cov}(\mathbf{t},\boldsymbol{\lambda})$, thus by eq. (5.2)

$$C = B^T B = m \left[\begin{array}{cc} (9\,n_0^2/4\,a_0^2)\,\mathrm{Var}(\mathbf{t}) & 0 \\ 0 & 1 \end{array}\right].$$

We have to start from a first guess $a = a_0$, $\lambda = \lambda_0$ with $a_0 > 0$ and $n_0 = n(a_0) > 0$: the residuals are $\boldsymbol{\xi} = \boldsymbol{\lambda} - n_0\,\mathbf{t} - \lambda_0\,\mathbf{1}$ and taking into account $\mathbf{t}\cdot\boldsymbol{\xi} = \mathbf{t}\cdot\boldsymbol{\lambda} - n_0\mathbf{t}\cdot\mathbf{t}$ the right-hand side of the normal equation is

$$D = -B^T\,\boldsymbol{\xi} = m \left[\begin{array}{c} -(3\,n_0/2\,a_0)\,(\mathrm{Cov}(\mathbf{t},\boldsymbol{\lambda}) - n_0\,\mathrm{Var}(\mathbf{t})) \\ \bar{\lambda} - \lambda_0 \end{array}\right].$$

The first differential correction, starting from the first guess (a_0,λ_0), is

$$a_1 = a_0 - \frac{2}{3}\frac{a_0}{n_0}\left[n_0 - \frac{\mathrm{Cov}(\mathbf{t},\boldsymbol{\lambda})}{\mathrm{Var}(\mathbf{t})}\right], \qquad \lambda_1 = \lambda_0 + (\bar{\lambda} - \lambda_0);$$

this first iteration needs to be compared with the results of the linear fit in the (n,λ) variables, which was $n^* = \mathrm{Cov}(\mathbf{t},\boldsymbol{\lambda})/\mathrm{Var}(\mathbf{t})$, $\lambda^* = \bar{\lambda}$. Thus the first iteration corrects λ at once to the right value, which is preserved in the following iterations. The correction to a can be interpreted as

$$a_1 = a_0 - \frac{n(a_0) - n^*}{dn/da},$$

i.e., one step of the Newton method to solve the equation $n(a) - n^* = 0$ with first guess $a = a_0$. In this simple case we can find whether the iterative differential corrections are convergent. If $a_1 = [(5/2)\,n_0 - n^*]/(3\,n_0/2\,a_0)$ is negative the second iteration is impossible: this occurs for $a_0 > (5/2)^{2/3}\,a^* \simeq 1.84\,a^*$, where a^* is the value such that $n(a^*) = n^*$. Else, if $a_0 < a^*$ the iterations take place where the convex decreasing function $n(a)$ has a value larger than n^*, and convergence is guaranteed. Else, if $(5/2)^{2/3}\,a^* > a_0 > a^*$ we have $0 < a_1 < a^*$ and the following iterations are in the region of guaranteed convergence. We can conclude that the differential corrections are convergent for $0 < a_0 < (5/2)^{2/3}\,a^*$, whatever λ_0.

We can see the effect of the nonlinearity also in the propagated confidence region (Figure 5.2). The propagation of the normal matrix C_0 to C_t at time t by eq. (5.4) uses the inverse of the state transition matrix

$$\frac{\partial(a,\lambda)}{\partial(a_0,\lambda_0)} = \left[\begin{array}{cc} 1 & 0 \\ -(3\,n_0/2\,a_0)\,t & 1 \end{array}\right], \quad \frac{\partial(a_0,\lambda_0)}{\partial(a,\lambda)} = \left[\begin{array}{cc} 1 & 0 \\ (3\,n_0/2\,a_0)\,t & 1 \end{array}\right]$$

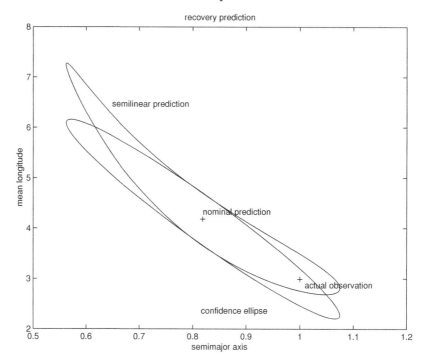

Fig. 5.2. Semilinear prediction compared with linear confidence ellipse for the model problem.

$$
C_t \;=\; m \left[\begin{array}{cc} (9\,n_0^2/4\,a_0^2)\,(\mathrm{Var}(\mathbf{t}) + t^2) & (3\,n_0/2\,a_0)\,t \\ (3\,n_0/2\,a_0)\,t & 1 \end{array} \right]
$$

which is the same as the normal matrix computed for time $t = 0$ when the average time is $\bar{t} = -t$. Thus it is no longer diagonal, but the determinant is still the same: that is, the propagated **confidence ellipse** has the same area, but it is skewed along the λ axis, and the ratio of the major semiaxis to the minor semiaxis increases. The image, by the full nonlinear integral flow, of the confidence ellipses for a_0, λ_0, that is of the ellipses defined by C_0, becomes increasingly banana shaped as the time t increases in absolute value (Figure 5.2). For large t such a **semilinear prediction** can become strongly incompatible with the linear one, see Section 7.4.

Observations of angle variables

The model problem can be made more realistic if we consider that λ is an **angle variable**, that is defined $mod\ 2\,\pi$. Indeed, when the satellite is observed at longitude λ_i at time t_i for $i = 1, 2$ the observations do not contain the information about whether the orbital arc covered in the interval $[t_1, t_2]$ has been $\lambda_2 - \lambda_1$ rather than $\lambda_2 - \lambda_1 + 2 p \pi$, with p integer.

As a simple example let us assume the $m = 2h + 1$ observations are at constant time intervals of length Δt, with $\bar{t} = 0$, that is $t_i = (i - h - 1) \Delta t$; let the best fit mean motion be n^*. Then $\lambda_{i+1} - \lambda_i \simeq n^* \Delta t$ if the satellite has done less than one revolution; but of course also $\lambda_{i+1} - \lambda_i \simeq n^* \Delta t + 2 \pi p$ for $p \in \mathbb{Z}$ provides a good fit, thus an alternative nominal solution with

$$n_p^* \simeq \frac{n^* \Delta t + 2 \pi p}{\Delta t} = n^* + \frac{2 \pi}{\Delta t} p.$$

There is no way to decide among these alternative solutions until a new observation, at time t^+ with $t^+/\Delta t$ not an integer, is available. In a simple numerical test, it is easy to get convergence to a value a_1^*, with $n(a_1^*) \simeq n_1^*$. Then for a time not an integer multiple Δt the predictions are completely off, which could result in a failed recovery.

This change in the topology of the problem, with multiple minima of the target function if the observable is an angle variable, is because the residuals are not $\lambda_i - \lambda(t_i)$ but $\xi_i = [\lambda_i - \lambda(t_i, a_0, \lambda_0) + \pi] \mod(2 \pi) - \pi$ where the shift by π before reducing to the principal value has the purpose of ensuring that the function Q is smooth at $\xi_i = 0$; however, it is not differentiable at $\xi_i = \pi$. Thus the target function $Q(a_0, \lambda_0)$ is not differentiable everywhere, and separate minima can be found in each domain of differentiability. If differential corrections are started with a first guess far from the true solution, the Newton method on a_0 can lead to the spurious values a_p^* (corresponding to n_p^* with $p = 1, 2, \ldots$). As we will see in Section 7.4, selecting an integer number of revolutions is a critical step of the identification problem.

5.7 Probabilistic interpretation

The probabilistic interpretation uses as source random variables the residuals themselves: the simplest assumption is that, after the best possible value has been found for the fit parameters $\mathbf{x}$, each residual ξ_i is a continuous random variable Ξ_i with zero mean and unit variance (in some appropriate unit), independent of the index i. It is also assumed that the error of each observation is a random variable independent (see Section 3.1) of those of the other observations. Under the additional hypothesis that the joint probability density is rotation invariant, thus it is a function depending only upon the target function, the only possible probability density is the Gaussian one $p_{\Xi_i}(\xi) = N(0, 1)(\xi)$ (see Section 3.2). Then the residuals random vector $\mathbf{\xi}$ has probability density $p_{\Xi}(\mathbf{\xi}) = N(\mathbf{0}, I)(\mathbf{\xi})$, with I

the $m \times m$ identity matrix. Under these conditions the solution for the fit parameters $\mathbf{x}$ can be seen as a set of jointly distributed random variables $\mathbf{X}$: the goal is to compute the probability density $p_{\mathbf{X}}(\mathbf{x})$, given the probability density $p_{\Xi}(\boldsymbol{\xi})$. The residuals are a function of the fit parameters

$$G : \mathbb{R}^N \longrightarrow \mathbb{R}^m, \quad \boldsymbol{\xi} = G(\mathbf{x})$$

obtained by subtracting from the observations the prediction function. Let $\mathbf{x}^*$ be the nominal solution and $\boldsymbol{\xi}^* = G(\mathbf{x}^*)$ the corresponding residuals. G is a differentiable function, thus we can linearize at the nominal solution

$$\boldsymbol{\xi} - \boldsymbol{\xi}^* = B(\mathbf{x}^*) (\mathbf{x} - \mathbf{x}^*) + \cdots$$

where $B(\mathbf{x}^*)$ is the design matrix, computed at convergence, and the dots stand for terms of order higher than 1 in $|\mathbf{x} - \mathbf{x}^*|$. The image of the fit parameters space $V = G(\mathbb{R}^N)$ is an N-dimensional submanifold of the residuals space $\mathbb{R}^m$. This manifold can have singularities, but the point $\boldsymbol{\xi}^*$ cannot be singular, because the matrix $B(\mathbf{x}^*)$ has rank N, otherwise differential corrections would fail and the nominal solution $\mathbf{x}^*$ could not be reached. Thus we can assume that the manifold V is smooth, at least in a neighborhood of $\boldsymbol{\xi}^*$. We need to compute the conditional density function of Ξ on V, as a step to compute the probability density of $\mathbf{X}$ on $\mathbb{R}^N$.

If we can neglect the higher order terms we can write a linearized equation

$$\Delta\boldsymbol{\xi} = B(\mathbf{x}^*) \Delta\mathbf{x}, \quad \Delta\boldsymbol{\xi} = \boldsymbol{\xi} - \boldsymbol{\xi}^*, \quad \Delta\mathbf{x} = \mathbf{x} - \mathbf{x}^*,$$

which is the tangent map between $\mathbb{R}^N$ and the linear N-plane $TV(\boldsymbol{\xi}^*)$ tangent to V at the point $\boldsymbol{\xi}^*$. To use this linearization is the same as considering the linear least squares problem with quadratic target function

$$Q(\mathbf{x}) = Q^* + \frac{1}{m} (\mathbf{x} - \mathbf{x}^*)^T C (\mathbf{x} - \mathbf{x}^*)$$

neglecting all higher order terms. Note that by using C instead of C_{new} we are neglecting the $\boldsymbol{\xi}^T H$ term in eq. (5.3).

By using a rotation R in the residuals space as in eq. (3.11), that is with coordinates $(\boldsymbol{\xi}', \boldsymbol{\xi}'')$ such that $\boldsymbol{\xi}''$ parameterizes $TV(\boldsymbol{\xi}^*)$, the conditional probability density of Ξ on $TV(\boldsymbol{\xi}^*)$ is Gaussian with the $N \times N$ identity matrix as covariance $p_{\Xi''} = N(\mathbf{0}, I)$. In these coordinates the linearized map $B(\mathbf{x}^*)$ has a simpler structure, since the $\boldsymbol{\xi}'$ component of the image is $\mathbf{0}$:

$$R\, B(\mathbf{x}^*) = \begin{bmatrix} \mathbf{0} \\ A \end{bmatrix}$$

with $A = A(\mathbf{x}^*)$ an invertible $N \times N$ matrix. Then the normal matrix

$C = C(\mathbf{x}^*)$ is

$$C = B^T\, B = B^T\, R^T\, R\, B = A^T\, A.$$

The inverse transformation from $TV(\boldsymbol{\xi}^*)$ to $\mathbb{R}^N$ is given by the matrix A^{-1}: by the Gaussian transformation formula (3.9) the probability density of $\mathbf{X}$

$$p_{\mathbf{X}}(\mathbf{x}) = N\left(\mathbf{x}^*, A^{-1}\, I\, [A^{-1}]^T\right)$$

is Gaussian with covariance matrix

$$\Gamma = A^{-1}\, [A^{-1}]^T = [A^T\, A]^{-1} = [B^T\, B]^{-1} = C^{-1}.$$

This is the fundamental result obtained by Gauss (1809): the solution of a linear least squares problem has a Gaussian probability density, with mean equal to the nominal solution and covariance matrix equal to the inverse of the normal matrix. $\Gamma = C^{-1}$ is the matrix solving the normal equation, thus connecting the probabilistic interpretation, the differential corrections "last step", and the optimization interpretation.

Additional connections between the two interpretations can be found by reviewing the results of Section 5.4 taking into account the probabilistic interpretation. The Gaussian probability density of the solution is

$$p_{\mathbf{X}}(\mathbf{x}) = \frac{\sqrt{\det C}}{(2\pi)^{N/2}}\, \exp\left[-\frac{1}{2}\,(\mathbf{x}-\mathbf{x}^*)^T\, C\,(\mathbf{x}-\mathbf{x}^*)\right].$$

Then the probability density contains $\mathbf{x}$ only through the penalty function[5] $\Delta Q = Q - Q^*$. The boundaries of the confidence ellipsoids are the level surfaces of the probability density function. The boundaries of the conditional and marginal confidence ellipsoids are the level surfaces of the conditional and marginal probability densities, respectively. In short, the computations required by the optimization and by the probabilistic interpretation are exactly the same. We can decide which interpretation to adopt after having computed the result, which is defined by $\mathbf{x}^*$, $C(\mathbf{x}^*)$, $\Gamma(\mathbf{x}^*)$, Q^*.

Normalization of the probability density

The hypotheses used above for the probabilistic interpretation are too restrictive to be applied in most cases. However, these hypotheses can be applied to the normalized residuals. For example, the variance of the true residuals ξ_i' could change with the index i, and even the mean could be non-zero:

---

[5] This applies in the approximation of neglecting the higher order terms; a fully nonlinear probability density could in principle be defined by the general transformation formula (3.8), but it cannot be explicitly computed by an analytical formula.

$p_{\Xi_i'}(\xi_i') = N(b_i, \sigma_i^2)(\xi_i')$. In this case the normalized residuals $\xi_i = (\xi_i' - b_i)/\sigma_i$, with **bias** b_i removed and the scale changed by the standard deviations σ_i, would all be random variables with the same probability density $N(0, 1)$.

Even the independence hypothesis can be relaxed: the joint random variables Ξ' could be a multivariate Gaussian with non-zero correlations $p_{\Xi'}(\xi') = N(\mathbf{b}, \Gamma_{\xi'})(\xi')$ and the normalization can be obtained with a square matrix P such that $P^T P = C_{\xi'} = \Gamma_{\xi'}^{-1}$. The normalized residuals are defined by $\xi = P(\xi' - \mathbf{b})$, thus $p_{\Xi}(\xi) = N(\mathbf{0}, I)(\xi)$ with I the $m \times m$ identity matrix. This procedure to normalize the residuals probability density function uses the same algorithms, giving the matrix square root P, to remove the weight matrix W from the weighted least squares problem; indeed, the residual normalization has a probabilistic interpretation with $W = C_{\xi'} = \Gamma_{\xi'}^{-1}$.

5.8 Gaussian error models and outlier rejection

That the errors in the observations, used for orbit determination, have a Gaussian distribution cannot be assumed a priori: it is an empirical fact to be confirmed by experience, by applying rigorous statistical tests. Experience shows that, whenever large data sets are processed, the error distribution cannot be described by a single normal distribution. There are two main reasons for this.

First, some observations are "wrong", that is contain errors which arise from unusual circumstances, including human error, software bugs, exceptionally difficult observing conditions: poor weather, too low **signal-to-noise** ratio (S/N), very fast moving objects, and so on. It is too difficult to include these cases in a statistical error model, and besides it would not be beneficial to the accuracy of the least squares solution. The best strategy is to decompose the residuals ξ_i into two subsets, one for which a probabilistic error model is attempted, the other containing the **outliers** which are removed from the fit (they are given weights $w_{ii} = 0$, see Section 5.3).

Second, even for the subset of residuals whose distribution is ruled by some probabilistic model, we cannot assume that the same probability density function, depending only upon the residuals values, can be applied to all. This is because there are hidden parameters, e.g., the S/N, the size of the pixels in the detector, the amount of trailing in the image, the accuracy of the star catalog used in the astrometric reduction, and more: the RMS error cannot be the same for different values of these parameters. The values of the hidden parameters are not always available, and they are not constant even for a given observing station: the accuracy of a given observatory often

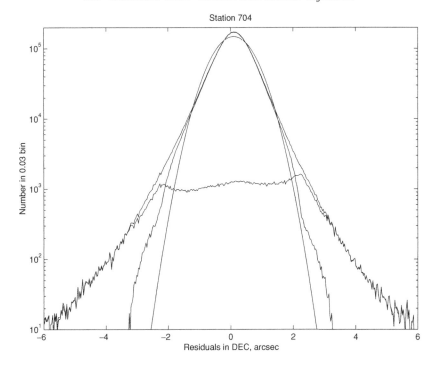

Fig. 5.3. Histograms of residuals for the declination of numbered asteroids as measured from the LINEAR observing station. For each of the 400 bins of width 0.03 arcsec, the four lines show, starting from the higher to the lower on the left side: all the residuals 1997–2003; only the outliers, automatically removed by the algorithm discussed in this section; the residuals actually used in the orbit fits; the Gaussian distribution with the same mean and the same RMS of the residuals actually used (the vertical scale is logarithmic, thus the Gaussian is represented by a parabola).

improves with time, as a result of the "learning curve" of the astronomers and of sporadic hardware, software, and star catalog upgrades.

To confirm this, we shall use as an example a large data set from population orbit determination; the collaborative case is discussed in Chapter 15.

In a typical population case, Figure 5.3 shows histograms of the residuals in one of the two angular coordinates, the declination δ, after fitting orbits of numbered asteroids, for which the orbit is over-determined: the residuals are essentially due to the astrometric errors, orbit errors giving a very small contribution. Only the residual from a single observatory, the most prolific producer of asteroid astrometry, are included: more than 9 million observations have been processed to produce this figure. The upper curve, showing the distribution of all the residuals, is clearly not a Gaussian, but has a significant tail of comparatively large residuals. For such a large data set, it is impossible to use human judgement to select the outliers, there needs to exist a fully automated outlier removal procedure, which is described below; these rejections remove the tail of large residuals. The other residuals have a distribution which is not a single Gaussian, but can be modeled as a combination of Gaussian distributions.

Outlier removal for weak fit

To select the outliers, to be discarded from the fit, it is not always enough to pick up the residuals with the largest absolute values. This would work for a strongly over-determined fit, when the average sampling time of the observations is shorter than the time-scale of the real signal: a single "wrong" observation cannot significantly change the nominal solution, as in Figure 5.4, top. This is normally the case in collaborative orbit determination, when the sampling time is decided as a design parameter of the tracking instruments, knowing the time-scale of the changes in the signal.

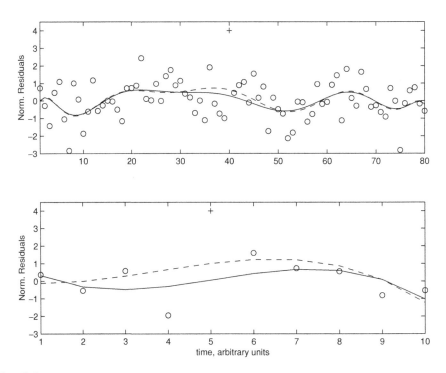

Fig. 5.4. The effect of an outlier on a fit. The points have a Gaussian distribution $N(0,1)$, the outlier (cross) is 4 RMS away from the mean. Above: for an over-determined fit with dense data (80 points, polynomial degree 10) the polynomial solution including the outlier (dashed) is different from the one without it (solid line), at least locally, but the outlier clearly has the largest residual. Below: for a weak fit with sparse data (10 points, polynomial degree 3) the fit with the outlier is bent towards it, thus making the residual much less prominent.

On the contrary, when the observations are sparse, the solutions with and without the "wrong" one are quite different, thus other residuals may become larger (see the residual at time 4 in Figure 5.4, bottom). This often happens in the population case, when the fits are only marginally over-determined.

To take this effect into account, we use eq. (5.1) for a linearized approximation of the post-fit residuals $\boldsymbol{\xi}$ as a function of the pre-fit residuals $\boldsymbol{\xi}_0$

$$\boldsymbol{\xi} = \left[I - B\,\Gamma_{\mathbf{x}}\,B^T\right]\,\boldsymbol{\xi}_0$$

where B is the design matrix, and compute the corresponding covariance by the propagation formula (5.5), with the uncertainty of the pre-fit residuals[6] given by the covariance matrix expressing the observation errors $\Gamma_{\boldsymbol{\xi}_0} = I$

$$\Gamma_{\boldsymbol{\xi}} = \left[I - B\,\Gamma_{\mathbf{x}}\,B^T\right]\,I\,\left[I - B\,\Gamma_{\mathbf{x}}\,B^T\right] = I - B\,\Gamma_{\mathbf{x}}\,B^T.$$

For an individual *post-fit residual* ξ_i, the variance is

$$\gamma_{\boldsymbol{\xi},i,i} = 1 - B_i\,\Gamma_{\mathbf{x}}\,B_i^T \qquad (5.6)$$

where $B_i = \partial \xi_i / \partial \mathbf{x}$ is the relevant row of the design matrix. In a similar way it is possible to compute the covariance matrix of a *discarded residual* ξ_k, namely the residual (with respect to the orbital solution) of an observation which has not been used in the fit: the contribution from the orbital uncertainty is just $B_k\,\Gamma_{\mathbf{x}}\,B_k^T$. In this case, however, the prediction error and the measurement error can be considered uncorrelated, thus the discarded residual has variance which is the simple sum of the two error sources

$$\gamma_{\boldsymbol{\xi},k,k} = 1 + B_k\,\Gamma_{\mathbf{x}}\,B_k^T. \qquad (5.7)$$

Thus, if a residual has been used in the fit, it has to be normalized by dividing by the square root of the variance (5.6), which is ≤ 1, else if it has already been discarded the normalized value is obtained by dividing by the square root of the variance (5.7), which is ≥ 1.

An algorithm for outlier rejection could be as follows: a residual ξ_i is discarded if its normalized absolute value is $\geq \chi_{disc}$; a discarded residual ξ_k is recovered if its normalized absolute value is $\leq \chi_{rec}$. The choice of the two parameters $\chi_{rec} < \chi_{disc}$ is delicate, although the standard χ^2 tables can be used to give an indication. Of course this is just a step of an iterative procedure, with the nominal solution recomputed between steps.

A practical implementation of a fully automated outlier rejection procedure needs to be complicated, to guarantee convergence of the differential corrections at each step and convergence of the subset of outliers: if there are very large residuals they have to be removed first, that is the values

---

[6] We are assuming that the residuals have already been normalized, as in Section 5.3. For the same formulae with a general weight matrix W, see (Carpino *et al.* 2003, Section 2.2.1).

of χ_{req}, χ_{disc} have to be adapted at each step: for a detailed description see
(Carpino *et al.* 2003, Section 2.2.2). Moreover, the observations are often
rejected by groups, as in the case of astrometric observations, where the
two-dimensional observations (α_i, δ_i) are rejected/recovered together: this
requires the analog of eqs. (5.6) and (5.7) for 2×2 covariance matrices.

Binning and local Gaussian models

Ideally, the weight matrix for the observations should be based upon complete information on the measurement process, including the assessment by
the observer, who is generally the best judge of the data quality. In practice the information is always incomplete, or anyway it has not been made
available to the orbit computer. A possible way out has been explored by
Carpino *et al.* (2003): by using all the information actually available, the
data are split into bins which should represent homogeneous observing and
measuring conditions.

The data are binned first by observatory, by observing technology,[7] and
by time. The data with low accuracy warnings, as in the primitive method
of storing less significant digits, are binned separately. They can be further
split by using the available information on the conditions which make the
measurement difficult, such as low S/N, fast proper motion, dense star fields
(e.g., observations close to the galactic plane). If the S/N value is not
available, the apparent magnitude can be used as a proxy (by assuming that
the exposure time is uniform for a given station in the same time period).

Once the data are binned, for each bin j a best fitting normal distribution
$N(b_j, \sigma_j^2)$ is computed (not using the outliers). It might be argued that the
bias b_j should be 0, otherwise the data should be re-calibrated: in practice,
this calibration has often to be done a posteriori. The kurtosis value can
be used as a control that the distribution is not too far from a Gaussian.
Then the error model is the list of bins (with their definitions, such that
each observation can be classified in one and only one bin) with their values
b_j, σ_j. Such a model already contains useful information because the RMS
values for different observatories and different years may differ by an order
of magnitude, from 0.2 to 2 arcsec, and the biases can be very significant
(Carpino *et al.* 2003, Figures 4–5); however, this is not enough.

---

[7] For example, the reduction technique for astrometric observations performed with CCD are
very different from those using photographic plates.

Correlations

Even after the residuals have been scaled with the appropriate weights, and shifted by the appropriate biases, it is not the case that their probability distributions are correctly represented by independent, unit variance Gaussian functions: they are found not to be independent. For Gaussian distributions, independence and zero correlation are equivalent, thus we can test the hypothesis of independence by measuring the correlation of a set of couples of residuals (ξ_i, ξ_j), with $(i, j) \in B$:

$$\mathrm{Corr}(\boldsymbol{\xi}, B) = \frac{1}{N_B} \sum_{(i,j) \in B} \xi_i \, \xi_j \tag{5.8}$$

where N_B is the total number of couples in the subset B, and the residuals are already shifted by the bias and scaled by the RMS of their bin.

The question is how to select the sets B of couples. Carpino *et al.* (2003) discuss several tests showing that the most significant correlations are *timewise*, appearing as a function of the time difference between the observations: they occur between observations of the same asteroid from the same station and with time differences of up to few weeks, with the largest values (typically in the range 0.2–0.4) for observations in the same night.

Correlations of this order cannot be safely neglected in the least squares fit: as the simplest example, let us suppose the covariance matrix of a set of m observations taken in the same night has 1 on the diagonal and $\alpha < 1$ outside. Such a matrix has the vector $\mathbf{1}$ as eigenvector with the largest eigenvalue $1 + (m - 1)\,\alpha$. For a tracklet containing $m = 5$ observations, correlations $\alpha = 0.2$–0.4 correspond to a longest semiaxis of the confidence ellipsoid $Z_L(1)$ of $\sqrt{1.8}$–$\sqrt{2.6}$, while neglecting correlations means replacing it with a sphere of radius 1. According to Carpino *et al.* (2003), spatial correlations, depending upon the position on the sky, did not need to be accounted for at the data quality level current at the time.

Thus the binning is done by time difference: from the values of the correlations (5.8) estimated for each time difference bin a linear model has to be fit, in such a way that the correlation $\mathrm{Corr}(\delta t, S)$ between two observations (from the same station S and of the same asteroid) can be represented as $\mathrm{Corr}(\delta t, S) = \sum_i c_i f_i(\delta t)$ for any time difference δt. However, we cannot use a linear combination of an arbitrary set of base functions f_i, because the covariance matrix of the observation errors has to be positive definite. It can be shown (Mussio 1984) that some functions of time have the property of ensuring that the correlation matrix is positive definite, so that also the covariance matrix is positive definite. One requirement is that all of these functions must decay to zero with time: $\lim_{\delta t \to +\infty} f_i(\delta t) = 0$. The list includes exponential decaying functions $\exp(-c\,\delta t)$, Gaussian-like functions $\exp(-c\,\delta t^2)$, quadratic times exponential functions of the form $(1 - d\,\delta t^2)\,\exp(-cT)$, and all their linear combinations. Carpino *et al.* (2003) provide tables of the coefficients c_i for these base functions, applicable to the observatories providing the largest data sets, thus allowing us to compute the non-diagonal weight matrix W to be used in the formulae of Section 5.3.

The above attempt to build an error model by processing incomplete information cannot give completely reliable results. Moreover, as the accuracy of the observations improves, the shortcomings of a "naive" error model (such as weighting all the observations at 1 arcsec) become more and more troublesome, especially for critical applications such as those of Chapter 12. A recent reanalysis of the set of residuals from all the astrometric observations of numbered asteroids (Baer *et al.* 2008) confirms the presence of significant biases and correlations reported by Carpino *et al.* (2003), but has also detected significant spatial correlations, such as declination biases depending upon the declination of the observations and found in the data of different observatories; the values of such biases can be as large as 0.2 arcsec. The most obvious interpretation is that there are regional systematic errors in the star catalogs mostly used by asteroid observers.[8] Whatever the cause, the effect of biases which are not constant for a given observatory cannot be removed by binning by observatory. They need to be removed by binning by region on the celestial sphere all the observations astrometrically reduced with the same star catalog. At the time this book is being written, the information on star catalogs used has not yet been made available to the orbit computers, thus there is no rigorous solution to the astrometric error model problem which is immediately applicable. This is a serious limitation to the reliability of a probabilistic interpretation of the orbit determination problem, which has nothing to do with the mathematical formulation but with the real-world limitations of an incomplete open data policy.

---

[8] It is not easy to find the cause of these regional errors, which should be investigated by the compilers of the catalogs.

6

RANK DEFICIENCY

This chapter discusses the cases in which the standard differential correction procedure for orbit determination fails. The worst failure is when the normal matrix is degenerate: this can result from the action of a group of symmetries leaving the residuals unchanged. The differential corrections as an iterative process may fail even when each individual step can be computed. The most common cause is an approximate degeneracy, which can result from an approximate symmetry. Different methods to constrain and stabilize the differential corrections in such difficult cases are discussed. This chapter systematizes and generalizes the results published in several papers (Milani and Melchioni 1989, Milani *et al.* 1995, Bonanno and Milani 2002, Milani *et al.* 2002, Milani *et al.* 2005d).

6.1 Complete rank deficiency

If, at some step $\mathbf{x}_j$ of the differential corrections, the normal matrix C is not invertible, then the correction solving

$$B^T B \left(\mathbf{x}_{j+1} - \mathbf{x}_j \right) = -B^T \boldsymbol{\xi}$$

cannot be computed by means of the covariance matrix Γ. Nevertheless solutions of the normal equation always exist (but are not unique). Let us introduce the **pseudo-inverse** C^*, defined as the matrix associated to the null map on the kernel of C times the inverse of C restricted to the subspace orthogonal to the kernel; C^* provides the solution of minimum norm

$$\mathbf{x}_{j+1} = \mathbf{x}_j - C^* B^T \boldsymbol{\xi}.$$

The pseudo-inverse C^* can be used as generalized covariance matrix for some purposes (see Section 8.3). However, corrections based on the pseudo-inverse are unlikely to converge towards a minimum of the target function.

Under these conditions, the **rank deficiency** order d is an integer such that the matrix C has rank $N - d$, that is a kernel of dimension d. Then the matrix B has the same rank $N - d$, its column m-vectors $\{\mathbf{b}_j\}, j = 1, N$, are linearly dependent: they span a linear subspace of dimension $N - d$, and there is a subspace $K \subset \mathbb{R}^N$ of dimension d, such that

$$\mathbf{v} \in K \Longrightarrow B\,\mathbf{v} = \mathbf{0}. \tag{6.1}$$

The effect on the residuals of a change in any one of the directions $\mathbf{v} \in K$ is of second order (with respect to the size of the change)

$$\boldsymbol{\xi}(\mathbf{x}_i + s\,\mathbf{v}) - \boldsymbol{\xi}(\mathbf{x}_i) = s\,B\,\mathbf{v} + \mathcal{O}(s^2) = \mathcal{O}(s^2). \tag{6.2}$$

The intuitive interpretation is the following: some linear combinations of the parameters $\mathbf{x}$ are uninfluential on the residuals, thus they cannot be constrained by the least squares optimization.

All the above discussion assumes *exact arithmetic*, e.g., that the inverses of the matrices are computed exactly when the determinant is non-zero, and are not computable when the determinant is zero. With a digital computer, rounding-off effects could result either in inverse matrices being computed even when the exact determinant is zero or in failure of the computation even for small but non-zero determinant. In practice, when rank deficiency occurs the iterative procedure fails disastrously, e.g., with increasingly large corrections until a meaningless value for some of the parameters $\mathbf{x}$ is reached.

Curing rank deficiency

The only solution in case of rank deficiency is to change the problem. Either fewer parameters are to be solved, or more observations are to be used.

If there is no way to increase the number of observations to be used, the only solution is **descoping**. That is, d parameters have to be removed from the fit list $\mathbf{x}$ (and added to the consider parameter list $\mathbf{k}$) in such a way that the matrices B and C remain of rank $N - d$. This implies that the values of the additional consider parameters are fixed at some nominal value, which are arbitrary unless a priori information is available (from other measurements). A more general procedure is as follows: change the basis in the space $\mathbb{R}^N$ of $\mathbf{x}$ in such a way that the first d vectors $\{\mathbf{v}_j\}, j = 1, d$, are a basis of K; the other $N - d$ coordinates can be selected as the new fit parameters. By using Gram–Schmidt orthonormalization, it is possible to do this with an orthogonal $N \times N$ matrix as coordinate change: with a suitable selection of signs, the coordinate change is a rotation.

As we shall see in the examples of this book, this procedure is natural when the rotation is done among parameters of the same type, such as the initial conditions $\mathbf{y_0}$, the dynamical parameters $\boldsymbol{\mu}$, the kinematical parameters $\boldsymbol{\nu}$. When we are forced to "rotate" a mix of variables with different dimensions and physical interpretations this solution is far from intuitive.

Another option is the use of **a priori observations**. This is equivalent to assuming that some information was available, on at least some of the variables $\mathbf{x}$, before processing the current set of observations. That is, we add to the set of observations constraints $\mathbf{x} = \mathbf{x}^P$ on the values of the parameters, with $\mathbf{x}^P \in \mathbb{R}^N$ a set of assumed values, and give to the a priori observations $x_i = x_i^P$ weights $1/\sigma_i, i = 1, N$ corresponding to the assumption that the a priori standard deviation is σ_i. This is equivalent to adding the *a priori* normal equation

$$C^P \mathbf{x} = C^P \mathbf{x}^P \tag{6.3}$$

with $C^P = \mathrm{diag}[\sigma_i^{-2}]$; a non-diagonal a priori normal matrix could be used if this is a better representation of the information already available. Thus an "a priori penalty" is added to the target function

$$Q(\mathbf{x}) = \frac{1}{N + m} \left[(\mathbf{x} - \mathbf{x}^P)^T C^P (\mathbf{x} - \mathbf{x}^P) + \boldsymbol{\xi}^T \boldsymbol{\xi} \right]$$

and the complete normal equation becomes

$$\left[C^P + B^T B \right] \Delta\mathbf{x} = -B^T \boldsymbol{\xi} + C^P (\mathbf{x}^P - \mathbf{x}_j)$$

with $\Delta\mathbf{x} = (\mathbf{x}_{j+1} - \mathbf{x}_j)$ as unknown. If the a priori uncertainties σ_i are small enough, the new normal matrix $C = C^P + B^T B$ has rank N and the problem is solved. The question is whether the a priori information used is reliable. For the parameters belonging to $\mathbf{y_0}$, this means assuming that some information on the orbit was already available. For the parameters belonging to $\boldsymbol{\mu}, \boldsymbol{\nu}$ the a priori information could be available from measurements having nothing to do with the orbit. For example, the coefficients of the gravity field could be available from ground-based gravimetry; the position of the observing stations could have been measured by previous missions.

The a priori observations could be applied to only $N' < N$ of the fit parameters x_i, with a minimum of d (at least as many a priori observations as the rank deficiency). If $N' = d$ and the a priori uncertainties tend to zero, the solution of the rank deficiency problem based upon a priori observations tends to the previous solution, based on assuming d fit parameters as known with some exact value. In practice, assuming a very strong constraint is the same as assuming exact values; this allows us to handle both methods to solve the rank deficiency problem with the same formulae. On the contrary, a weak

constraint is a way to introduce a non-exclusive preference for some portions of the parameter space: for example, in solving for an orbit in
Keplerian elements, we can introduce an a priori observation of the eccentricity of the form $e = 0 \pm 1$ to force the solution to be elliptic, still allowing
for **hyperbolic orbits** only if there are significantly larger residuals for
all the $e < 1$ solutions. This corresponds to the a priori information that
strongly hyperbolic orbits are very rare in the Solar System (indeed, none
is known), thus assuming an $e < 1$ orbit is better in almost all cases, nevertheless the orbit determination algorithm should not be strongly biased
against discovering objects belonging to a new orbital class.

The use of a priori information can take a more general form, as **a priori constraints** to the least squares solution: the search for a minimum of
the target function is restricted to the set of parameters **x** fulfilling a set
of k equations $\mathbf{f}(\mathbf{x}) = \mathbf{0}$. If $k \geq d$ functional constraints are suitably chosen, the rank deficiency can be removed. The general theory of constrained
optimization is beyond the aim of this book; see, e.g., (Conn *et al.* 1992).
We linearize the constraint by taking the differential of **f** at the current
value $\mathbf{x}_k$ of the parameters (at a given iteration of the differential correc-
tions); let $A = \partial \mathbf{f}/\partial \mathbf{x}(\mathbf{x}_k)$ be the Jacobian matrix. Then the constraint can
be described by adding to the target function a quadratic function of the
correction $\mathbf{x}_{k+1} - \mathbf{x}_k$, defined by using A (see Section 15.5).

Model problem with degeneracy

The general solution of the model problem of Section 5.6 depends upon the
parameter k, proportional to the square root of the mass of the central body

$$a(t) = a_0, \quad \lambda(t) = \frac{k}{a_0^{3/2}} t + \lambda_0, \tag{6.4}$$

the derivatives of the residuals with respect to the parameters (a_0, λ_0, k) are

$$B = \begin{bmatrix} \dfrac{3}{2} \dfrac{n_0}{a_0} \mathbf{t} & -1 & -\dfrac{n_0}{k} \mathbf{t} \end{bmatrix};$$

the first and the last column are proportional. For $\bar{t} = 0$ the normal matrix

$$C = B^T B = m \begin{bmatrix} \dfrac{9}{4} \dfrac{n_0^2}{a_0^2} \operatorname{Var}(\mathbf{t}) & 0 & -\dfrac{3}{2} \dfrac{n_0^2}{k\,a_0} \operatorname{Var}(\mathbf{t}) \\[2ex] 0 & 1 & 0 \\[2ex] -\dfrac{3}{2} \dfrac{n_0^2}{k\,a_0} \operatorname{Var}(\mathbf{t}) & 0 & \dfrac{n_0^2}{k^2} \operatorname{Var}(\mathbf{t}) \end{bmatrix}$$

has determinant zero. In the normal equation

$$
C \left[\begin{array}{c} \Delta a_0 \\ \Delta \lambda_0 \\ \Delta k \end{array} \right] = D = -B^T \, \boldsymbol{\xi} = m \left[\begin{array}{c} -\frac{3}{2} \frac{n_0}{a_0} \left[\mathrm{Cov}(\mathbf{t}, \boldsymbol{\lambda}) - n_0 \, \mathrm{Var}(\mathbf{t}) \right] \\ \overline{\lambda} - \lambda_0 \\ \frac{n_0}{k} \left[\mathrm{Cov}(\mathbf{t}, \boldsymbol{\lambda}) - n_0 \, \mathrm{Var}(\mathbf{t}) \right] \end{array} \right]
$$

there are only two independent equations

$$
\Delta \lambda_0 = \overline{\lambda} - \lambda_0, \quad \frac{n_0}{2} \left[-3 \frac{\Delta a_0}{a_0} + 2 \frac{\Delta k}{k} \right] = n^* - n_0 \tag{6.5}
$$

where n^* is the solution of the linear model problem. Thus there are infinite solutions, with all the combinations of Δa_0 and Δk satisfying the second equation. The condition to be satisfied by the solution is $n^* = n(a_0) = k/a_0^{3/2}$; the equation (6.5) above constraining Δa_0 and Δk is obtained by expanding and neglecting second-order terms. Fixing the value of either a_0 or k removes the problem. An a priori observation of either a_0 or k removes the rank deficiency, although it may leave an approximate one, see below.

6.2 Exact symmetries

An **exact symmetry** of an orbit determination problem is the action of a group on the space of the fit parameters $\mathbf{x}$, such that all the residuals are invariant. Let G be a group of transformations $g[\mathbf{x}]$ of $\mathbb{R}^N$: if for every $g \in G$

$$
\boldsymbol{\xi}(\mathbf{x}) = \boldsymbol{\xi}(g[\mathbf{x}]) \tag{6.6}
$$

then G is a group of *exact* symmetries of the orbit determination. The simplest case is a one-parameter group of symmetries of the orbit determination: G is either $\mathbb{R}$ or $\mathbb{R}/(2\pi\mathbb{Z})$, that is $g(s) \in G$ is parameterized by either a real number or an angle variable; the internal operation in G corresponds to the sum of the parameters. Moreover, we assume that there is a *differentiable action*, that is the map $(s, \mathbf{x}) \mapsto g(s)[\mathbf{x}]$ is differentiable, and that G has *no isotropy*, that is $g(s)[\mathbf{x}] \neq \mathbf{x}$ for every $\mathbf{x}$ unless $s = 0$ applies.[1] As for the other form of symmetries discussed in Section 4.1, the same results apply to a *local one-parameter group of symmetries*, with the same properties only for s in a neighborhood of 0.

If there is a (local) one-parameter group of exact symmetries the normal equation has rank deficiency ≥ 1. The residuals do not change with s:

$$
\mathbf{0} = \frac{\partial \boldsymbol{\xi}(s)[\mathbf{x}]}{\partial s} = \frac{\partial \boldsymbol{\xi}}{\partial \mathbf{x}} \frac{\partial g(s)[\mathbf{x}]}{\partial s} = B \frac{\partial g(s)[\mathbf{x}]}{\partial s}
$$

---

[1] If s is an angle variable, unless $s = 0 \pmod{2\pi}$.

and the hypothesis that the group has no isotropy implies that

$$0 \neq \frac{\partial g(s)[\mathbf{x}]}{\partial s}\bigg|_{s=0} = \mathbf{v}_1 \in \mathbb{R}^N.$$

The vector $\mathbf{v}_1$ plays the same role as one of the $\mathbf{v}_k$ in the previous section: it is orthogonal to each of the rows of B, thus it belongs to the kernel of $C = B^T B$. The hypothesis of no isotropy can be relaxed as follows: if

$$0 \neq \frac{\partial g(s)[\mathbf{x}^*]}{\partial s}\bigg|_{s=0}$$

the rank deficiency occurs for the normal matrix at the nominal solution $\mathbf{x}^*$.

If there are d groups of symmetries with one parameter and the vectors $\mathbf{v}_k$, defined by the derivative with respect to s of each group action, are linearly independent, then the rank deficiency is d.

The symmetries could be organized in higher dimension groups, such as the groups of translations and rotations. In such case we need to exploit the differentiable structure of the groups, that is they have to be **Lie groups** with differentiable actions on the parameter space $\mathbb{R}^N$. The no-isotropy condition cannot in general be satisfied and the group internal operation may not be commutative (the rotation group $SO(3)$ acting on $\mathbb{R}^3$ is an example of both difficulties). A theory of the Lie groups of symmetries would require some mathematical background, which is not worth presenting here. Thus in this book we shall replace, e.g., the symmetry group $SO(3)$, with three one-parameter symmetry groups, corresponding to the rotations around three orthogonal axes; this is analogous to what was done in Section 4.1.

If there is a symmetry group, then there is also a corresponding rank deficiency. If there is a rank deficiency, by (6.2) the one-parameter group of translations $\mathbf{x} \mapsto \mathbf{x} + s\,\mathbf{v}$ is such that the residuals change by quantities $\mathcal{O}(s^2)$, of higher order with respect to the change $s\,\mathbf{v}$, that is a "first order symmetry". Is there also an exact symmetry, defined by a one-parameter group operating by transformations other than translations? Under somewhat more restrictive hypotheses, this can be guaranteed. If the normal matrix $C(\mathbf{x})$, as computed at each value of the fit parameters $\mathbf{x}$, always has rank $N-1$, then the eigenvector with zero eigenvalue defines a smooth vector field.[2] The integral flow of such a vector field provides the symmetry group.

In conclusion, symmetries imply rank deficiencies, and (under some additional hypotheses) rank deficiencies imply symmetries. The two phenomena occur in the same cases: the examples in this and in the following chapters show that they occur often, and their understanding is critical.

---

[2] For the details of this construction, see Section 10.1.

Model problem with scaling

In the model problem with integral flow (6.4) there is a symmetry with a multiplicative parameter $w \in \mathbb{R}^+$

$$k \mapsto w^3 k, \qquad a_0 \mapsto w^2 a_0$$

leaving n invariant, thus the general solution of the equation of motion is also invariant, and so is the prediction function (the observation function being just a projection onto the second component of the state vector). This is a change of scale (see Section 4.1), e.g., could be obtained by changing the unit of length by a factor $L = w^2$ without changing the unit of time, then k^2 would change by a factor $L^3 = w^6$. The symmetry can be represented with an additive parameter s by setting $w = \exp(s)$. The derivative of the symmetry group action with respect to s is

$$\frac{da_0}{ds} = 2\,w^2\,a_0, \qquad \frac{dk}{ds} = 3\,w^3\,k$$

and for $s = 0, w = 1$ this gives a vector $(2\,a_0, 3\,k)$ orthogonal to $(-3/a_0, 2/k)$ which is the vector of coefficients of the equation constraining Δa_0 and Δk.

6.3 Approximate rank deficiency and symmetries

An **approximate rank deficiency** of order d means there is a subspace K of dimension d in $\mathbb{R}^N$, and a constant ϵ with $0 < \epsilon \ll 1$, such that

$$\mathbf{v} \in K, \ |\mathbf{v}| = 1 \Longrightarrow |B\,\mathbf{v}| \le \epsilon, \qquad (6.7)$$

a generalization of (6.1). Moreover (6.7) must not apply to a subspace of dimension $> d$. Equation (6.7) implies that the quadratic form defined by the normal matrix C, restricted to K, has values of order ϵ^2 on unit vectors:

$$\mathbf{v} \in K, \ |\mathbf{v}| = 1 \Longrightarrow \mathbf{v}^T\,C\,\mathbf{v} = (B\,\mathbf{v})^T\,B\,\mathbf{v} \le \epsilon^2.$$

We shall now study the properties of the eigenvalues of C. Let $\mathbf{v}_j$, for $j = 1, N$, be unit eigenvectors of the normal matrix C, with non-negative eigenvalues $0 \le \lambda_1 \le \lambda_2 \le \cdots \le \lambda_N$. The values of the quadratic form on the unit sphere are constrained by the spectrum of the matrix C:

$$\min_{|\mathbf{x}|=1} \mathbf{x}^T\,C\,\mathbf{x} = \lambda_1, \qquad \max_{|\mathbf{x}|=1} \mathbf{x}^T\,C\,\mathbf{x} = \lambda_N,$$

thus $\lambda_1 \le \epsilon^2$. In the simple case $d = 1$ there is one eigenvalue of the normal matrix $\le \epsilon^2$, one eigenvalue of the covariance matrix $\ge 1/\epsilon^2$, and one semiaxis of the confidence ellipsoid $Z_L(1)$ of length $\ge 1/\epsilon$.

Approximate rank deficiency of order $d > 1$

For $d > 1$ a similar result holds: if the subspace K has dimension d, and the quadratic form defined by C, restricted to the unit vectors of K, is $\leq \epsilon^2$, then the matrix C has at least d eigenvalues $\leq \epsilon^2$.

This can be proven by recursion on d. The case $d = 1$ has been proven above. Let us prove the result for d, assuming it applies to $d - 1$.

Let Z be the linear subspace of $\mathbb{R}^N$ orthogonal to the eigenvector $\mathbf{v}_1$. The intersection $K' = K \cap Z$ does not contain $\mathbf{v}_1 \in K$ and has dimension $d - 1$. Thus we can apply the same result to $K' \subset Z$: the quadratic form defined by C restricted to Z has values $\leq \epsilon^2$ on the unit vectors of K'. Thus the restriction of C to Z has $d - 1$ eigenvalues less than ϵ^2, but these eigenvalues are $\lambda_2 \leq \lambda_3 \leq \cdots \leq \lambda_N$: we conclude $\lambda_d \leq \epsilon^2$.

This has implications on the uncertainty of the results: in an approximate rank deficiency of order d, there are d eigenvalues of the covariance matrix $\Gamma = C^{-1}$ larger than $1/\epsilon^2$ and the confidence ellipsoid $Z_L(1)$ has d semiaxes longer than $1/\epsilon$. The converse is also true: if there are d semiaxes longer than $1/\epsilon$, then there are d eigenvalues of C smaller than ϵ^2, and the subspace K generated by the related eigenvectors $\mathbf{v}_1, \mathbf{v}_2, \ldots, \mathbf{v}_d$ fulfills the definition for approximate rank deficiency of order d, with small parameter ϵ.

This implies that approximate rank deficiency can be found a posteriori by **principal components analysis**, that is, after computing the covariance matrix Γ and its eigenvalues, by selecting the ones larger than some value, or anyway the few largest ones. As we shall see in the applications of this method, see Section 16.5, if the approximate rank deficiency is found a posteriori from the spectrum of the covariance matrix, we need additional effort to understand what is the source of the problem and whether it can be fixed or just needs to be swept under the carpet by descoping.

Approximate symmetries

An **approximate symmetry** is a differentiable group action changing the residuals by $\mathcal{O}(\epsilon)$, where ϵ is a small parameter. If G is a one-parameter group with a differentiable action on the fit parameters $\mathbf{x}$, such that for each $g(s) \in G$

$$\boldsymbol{\xi}\left(g(s)[\mathbf{x}]\right) = \boldsymbol{\xi}(\mathbf{x}) + \epsilon\, s\, \mathbf{a} + \mathcal{O}(s^2), \qquad \mathbf{a} \in \mathbb{R}^m;\ |\mathbf{a}| = 1 \qquad (6.8)$$

and the no-isotropy condition applies, at least locally near $\mathbf{x}^*$, then

$$\epsilon\, \mathbf{a} = \left.\frac{d\boldsymbol{\xi}\left(g(s)[\mathbf{x}]\right)}{ds}\right|_{s=0} = \frac{\partial\boldsymbol{\xi}}{\partial\mathbf{x}}(\mathbf{x})\left.\frac{\partial g(s)[\mathbf{x}]}{\partial s}\right|_{s=0} = B\,\mathbf{v}.$$

Thus, if $|\mathbf{v}| = v$ for the corresponding unit vector

$$B\,\hat{\mathbf{v}} = \frac{\epsilon}{v}\,\mathbf{a}$$

and the small parameter is ϵ/v. We conclude that there is an eigenvalue of the covariance matrix $\geq v^2/\epsilon^2$.

The case of a higher dimension group of symmetries is more delicate. We proceed as in Section 6.2 and assume that there are d local one-parameter groups of approximate symmetries, resulting in d weak directions $\mathbf{v}_i$:

$$B\,\mathbf{v}_i = \epsilon\,\mathbf{a}_i, \qquad |\mathbf{a}_i| \leq 1.$$

To simplify the discussion, we assume the symmetry groups have been re-parameterized in such a way that $|\mathbf{v}_i| = v_i = 1$. Let K be the subspace of dimension d generated by the $\mathbf{v}_i$. Then the quadratic form defined by C, restricted to K, has values $\mathcal{O}(\epsilon^2)$. However, if we need an explicit estimate we have to find the value p such that

$$\mathbf{x} \in K, |\mathbf{x}| = 1 \Longrightarrow \mathbf{x}^T\,C\,\mathbf{x} \leq p\,\epsilon^2.$$

In general $p \leq d^2$, and a better estimate can be obtained in special cases. Then, by applying the same recursive argument as before, there are d eigenvalues of C which have to be $\leq p\,\epsilon^2$. Thus there are d semiaxes of the ellipsoid $Z_L(1)$ longer that $1/(\epsilon\,\sqrt{p})$.

Taking also into account the re-parameterization problem, the above result is weaker than the statement "d one-parameter approximate symmetry groups imply an approximate rank deficiency of order d", because the small parameter is not the same. Still, symmetries can be an effective heuristic method to find and explain rank deficiencies, even approximate ones.

This method can be effectively used in case of **symmetry breaking**, if the residuals can be expanded in power series of some small parameter ϵ:

$$\boldsymbol{\xi} = \boldsymbol{\xi}_0 + \epsilon\,\boldsymbol{\xi}_1 + \mathcal{O}(\epsilon^2).$$

This can occur for different reasons, e.g., $\boldsymbol{\xi}_0$ could be the residuals obtained for some unperturbed equation of motion, $\epsilon\,\boldsymbol{\xi}_1$ could be the first-order change due to a minor perturbation containing ϵ. If a group G is an exact symmetry of the unperturbed problem obtained for $\epsilon = 0$, but the perturbation $\epsilon\,\boldsymbol{\xi}_1$ is not symmetric, then G is a group of approximate symmetries, although the explicit computation of the factor p is far from trivial. We shall see many examples of this in Section 6.5, and in Chapters 15, 16, and 17.

6.4 Scaling and approximate rank deficiency

Rank deficiency and exact symmetries are topological properties, that is their definitions (6.1) and (6.6) are independent of the choice of a metric in the $\mathbf{x}$ space of fit parameters. Approximate rank deficiency (and approximate symmetries) are metric properties, that is, (6.7) and (6.8) are conditions on the Euclidean lengths of the vectors $\mathbf{v}$ and $B\mathbf{v}$, with $\mathbf{v}$ in some subspace. If the Euclidean norms $|\mathbf{v}|, |B\mathbf{v}|$ are replaced by some other norms, such as $||\mathbf{v}||_W = \sqrt{\mathbf{v}^T\,W\,\mathbf{v}}$, with a symmetric positive definite weight matrix W, the definitions for the same ϵ are changed.

Let P be a diagonal matrix $P = \text{diag}[p_1, p_2, \ldots, p_N]$ with all $p_i > 0$. By **scaling** we mean a linear transformation in the space $\mathbb{R}^N$ of the fit parameters $\mathbf{x} = P\,\mathbf{y}$. A scaling changes the metric, that is $|\mathbf{x}| = ||\mathbf{y}||_W$ with $W = P^2$, and has the following effects on the normal equation

$$B_\mathbf{y} = \frac{\partial\xi}{\partial\mathbf{y}} = B\,P, \quad C_\mathbf{y} = B_\mathbf{y}^T\,B_\mathbf{y} = P\,C\,P, \quad \Gamma_\mathbf{y} = P^{-1}\,\Gamma\,P^{-1},$$

thus also the eigenvalues of C, Γ are changed by scaling. Scaling is implicit in the choice of units to convert the quantities to be determined into the numeric fit vector $\mathbf{x}$. The problem is, the units used to measure dimensionally different quantities are not easy to be compared. Are the initial conditions measured in astronomical units, rather than in centimeters, more or less suitable to be compared with the normalized[3] harmonic coefficients of the gravity of some planet? This question has no firm answer, still the change introduces a factor $\simeq \left[1.5 \times 10^{13}\right]^2$ in the matrix W.

We would like to define approximate rank deficiency in a way independent of the units used in the $\mathbf{x}$ space, that is invariant with respect to the choice of an arbitrary diagonal weight matrix. To this goal we need to define some standard metric, applicable to all orbit determination problems. There are two meaningful ways to do this.

A posteriori scaling

The *a posteriori* scaling of the normal matrix C is obtained by reducing to unitary length the columns of the design matrix B

$$b_i = \left|\frac{\partial\xi}{\partial x_i}\right| = \sqrt{c_{ii}}, \quad p_i = 1/b_i \quad \text{for } i = 1, N, \tag{6.9}$$

with $C = (c_{ij})$. Then the scaled normal matrix $C_\mathbf{y}$ has all the main diagonal coefficient equal to 1, all the others less than 1 in absolute value. This is not possible if some $b_i = 0$, but then there would be exact rank deficiency.

---

[3] Another example of scaling is the normalization of the harmonic coefficients, see Section 13.2.

A posteriori scaling is also used to increase the numerical stability of the matrix inversion $C^{-1} = \Gamma$, by computing $\Gamma_\mathbf{y} = C_\mathbf{y}^{-1}$ and then rescaling it to $\Gamma = P\Gamma_y P$. The reason is that the conditioning number of $C_\mathbf{y}$ is often much smaller than that of C. For example, if the matrix C is diagonal, whatever cond(C) the inversion is trivial, indeed $C_\mathbf{y} = I$ has conditioning number 1; if C is diagonal dominant[4] then $C_\mathbf{y}$ is also diagonal dominant and the inversion is numerically stable, even when cond(C) is very large. The scaled cond($C_\mathbf{y}$) indeed characterizes the difficulty of inversion and the stability of the result: even if there are methods (discussed in Section 5.3) to perform the inversion for very large cond($C_\mathbf{y}$), the iterations of differential corrections may diverge if this is the case.

The approximate rank deficiencies could be analyzed for the a posteriori scaled matrix $C_\mathbf{y}$. They indicate **aliasing**, that there are some fit parameters **x** which cannot be accurately entangled, having equivalent effect on the residuals (to first order). They also measure the numerical difficulty of inversion, even with rescaling: since the eigenvalues of $C_\mathbf{y}$ cannot be very large (they are anyway $\leq N$), if there is a large cond($C_\mathbf{y}$) there needs to be at least one eigenvalue of $C_\mathbf{y}$ smaller than $\epsilon = N/\text{cond}(C_\mathbf{y})$, thus an approximate rank deficiency for C_y of order $d \geq 1$, with ϵ as small parameter.

The matrices $C_\mathbf{y}, \Gamma_\mathbf{y}$ have no obvious probabilistic interpretation. If we scale the covariance matrix so that the main diagonal has coefficients 1

$$\text{Corr}(\mathbf{x}) = (r_{ij}), \quad r_{ij} = \frac{\gamma_{ij}}{\sqrt{\gamma_{ii}\,\gamma_{jj}}}$$

we obtain the **correlation matrix**: $r_{ij}(\mathbf{x}) = \text{Corr}(x_i, x_j)$ measures the dependence of the random variables x_i and x_j; if only these two were considered, there would be a simple interpretation in terms of regression line and of the size of the residuals (Section 5.1). Unfortunately, $\text{cond}(\text{Corr}(\mathbf{x})) \neq \text{cond}(\Gamma_\mathbf{y})$, although numerical tests with random design matrices B show that they are of the same order of magnitude.

A priori scaling

The **a priori scaling** of the normal matrix C is obtained by considering an a priori diagonal normal matrix C_0. It is not meant to represent a formal a priori knowledge, which could be handled with the addition of a priori observations according to eq. (6.3). The matrix C_0 could be built from upper bounds, derived from theory and/or very weak observational constraints, e.g., of the form $|x_i| \leq 1/\sqrt{c_{ii}}$. Then the a priori scaling would

---

[4] There are different possible definitions, e.g., with the coefficients on the main diagonal larger than the sum of the absolute values of the others on the same row or column.

be given by the diagonal matrix P such that $P\,C_0\,P = I$; i.e., each parameter x_i would be measured as a fraction of its upper bound.

The approximate rank deficiencies could be analyzed for the a priori scaled matrix $C_{\mathbf{y}}$. They are useful to identify the parameters, or group of parameters, which cannot be measured at all with the current experiment, to the point that the values formally determined may be absurd.

As the simplest example, if there is just one parameter such that the corresponding column of the a priori scaled design matrix $B_{\mathbf{y}} = B\,P$ has a norm $< \epsilon$, then the unit vector $\hat{\mathbf{v}}$ along the corresponding axis is such that $\hat{\mathbf{v}}^T\,C_{\mathbf{y}}\,\hat{\mathbf{v}} < \epsilon^2$ and the confidence ellipsoid has at least one semiaxis larger than $1/\epsilon$. In this case it does not matter whether the matrix inversion is stable, anyway the result is going to be absurd, and the following iterations of differential corrections may destabilize the result for other parameters. In such cases either descoping or constraining needs to be applied.

In conclusion, approximate rank deficiencies and approximate symmetries are extremely valuable as heuristic tools, either to debug the design of an orbit determination experiment or to warn about the unreliable results of one actual test. However, in most cases they provide only order of magnitude estimates, and have to be used with a bit of common sense.

6.5 Planetary systems: extrasolar planets

Planets orbiting around a star different from the Sun are one of the most important astronomical discoveries of recent years. Most such **extrasolar planets** have been discovered by measuring the radial velocity of the star with respect to the Solar System. As of October 2007, this was the case for 245 extrasolar planets[5] out of a confirmed total of 279. In most cases the radial velocity of the star was measured by spectrometry, with accuracies currently at a few m/s; in a few cases planets have been discovered around pulsars, where the radial velocity is measured by the Doppler shift of the pulsar. Anyway, the radial velocity needs to be measured with respect to the barycenter of the Solar System, thus the orbit of the Earth from which we observe must also be accurately computed in barycentric coordinates (see Section 4.3) and its radial velocity component along the line of sight has to be subtracted.[6] In this section we will outline the procedure for orbit determination of an extrasolar planetary system.

---

[5] See http://vo.obspm.fr/exoplanetes/encyclo/catalog.php for an up-to-date list; occultation is currently the second most effective method.

[6] If this subtraction were not accurate enough, it would lead to a pretended discovery of a planet with period an integer fraction of a year; in fact this happened once.

One planet

If there is only one planet, the dynamical model is the two-body problem with m_0, the mass of the observed star, significantly larger than m_1, the mass of the planet. Neglecting the interaction with the other stars (and with the galaxy) the center of mass of the system of two bodies with position vectors $\mathbf{r}_0$, $\mathbf{r}_1$ has a constant velocity in any inertial system, such as the one we adopt, which is moving with the center of mass of the Solar System. We are also neglecting stellar and galactic perturbations on the Solar System barycenter motion. The acceleration of the Sun towards the galactic center is just a few m/day^2, and if the star is much closer than the galactic center, the differential acceleration is much less.

With the standard reduction of the two-body problem, eq. (4.9),

$$\mathbf{b}_0 = \mu_1\, \mathbf{r}_1 + (1 - \mu_1)\mathbf{r}_0, \quad \mu_1 = \frac{m_1}{m_0 + m_1}, \quad \mathbf{b}_1 = \mathbf{r}_1 - \mathbf{r}_0$$

the coordinate of the star is $\mathbf{s} = \mathbf{r}_0 - \mathbf{b}_0 = -\mu_1\mathbf{b}_1$ in the barycentric system, and $\mathbf{r}_0 = \mathbf{b}_0 + \mathbf{s} = \mathbf{b}_0 - \mu_1\mathbf{b}_1$, with respect to the center of mass of our Solar System. Let (x, y, z) be the Cartesian coordinates of $\mathbf{b}_1$ in a reference system with the z axis in the direction $\hat{\mathbf{z}}$ of the Earth, (x_0, y_0, z_0) the corresponding coordinates of the center of mass, and let $(a, e, I, \Omega, \omega, v)$ be the Keplerian orbital elements of the two-body orbit of $\mathbf{b}_1$; the mean motion is $n = \sqrt{G\, M_0/a^3}$, with $M_0 = m_0 + m_1$.

In an orthogonal reference system centered in $\mathbf{b}_0$, with the x_1 axis directed along the Laplace–Lenz vector $\mathbf{e}$ and the z_1 axis along the direction $\hat{\mathbf{c}}_1$ of the angular momentum of $\mathbf{b}_1$, the coordinates of the Keplerian orbit are

$$\begin{bmatrix} x_1 \\ y_1 \\ z_1 \end{bmatrix} = \begin{bmatrix} r\cos v \\ r\sin v \\ 0 \end{bmatrix}, \quad r = \frac{a\,(1 - e^2)}{1 + e\cos v}.$$

By rotating this vector around the angular momentum axis by an angle ω

$$\begin{bmatrix} x_2 \\ y_2 \\ z_2 \end{bmatrix} = R_{\omega\,\hat{\mathbf{c}}_1} \begin{bmatrix} x_1 \\ y_1 \\ z_1 \end{bmatrix} = \begin{bmatrix} r\cos(v + \omega) \\ r\sin(v + \omega) \\ 0 \end{bmatrix}$$

we have the x_2 axis along the ascending node $\hat{\mathbf{N}} = (\hat{\mathbf{z}} \times \hat{\mathbf{c}}_1)/\sin I$. Rotating by an angle I around $\hat{\mathbf{N}}$

$$\begin{bmatrix} x_3 \\ y_3 \\ z_3 \end{bmatrix} = R_{I\,\hat{\mathbf{N}}} \begin{bmatrix} x_2 \\ y_2 \\ z_2 \end{bmatrix} = \begin{bmatrix} r\cos(v + \omega) \\ r\sin(v + \omega)\,\cos I \\ r\sin(v + \omega)\,\sin I \end{bmatrix}$$

and finally, rotating by an angle Ω around the $\hat{z}$ axis, we obtain the vector $\mathbf{b}_1$, but this is not necessary because $z = z_3$. In conclusion the observable radial velocity of the star is

$$\dot{\mathbf{r}}_0 \cdot \hat{z} = \dot{\mathbf{b}}_0 \cdot \hat{z} - \mu_1 \dot{z} = \dot{z}_0 - \mu_1 \sin I \, \frac{d}{dt}[r \, \sin(v + w)],$$

with time derivative

$$\frac{d}{dt}[r \, \sin(v + w)] = \dot{r} \, \sin(v + w) + r \, \dot{v} \, \cos(v + w)$$

where $\dot{v}$ is obtained from the angular momentum c_1

$$\dot{v} = \frac{c_1}{r^2} = \sqrt{G \, M_0 \, a \, (1 - e^2)} \, \frac{(1 + e \, \cos v)^2}{[a \, (1 - e^2)]^2} = \frac{n}{(1 - e^2)^{3/2}} \, (1 + e \, \cos v)^2$$

$$r \, \dot{v} = \frac{n \, a}{\sqrt{1 - e^2}} \, (1 + e \, \cos v)$$

and $\dot{r}$ is obtained from the derivative of the conic section equation

$$\dot{r} = \frac{a \, (1 - e^2)}{(1 + e \, \cos v)^2} \, e \, \sin v \, \dot{v} = \frac{n \, a}{\sqrt{1 - e^2}} \, e \, \sin v.$$

Then

$$\frac{d}{dt}[r \, \sin(v + w)] = \frac{n \, a}{\sqrt{1 - e^2}} \, [\cos(v + w) + e \, \cos w],$$

$$\dot{\mathbf{r}}_0 \cdot \hat{z} = \dot{z}_0 - \frac{\mu_1 \, \sin I \, n \, a}{\sqrt{1 - e^2}} \, [\cos(v + w) + e \, \cos w] \qquad (6.10)$$

is the observable as an explicit function of the orbital elements, dependent upon time only through the true anomaly v.

Circular approximation

If we assume a circular orbit ($e = 0$) then $\dot{v} = n$, w is just a phase, and

$$\dot{\mathbf{r}}_0 \cdot \hat{z} = \dot{z}_0 - K \, \cos(v + w) = \dot{z}_0 + (-K \, \cos w) \, \cos v + (K \, \sin w) \sin v, \quad (6.11)$$

where
$$K = \mu_1 \, \sin I \, n \, a = \frac{G^{1/3} \, n^{1/3} \, m_1 \, \sin I}{M_0^{2/3}}$$

is a constant depending upon the masses m_0, m_1, the mean motion n, and the inclination I. If we have measured the mean motion n as the main frequency of the signal, then eq. (6.11) is a linear model (Section 5.1) with three constants $k_1 = \dot{z}_0$, $k_2 = -K \, \cos w$, $k_3 = K \, \sin w$ to be fitted.

If the observations were available for every time t this would be the simplest problem of Fourier analysis, but in practice the observations are given at times sampled in a non-uniform way (only during the nights) and over a finite time span. Let the m observations $\dot{\mathbf{r}} = (\dot{r}_i)$ be at times $\mathbf{t} = (t_i)$ and let us assemble the vectors of values $\cos \mathbf{v} = \cos(v(t_i))$, $\sin \mathbf{v} = \sin(v(t_i))$:

$$\boldsymbol{\xi} = \dot{\mathbf{r}} - k_1 \mathbf{1} - k_2 \cos \mathbf{v} - k_3 \sin \mathbf{v},$$

$$B = \frac{\partial \boldsymbol{\xi}}{\partial (k_1, k_2, k_3)} = -\begin{bmatrix} \mathbf{1} & \cos \mathbf{v} & \sin \mathbf{v} \end{bmatrix}.$$

If $\cos \mathbf{v} \cdot \mathbf{1} = \sin \mathbf{v} \cdot \mathbf{1} = \cos \mathbf{v} \cdot \sin \mathbf{v} = 0$ the normal matrix is diagonal and the solution is

$$k_1 = \bar{\dot{r}} = \frac{1}{m} \sum_{i=1}^{m} \dot{r}_i, \qquad k_2 = \frac{\dot{\mathbf{r}} \cdot \cos \mathbf{v}}{\cos \mathbf{v} \cdot \cos \mathbf{v}}, \qquad k_3 = \frac{\dot{\mathbf{r}} \cdot \sin \mathbf{v}}{\sin \mathbf{v} \cdot \sin \mathbf{v}}.$$

Moreover, if the data points are evenly distributed with respect to the phase of the signal sinusoidal terms, then $\cos \mathbf{v} \cdot \cos \mathbf{v} \simeq \sin \mathbf{v} \cdot \sin \mathbf{v} \simeq m/2$ and the least squares solution is given by the classical Fourier analysis formulae, the integrals being replaced by finite sums over the available data points. In practice the base vectors are not orthogonal, the normal matrix is full and it has to be inverted. A solution could be obtained by Gram–Schmidt orthonormalization of the three vectors $\mathbf{1}, \cos \mathbf{v}, \sin \mathbf{v}$ (Ferraz-Mello 1981). If the time span of observations is not much longer than the period $2\pi/n$, the correlations between the constant and the sine wave are large and the solution is significantly different from the one obtained by the Fourier analysis formulae.

Anyway the result is that the radial velocity $\dot{z}_0$ of the center of mass, the phase w, and the amplitude K can be independently determined. The problem is that, even assuming n is known, K contains three parameters $m_1, m_0, \sin I$. Let us assume the mass of the star is measured independently from the spectrum, using the Hertzsprung–Russel relationship between color and mass, and let us approximate $M_0 \simeq m_0$.[7] Then only the product $m_1 \sin I$ appears in the signal. Inspection of eq. (6.10) shows that this has nothing to do with the circular orbit approximation, but is an intrinsic property of the observable radial velocity.

---

[7] A more rigorous procedure would assign an uncertainty to this estimate and add an a priori observation to account for this. Also, if the planetary mass m_0 is not negligible with respect to that of the star, the nonlinear effects are stronger.

First order in eccentricity

A solution applicable to eccentric orbits, but approximated by neglecting the terms of degree two in eccentricity, can be obtained by truncation from the formula for the time derivative of the true anomaly v

$$\frac{(1-e^2)^{3/2}}{(1+e\cos v)^2}\, \dot{v} = n \Longrightarrow l = n\,(t-t_0) = \int_0^v (1-2\,e\cos v)\,dv + \mathcal{O}(e^2)$$

providing an approximate expression for the **mean anomaly** l

$$v - 2\,e\sin v + \mathcal{O}(e^2) = l$$

and the inverse formula

$$v = l + 2\,e\sin l + \mathcal{O}(e^2)$$

implying

$$\sin v = \sin l + e\sin(2\,l) + \mathcal{O}(e^2), \qquad \cos v = \cos l + e\cos(2\,l) - e + \mathcal{O}(e^2).$$

Substituting into $\cos(v+\omega)$ we obtain, neglecting $\mathcal{O}(e^2)$ terms,

$$\dot{\mathbf{r}}_0 \cdot \hat{\mathbf{z}} = \dot{z}_0 - \mu_1 \sin I\, n\, a\, [\cos(l+\omega) + e\cos(2\,l+\omega)]$$

and the signal appears as a constant plus two trigonometric terms, with frequencies n and $2\,n$ and phase constants $-n\,t_0+\omega$, $-2\,n\,t_0+\omega$, respectively. Assuming n is known from frequency analysis, this suggests using a linear model with five base functions and five coefficients k_i, $i = 1,5$:

$$\dot{\mathbf{r}}_0 \cdot \hat{\mathbf{z}} = k_1 + k_2\,\cos(n\,t) + k_3\,\sin(n\,t) + k_4\,\cos(2\,n\,t) + k_5\,\sin(2\,n\,t).$$

Thus it is possible to find a linear least squares solution for these parameters, determining $e, \omega, m_1\, \sin I, t_0$ and $\dot{z}_0$; a is determined from n (again assuming m_0 is known, and approximating $M_0 \simeq m_0$). The orbital element Ω cannot be determined at all, and the mass m_1 cannot be entangled from the orbit plane parameter $\sin I$. Thus, if the problem were posed with eight parameters to be solved $(a, e, I, \Omega, \omega, t_0, \dot{z}_0, m_1)$ there would be a rank deficiency of 2. Again, this is not due to the first-order truncation in e.

Rank deficiency of the exoplanet problem

In this problem there is an exact symmetry, even without any truncation to some power of e: the rotations of the exoplanetary system around the z axis $R_{s\,\hat{z}}$, for each rotation angle s, leave the z component of $\dot{\mathbf{r}}_0$ unchanged, thus the residuals are exactly the same. The effect of such rotation in the orbital elements, given that the inclination is measured with respect to the plane parallel to (x, y) passing through the star, is simply $\Omega \to \Omega + s$. Thus $\partial \xi / \partial \Omega = \mathbf{0}$, that is, if Ω is included in the list of parameters to be solved, the matrix B has one column of zeros, and the normal matrix $C = B^T B$ has one row and one column of zeros. This problem is easily solved by descoping, that is by determining a list of parameters not including Ω.

In the approximation $M_0 \simeq m_0$ the observables of eq. (6.10) contain m_1 and $\sin I$ only through their product $m_1 \sin I$

$$m_1 \frac{\partial \xi}{\partial m_1} = \sin I \frac{\partial \xi}{\partial (\sin I)},$$

thus two rows of the matrix B are always linearly dependent and there is one additional rank deficiency. A vector $\mathbf{v}$ satisfying $B \mathbf{v} = \mathbf{0}$ is

$$\mathbf{v} = \sin I \, \nabla(m_1) + m_1 \, \nabla(\sin I)$$

where $\nabla(m_1)$, $\nabla(\sin I)$ are vectors pointing in the direction of change of m_1, $\sin I$, respectively.

Again descoping is the only solution, that is the parameter to be solved is just the product $m_1 \sin I$. This has a serious implication due to the fact that an extrasolar planet is, by definition, not a companion star. Thus to prove that an extrasolar system contains a planet it is necessary to provide an upper limit for the mass (e.g., 13 times the mass of Jupiter is estimated to be the lower mass limit for deuterium fusion, thus for a body shining by its own radiation). If only the combination $m_1 \sin I$ is constrained by the observations, there is often no way to exclude a very low value of I and a large mass. In some cases it is possible to exclude a low I because partial occultations of the star by the planet are observable.

Exoplanetary systems

The rank deficiencies described above can be found also in an **extrasolar planetary system** with more than one planet. Let us suppose there is a star and two planets, with masses m_0, m_1, m_2 and let $\mathbf{b}_0, \mathbf{b}_1, \mathbf{b}_2$ be their positions in barycentric coordinates. Then the observable is

$$\dot{\mathbf{r}}_0 \cdot \hat{\mathbf{z}} = [\dot{\mathbf{b}}_0 + \dot{\mathbf{s}}] \cdot \hat{\mathbf{z}}$$

with the barycentric star position

$$\mathbf{s} = -\frac{m_1}{M_0} \mathbf{b}_1 - \frac{m_2}{M_0} \mathbf{b}_2,$$

where M_0 is the total mass. The rotations $R_{s\hat{\mathbf{z}}}$ around the line of sight are exact symmetries of the equation of motion. If the two planets are parameterized by Keplerian orbital elements $(a_j, e_j, I_j, \Omega_j, \omega_j, v_j)$, $j = 1, 2$, the effect of $R_{s\,\hat{z}}$ is $\Omega_j \to \Omega_j + s$. Thus a rank deficiency occurs with

$$\frac{\partial \xi}{\partial \Omega_1} + \frac{\partial \xi}{\partial \Omega_2} = 0.$$

The solution is to remove Ω_1 and Ω_2 from the list of parameters, replacing them with $\Omega_2 - \Omega_1$. However, even after this descoping by one parameter the system still has some approximate rank deficiencies, resulting from the smallness of the mass ratios m_1/m_0, m_2/m_0. If we were to neglect the mutual perturbations of the two planets (both direct and indirect), by using eq. (6.10) for each of the two planets we obtain

$$\dot{\mathbf{r}}_0 \cdot \hat{\mathbf{z}} = \dot{z}_0 - \sum_{j=1}^{2} \frac{m_j \sin I_j \, n_j \, a_j}{(m_0 + m_j)\,\sqrt{1 - e_j^2}} \; [\cos(v_j + \omega_j) + e_j \, \cos \omega_j]$$

with $n_j = \sqrt{G\,m_0/a_j^3}$. Then both angles Ω_j do not appear and the inclinations appear only in the combinations $m_j/(m_0 + m_j) \sin I_j$. If we were trying to solve for 15 parameters (2×6 orbital elements plus $m_1, m_2, \dot{z}_0$) the normal matrix would have rank 11.

The equation of motion for a three-body system in barycentric coordinates can be described as two two-body systems with perturbing accelerations containing the planetary masses m_1, m_2. We define an adimensional small parameter $\mu = (m_1 + m_2)/m_0$, which is contained in all perturbation terms. Then the short periodic perturbations of the two two-body subsystems are $\mathcal{O}(\mu)$ and the perturbations in the observable due to these perturbations are $\mathcal{O}(\mu^2)$. Hence the observable is $\dot{z}_0 + \mathcal{O}(\mu) + \mathcal{O}(\mu^2)$, the second-order part depending only upon the perturbations. Thus the coupling between two (or more) planets is an example of symmetry breaking with small parameter μ, and there is an approximate rank deficiency of order 3, with $\epsilon = \mathcal{O}(\mu)$; however, an explicit computation of ϵ is not easy. The practical rule of thumb is that an observation time span of the order of $1/\mu$ periods of the planets is required to determine m_1, m_2, I_1, I_2 in a robust way.[8]

6.6 Planetary systems: the Solar System

The orbits of the major planets of our Solar System can be the subject of orbit determination. From the discussion in Section 4.6, at least the eight major planets have significant interactions, thus their orbits need to be solved at

---

[8] In systems with two planets in a mean motion resonance a time span of the order of $1/\sqrt{\mu}$ could be enough.

once, with all their masses among the fit parameters. Some of the major satellites (especially the Moon) and the largest asteroids need to be taken into account. To control the complexity of the problem, it is possible to decouple the orbit determination for some satellites and asteroids.

Symmetries

This orbit determination problem has symmetries and rank deficiencies depending upon the set of available observations. If the observations are only relative, between planets of our Solar System (e.g., if all the observations are from Earth), then all observables, be they either angles or distances between planets or their derivatives, are exactly symmetric with respect to the rotation group $SO(3)$. Thus it is essential to add three a priori constraints, that is to select an inertial reference system (see Appendix C).

If the observations are absolute, connected to some extrasolar inertial reference system, e.g., angles relative to "fixed stars" (in practice, a non-rotating astrometric catalog of stars), then the $SO(3)$ symmetry disappears. If there are only angular observations the change of scale defined in Section 4.1 is an exact symmetry and rank deficiency of order 1 occurs. Thus the classical results on the orbit determination of the planets were only expressed in terms of a unit of length, the **astronomical unit** (AU), whose value in terrestrial units could not be accurately determined.[9]

The current state-of-the-art is more complex. Both absolute angular observations of the planets (with respect to star catalogs) and relative range/range-rate observations are available. The latter are obtained by planetary radar, by Lunar laser ranging, by tracking landers on other planets, by the orbit determination of artificial satellites of other planets. In general, range observations are more accurate than angular observations, although there has been progress also in the latter. For example, in the case study of Chapter 17 the accuracy in range is expected to be 10 cm over a distance $\simeq 1$ AU, with a relative accuracy $< 10^{-12}$. Thus there is an approximate $SO(3)$ symmetry, in that the angular observations are by far less accurate.

The translations of the barycenter $\mathbf{b}_0$ (see Section 4.3) do not affect observations between Solar System bodies. To remove the corresponding exact rank deficiency we can use an equation of motion in barycentric coordinates (4.16) with a priori constraints $\mathbf{b}_0 = \dot{\mathbf{b}}_0 = \mathbf{0}$: the coordinates of the Sun are eliminated, being computed from (4.15).

---

[9] The first accurate value of the AU was obtained from the parallax of the asteroid Eros in 1898.

Relativistic effects

A full discussion of the complex problem of the **planetary ephemerides** is beyond the scope of this book. We would like to mention one key feature, which will be needed in Chapter 17. The equation of motion for the planets to be used for orbit determination, given the current observational accuracy, is not just that of the N-body problem, but has to be fully compliant with **general relativity**. Also the observation functions have to be fully relativistic, taking into account the non-Newtonian properties of light propagation in a curved space-time, and the space-time coordinates have to be selected with care and computed with accurate transformation equations. We have used two simulations of an orbit determination experiment for a

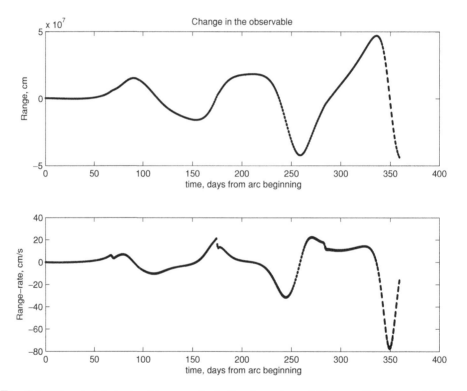

Fig. 6.1. The relativistic signal in range (including dynamics and Shapiro effect) from a Mercury orbiter over a 1-year mission has a peak-to-peak amplitude of $\simeq 900$ km, with S/N $\simeq 9 \times 10^6$ (assuming the accuracy of the experiment described in Chapter 17). In range-rate the peak-to-peak signal is $\simeq 1$ m/s with S/N $\simeq 3 \times 10^5$.

Mercury orbiter, one with a pure Newtonian N-body model for the orbits of the planets, the other with a fully relativistic model. The differences plotted in Figure 6.1 show that, with state-of-the-art tracking systems, there is a very large signal-to-noise ratio in the relativistic effects.

There is no way to really explain general relativity in short. We shall give the equation of motion by using the **parametric post-Newtonian** approach: the relativistic equation of motion is linearized with respect to the small parameters v_i^2/c^2 and Gm_i/r_{ik}, where v_i is the barycentric velocity for each of the bodies of mass m_i, c is the speed of light, and r_{ik} is a mutual distance, appearing in the metric of the curved space-time, hence in the equations for geodesic motion. This can be formalized by adding to the Lagrangian L_{NEW} of the N-body problem, given in eq. (4.2), some corrective terms of post-Newtonian (PN) order 1 in the small parameters

$$L = L_{NEW} + L_{GR}. \tag{6.12}$$

By using the notation (Moyer 2003)

$$\mathbf{r}_{ij} = \mathbf{r}_j - \mathbf{r}_i, \quad r_{ij} = |\mathbf{r}_{ij}|$$
$$\mathbf{v}_{ij} = \dot{\mathbf{r}}_j - \dot{\mathbf{r}}_i = \mathbf{v}_j - \mathbf{v}_i, \quad v_{ij} = |\mathbf{v}_{ij}|,$$

L_{GR} can be written in a synthetic way as

$$L_{GR} = \frac{1}{8\,c^2} \sum_i m_i v_i^4 - \frac{1}{2\,c^2} \sum_i \sum_{j\neq i} \sum_{k\neq i} \frac{G^2\,m_i\,m_j\,m_k}{r_{ij}\,r_{ik}} \tag{6.13}$$

$$+ \frac{1}{2\,c^2} \sum_i \sum_{j\neq i} \frac{G\,m_i\,m_j}{r_{ij}}$$

$$\times \left[\frac{3}{2}(v_i^2 + v_j^2) - \frac{7}{2}(\mathbf{v}_i \cdot \mathbf{v}_j) - \frac{1}{2\,r_{ij}^2}(\mathbf{r}_{ij} \cdot \mathbf{v}_i)(\mathbf{r}_{ij} \cdot \mathbf{v}_j) \right].$$

If the observations are between bodies of the Solar System, including space probes, the position and velocity of the Sun have to be constrained by the barycenter integrals to avoid an exact rank deficiency of order 6. However, the integrals for the Lagrangian system defined by L are different from the ones of L_{NEW}: by using the translations group of symmetries and Noether's theorem we get a relativistic total linear momentum $\mathbf{p}$:

$$\mathbf{p} = \sum_i \frac{\partial L}{\partial \mathbf{v}_i} = \sum_i m_i \mathbf{v}_i \left[1 + \frac{v_i^2}{2\,c^2} - \frac{U_i}{2\,c^2} \right] - \frac{1}{2\,c^2} \sum_i \sum_{j\neq i} \frac{G\,m_i\,m_j}{r_{ij}^3} (\mathbf{r}_{ij} \cdot \mathbf{v}_j)\,\mathbf{r}_{ij}$$

$$\tag{6.14}$$

where $U_i = \sum_{k\neq i} G m_k/r_{ik}$ is the potential at the i-th body, neglecting terms of PN order 2. We have $\dot{\mathbf{p}} = \mathbf{0}$, thus $\mathbf{p}$ is a vector integral. The vector

$$\mathbf{P} = \sum_i m_i \mathbf{r}_i \left[1 + \frac{v_i^2}{2\,c^2} - \frac{U_i}{2\,c^2} \right] \tag{6.15}$$

has the property, again neglecting $\mathcal{O}(v^4/c^4)$, that $\dot{\mathbf{P}} = \mathbf{p}$, thus it moves in a linear uniform way like the Newtonian center of mass. A relativistic center of mass can be defined as

$$
\mathbf{b}_0 = \frac{\sum_i m_i \mathbf{r}_i \left[1 + \frac{v_i^2}{2\,c^2} - \frac{U_i}{2\,c^2} \right]}{\sum_i m_i \left[1 + \frac{v_i^2}{2\,c^2} - \frac{U_i}{2\,c^2} \right]}
\tag{6.16}
$$

where the denominator, neglecting PN order 2 terms, is $\sum_i m_i + H/c^2$, with H the Hamiltonian, and it is also an integral. Thus we can use the constraint $\mathbf{b}_0 = \dot{\mathbf{b}}_0 = \mathbf{0}$ to reduce the dimensionality of the vector of parameters to be solved: the position and velocity of the Sun can be eliminated from the equation of motion and computed from those of the other bodies by

$$
\mathbf{r}_0 = \frac{-\sum_{i \neq 0} m_i \mathbf{r}_i \left[1 + \frac{v_i^2}{2\,c^2} - \frac{U_i}{2\,c^2} \right]}{m_0 \left(1 + \frac{v_0^2}{2\,c^2} - \frac{U_0}{2\,c^2} \right)}.
\tag{6.17}
$$

With this Lagrangian formalism, the relativistic equation of motion at the post-Newtonian order 1 are well defined and can be used for orbit determination of the planets, also of interplanetary space probes. Moreover, with state-of-the-art interplanetary tracking data, given the very large S/N ratio of the relativistic effects as shown in Figure 6.1, it is possible to test general relativity to great accuracy. The same formalism allows us to parameterize the equation of motion (and other relativistic effects) with constants having fixed values in Einstein's theory, and to solve for their value, together with the initial conditions and instrumental parameters, in the orbit determination procedure. One such post-Newtonian parameter γ, with value 1 in general relativity, controls how the space-time curvature depends upon the gravitational potential energy. A deviation from general relativity can be introduced with velocity-dependent terms

$$
L_{\bar{\gamma}} = \frac{\bar{\gamma}}{2c^2} \sum_i \sum_{j \neq i} \frac{G m_i m_j}{r_{ij}} v_{ij}^2,
$$

where $\bar{\gamma} = \gamma - 1$. The Eddington parameter β, equal to 1 in general relativity, appears in the three-body interactions, thus a violation can be introduced with

$$
L_{\bar{\beta}} = -\frac{\bar{\beta}}{c^2} \sum_i \sum_{j \neq i} \sum_{k \neq i} \frac{G^2 \, m_i \, m_j \, m_k}{r_{ij} \, r_{ik}},
$$

where $\bar{\beta} = \beta - 1$. In the general relativity theory of Einstein the only free parameter is G, which is constant ($\dot{G} = 0$); nevertheless, the product $G m_0$ changes because of the mass shed by the Sun as radiation and charged particles, see Section 4.6. This effect can be included in the dynamical model by the Lagrangian term

$$L_\zeta = (t - t_0)\,\zeta \sum_{i \neq 0} \frac{G\,m_0\,m_i}{r_{i0}}, \qquad \zeta = \frac{d(G\,m_0)/dt}{G\,m_0},$$

where t_0 is a reference epoch for m_0. For an accurate orbit determination of Mercury, the non-spherical shape of the rotating Sun does matter, since $r_{10}/R_\odot \simeq 900$: the corresponding Lagrangian term $L_{J2\odot}$ is the zonal spherical harmonic of degree 2 of the Sun, with respect to the rotation axis of the Sun, see Section 13.2. Thus the equation of motion has to be deduced from the Lagrangian

$$L = L_{NEW} + L_{GR} + L_{\bar\gamma} + L_{\bar\beta} + L_\zeta + L_{J2\odot}. \tag{6.18}$$

Equation (6.14) for the total linear momentum does not change for the added terms, because the $\partial L_\gamma/\partial \mathbf{v}_i$ cancel in the sum ($\sum_i \partial L_\gamma/\partial \mathbf{v}_i = \mathbf{0}$), the other three terms do not depend upon the velocities, thus the equation (6.17) for the Sun is not changed.

Lagrangian terms can describe other violations, e.g., the violations of the strong equivalence principle and preferred frame effects (see Section 17.5). A violation of the **equivalence principle** is obtained by assuming that in the Newtonian Lagrangian L_{NEW} the gravitational masses m_i, m_j, as they appear in the potential terms $G\,m_i\,m_j/r_{ij}$, are not the same as the inertial masses m_j^I appearing in the kinetic energy terms $m_j^I\,v_j^2/2$. The difference may depend upon the composition of the mass m_j, e.g., the fraction of the mass which results from the rest mass, from nuclear binding energy, from gravitational self-energy. While the dependence upon the nuclear binding energy has been excluded by laboratory experiments to very great accuracy (better than 10^{-12}), the dependence on gravitational self-energy is difficult to test because this fraction is very small for all bodies we can use: even for the Sun the fraction is $\Omega_0 \simeq -3.52 \times 10^{-6}$. If we assume $m_0^I = m_0\,[1 - \eta\,\Omega_0]$, the parameter η can be tested by orbit determination. However, in the equation of motion for the planets in barycentric coordinates the inertial mass of the Sun m_0^I does not appear directly. The change occurs in the integral of the center of mass, where m_0 is replaced by m_0^I, resulting in a modified equation for the coordinates of the Sun

$$\mathbf{r}_0 = \frac{-\sum_{i \neq 0} m_i \mathbf{r}_i \left[1 + \frac{v_i^2}{2\,c^2} - \frac{U_i}{2\,c^2}\right]}{m_0\,[1 - \eta\,\Omega_0]\left(1 + \frac{v_0^2}{2\,c^2} - \frac{U_0}{2\,c^2}\right)}.$$

The indirect perturbation from this displacement of the Sun affects the orbits of the other bodies: $\partial \ddot{\mathbf{r}}_j / \partial \eta \neq \mathbf{0}$ contains $\Omega_0 \, m_k$ for each $k \neq 0$, the contribution of Jupiter is as large as that of all the other planets together.

To model **preferred frame** effects requires us to add to the Lagrangian

$$L_\alpha = \frac{\alpha_2 - \alpha_1}{4 \, c^2} \sum_j \sum_{i \neq j} \frac{G \, m_i \, m_j}{r_{ij}} (\mathbf{v}_i + \mathbf{w}) \cdot (\mathbf{v}_j + \mathbf{w})$$

$$- \frac{\alpha_2}{4 \, c^2} \sum_j \sum_{i \neq j} [\mathbf{r}_{ji} \cdot (\mathbf{v}_j + \mathbf{w})] \, [\mathbf{r}_{ji} \cdot (\mathbf{v}_i + \mathbf{w})] \frac{G \, m_i \, m_j}{r_{ij}^3}$$

with two additional post-Newtonian parameters α_1, α_2 and with $\mathbf{w}$ the velocity of the Solar System barycenter with respect to the preferred frame, usually assumed to be that of the cosmic microwave background, thus $|\mathbf{w}| = 370 \pm 10$ km/s in the direction $(\alpha, \delta) = (168°, 7°)$.

The problem arises from the presence of additional terms in the total linear momentum integral $\mathbf{p}$: the Lagrangian $L + L_\alpha$, where L is from eq. (6.18), is still invariant by translation. The integral from Noether's theorem is

$$\sum_i \frac{\partial (L + L_\alpha)}{\partial \mathbf{v}_i} = \mathbf{p} + \sum_i \frac{\partial L_\alpha}{\partial \mathbf{v}_i}$$

with $\mathbf{p}$ from eq. (6.14). This integral is not the derivative of $\mathbf{P}$ and this cannot be fixed by changing the definition, that is, a center of mass integral does not exist (Will 1981, Section 4.4). A possible solution to formulate in a consistent way the equation of motion with preferred frame effects is still to use a reference system centered in $\mathbf{b}_0$ as defined in eq. (6.15), which is however accelerated by

$$\ddot{\mathbf{b}}_0 = - \frac{\frac{d}{dt} \sum_i \frac{\partial L_\alpha}{\partial \mathbf{v}_i}}{\sum_i m_i \left[1 + \frac{v_i^2}{2 \, c^2} - \frac{U_i}{2 \, c^2} \right]},$$

and the equation of motion is the Lagrange equation with Lagrangian $L + L_\alpha$, with the additional "apparent acceleration" $-\ddot{\mathbf{b}}_0$.

There are many other possibilities of violations of the fundamental laws of gravitation and of inertia, including violations of the conservation of total linear momentum, of total angular momentum, and of total energy, and also violations of the action–reaction law, but most of these appear unlikely. Thus a list of parameters to be solved including $\gamma, \beta, \zeta, J_{2\odot}, \eta, \alpha_1, \alpha_2$ can be appropriate for a test of the theory of gravitation based on Solar System orbit determination. More on this subject is explained in Chapter 17.

Part III
Population Orbit Determination

7

THE IDENTIFICATION PROBLEM

The **identification** problem is the attempt to find, among independent detections of celestial bodies, those belonging to the same physical object. The problem becomes more difficult as the population of observed objects increases (see Chapter 11). This chapter is based on (Milani 1999, Milani *et al.* 2000a, Milani *et al.* 2001a) and ongoing research. The main example is the population of small Solar System bodies. Most of the observable ones are asteroids, although the observed population contains a smaller fraction of comets and others. We will use the word **asteroid** in the following discussion, although it applies also to the other populations.

An asteroid is typically observed only over a time span of a few hours to a few weeks, and is bright enough to be visible only over the **apparition**, a time interval spanning at most a few months. If this time span is not exploited in full, the single apparition orbit determination either is impossible or results in a rapidly growing prediction uncertainty: by the time of the next apparition the asteroid could be in a portion of the sky larger than the field of view of the telescopes available for the recovery. Thus we have a **lost asteroid**, that is, it is more likely to be rediscovered by chance than by looking at the predicted position. The databases of detected Solar System objects contain many single apparition arcs: the goal is to join together those of the same object, allowing for an accurate orbit determination.

7.1 Classification of the problem

The identification problem deals with separate sets of observations, which might, and might not, belong to the same object. As a basic form of the problem, we assume that these observations are partitioned into exactly two arcs, and that the observations of the same arc are of the same object.[1]

---
[1] The latter assumption may also fail, as discussed in Chapter 11.

Orbit identification

The problem can be classified as **orbit identification** when the observations of both arcs are sufficient to separately solve for two least squares orbits, one for each arc: then the input data include two sets of orbital elements, with their covariance matrices and residuals. The identification is confirmed if the observations from both arcs can be fitted to a single orbit.

To test two given orbits for possible identification is not simple, because of the strong nonlinearity of the orbit determination problem: we need a first guess orbit to start the differential corrections procedure. Nevertheless, this basic problem is much less difficult than the global problem: given a catalog containing N short arc orbits, we want to know which of the $N(N-1)/2$ couples belong to the same object, and how to compute a catalog of all the orbits of the physically distinct objects. With the modern catalogs including hundreds of thousands of orbits, and the next-generation surveys expected to discover tens of millions of objects, such a problem could lead to unacceptable computational complexity, unless it is tackled with a smart algorithm. Thus there are three steps in the orbit identification problem:

(i) to propose identifications, by selecting a small subset of couples;

(ii) to compute a preliminary orbit as first guess for each couple in (i);

(iii) to iterate differential corrections for each couple of arcs together, checking convergence to an orbit solution with acceptable residuals.

Attribution

The identification problem can be classified as **attribution** when an amount of data insufficient to compute a unique orbit for one arc (e.g., two two-dimensional observations, that is $m = 4$) is compared to an orbit already computed for the other arc. Not enough information is available in the orbit space, thus we need to compare the data in the observation space: the predictions from the orbit with the observations from the other arc.

If there is a catalog of N orbits and M observed arcs, each one too short to compute an orbit, this global problem has to be decomposed into three steps similar to those of the orbit identification case. The number of proposed attributions in step (i) must be much smaller than $M \times N$. Step (ii) can be less difficult in that the original orbit from the better arc could be good enough to serve as preliminary orbit, but this is not always the case. Step (iii) is the same as above, but the quality control to be applied to the residuals can take into account the asymmetry between the two arcs.

Recovery and precovery

This is the procedure to search for other observations belonging to the same physical object, assuming they are not already in the databases of past observations. It can take two forms: **recovery** in the future and in the sky, by pointing a telescope at one or more predicted positions of an already known object, and **precovery** in the past and in the archives of images of the sky, looking for observations which were either not measured or not included in the observation databases.

The main problem of recovery/precovery is that the resources needed (telescope time for recovery, human labor, and/or computational resources for precovery) depend upon the uncertainty of the prediction. When recovery observations are performed, often intruder asteroids are found along with (sometimes instead of) the wanted one; the same for precovery. Thus an attribution problem has to be solved after obtaining the observations.

Linkage

The most difficult kind of identification problem is **linkage**, when two arcs of observations, both too short to perform orbit determination, are to be joined into an arc good enough to compute an orbit. In this case there is no way to directly compare quantities of the same nature, such as orbits with orbits, observations with observations: orbits are not available, and observations at different times cannot be directly compared (unless the time difference is very short). Thus the sequence of steps has to be different:

(i) to compute one or more hypothetical orbit, compatible with the observations of the first arc, together with some replacement of the covariance matrix to assess uncertainty;

(ii) to compare predictions of the observations from the hypothetical orbit(s) with the observations of the other arcs, selecting the couples proposed for identification;

(iii) to compute a preliminary orbit compatible with both arcs;

(iv) to check the convergence of differential corrections, with the data of both arcs, and the quality of the residuals.

Linkage may be a difficult problem even when there are just a few observed arcs. Thus, when dealing with a large database of observed arcs too short for orbit determination, it is especially necessary to keep under control the computational complexity of the global linkage problem. Since linkage is a more difficult problem than the other classes of identifications, it will be discussed in the dedicated Chapter 8.

7.2 Linear orbit identification

The starting point for the basic orbit identification problem is a set of two nominal orbits, obtained by convergent differential corrections, as described in Chapter 5, with the initial conditions as the only fit parameters. Let $\mathbf{x}_1, \mathbf{x}_2 \in \mathbb{R}^6$ be two separately determined vectors of initial conditions, and $C_1, C_2, \Gamma_1, \Gamma_2$ be the normal and covariance matrices computed at convergence, that is at $\mathbf{x}_1, \mathbf{x}_2$, respectively. We assume that these initial conditions are at the same epoch; if this is not the case, the orbits and the matrices have to be propagated to some common epoch (see Section 5.5). To determine $\mathbf{x}_1, \mathbf{x}_2$ we have used two separate sets of observations

$$(t_i, r_i), \ i = 1, m_1, \qquad (t_i, r_i), \ i = m_1 + 1, m_1 + m_2$$

with m_1 observations in the first arc and m_2 in the second arc; they have resulted in the residuals, with respect to the nominal solutions,

$$\boldsymbol{\xi}_1 = (\xi_i), \ i = 1, m_1, \qquad \boldsymbol{\xi}_2 = (\xi_i), \ i = m_1 + 1, m_1 + m_2.$$

We can compute the two separate target functions for $i = 1, 2$

$$Q_i(\mathbf{x}) = \frac{1}{m_i} \boldsymbol{\xi}_i \cdot \boldsymbol{\xi}_i = Q_i(\mathbf{x}_i) + \Delta Q_i(\mathbf{x}) = Q_i(\mathbf{x}_i) + \frac{1}{m_i} (\mathbf{x} - \mathbf{x}_i) \cdot C_i (\mathbf{x} - \mathbf{x}_i) + \cdots$$

where the dots contain the terms of degree 3 in $(\mathbf{x} - \mathbf{x}_i)$ and those of degree 2 containing the residuals, see Section 5.2. The two penalties ΔQ_i would be zero if the nominal orbits could be assumed, but if a single physical body has been observed, there must be a single orbit fitting both sets of observations, and we cannot assume $\mathbf{x} = \mathbf{x}_1$ *and* $\mathbf{x} = \mathbf{x}_2$. Then the joint target function Q contains a linear combination Q_0 of the two separate minima $Q_1(\mathbf{x}_1), Q_2(\mathbf{x}_2)$ plus a penalty ΔQ measuring the increase of the target function resulting from the hypothesis that the two objects are the same: with $m = m_1 + m_2$

$$
\begin{aligned}
m\,Q(\mathbf{x}) &= \boldsymbol{\xi}_1 \cdot \boldsymbol{\xi}_1 + \boldsymbol{\xi}_2 \cdot \boldsymbol{\xi}_2 = m_1 Q_1(\mathbf{x}) + m_2 Q_2(\mathbf{x}) = mQ_0 + m\Delta Q(\mathbf{x}) \\
m\,Q_0 &= [m_1 Q_1(\mathbf{x}_1) + m_2 Q_2(\mathbf{x}_2)] \\
m\,\Delta Q(\mathbf{x}) &= m_1 \Delta Q_1(\mathbf{x}) + m_2 \Delta Q_2(\mathbf{x}) \\
&= (\mathbf{x} - \mathbf{x}_1) \cdot C_1 (\mathbf{x} - \mathbf{x}_1) + (\mathbf{x} - \mathbf{x}_2) \cdot C_2 (\mathbf{x} - \mathbf{x}_2) + \cdots
\end{aligned}
$$

Linear theory

The linear algorithm to solve the problem is obtained when the linear approximation can be used, not only locally, in the neighborhood of the two separate solutions $\mathbf{x}_1$ and $\mathbf{x}_2$, but even globally for the joint solution. This is a strong assumption, because we cannot assume that the two separate solutions are near to each other. However, if the assumption is true, we can use the quadratic approximation for both penalties ΔQ_i, and obtain an explicit formula for the solution of the identification problem (Milani *et al.* 2000a). Neglecting all the higher order terms (the dots in the previous formula)

$$m\,\Delta Q(\mathbf{x}) \simeq (\mathbf{x} - \mathbf{x}_1) \cdot C_1\,(\mathbf{x} - \mathbf{x}_1) + (\mathbf{x} - \mathbf{x}_2) \cdot C_2\,(\mathbf{x} - \mathbf{x}_2)$$
$$= \mathbf{x} \cdot (C_1 + C_2)\mathbf{x} - 2\mathbf{x} \cdot (C_1\,\mathbf{x}_1 + C_2\,\mathbf{x}_2) + \mathbf{x}_1 \cdot C_1\,\mathbf{x}_1 + \mathbf{x}_2 \cdot C_2\,\mathbf{x}_2\,.$$

Then the minimum of the penalty ΔQ can be found by minimizing the non-homogeneous quadratic form of the formula above. If the new joint minimum is $\mathbf{x}_0$, then by expanding around $\mathbf{x}_0$ we have

$$m\,\Delta Q(\mathbf{x}) \simeq (\mathbf{x} - \mathbf{x}_0) \cdot C_0\,(\mathbf{x} - \mathbf{x}_0) + K$$

and by comparing the last two formulae we find:

$$
\begin{aligned}
C_0 &= C_1 + C_2, \\
C_0\,\mathbf{x}_0 &= C_1\,\mathbf{x}_1 + C_2\,\mathbf{x}_2, \\
K &= \mathbf{x}_1 \cdot C_1\,\mathbf{x}_1 + \mathbf{x}_2 \cdot C_2\,\mathbf{x}_2 - \mathbf{x}_0 \cdot C_0\,\mathbf{x}_0.
\end{aligned}
$$

If the matrix C_0, which is the sum of the two separate normal matrices C_1 and C_2, is positive definite, then it is invertible and we can solve for the new minimum point by using the covariance matrix $\Gamma_0 = C_0^{-1}$:

$$\mathbf{x}_0 = \Gamma_0\,(C_1\,\mathbf{x}_1 + C_2\,\mathbf{x}_2). \tag{7.1}$$

This has a simple interpretation in terms of differential corrections: at convergence in each of the two iterations, $\mathbf{x} \to \mathbf{x}_i$ with $C_i = C_i(\mathbf{x}_i)$ and the right-hand side of the normal equation $D_i = D_i(\mathbf{x}_i) = C_i\,\Delta\mathbf{x}_i$ is $\mathbf{0}$. Thus

$$C_1\,(\mathbf{x} - \mathbf{x}_1) = \mathbf{0} \ \text{ and } \ C_2\,(\mathbf{x} - \mathbf{x}_2) = \mathbf{0} \Longrightarrow (C_1 + C_2)\,\mathbf{x} = C_1\,\mathbf{x}_1 + C_2\,\mathbf{x}_2.$$

By the linearity assumption C_1, C_2 have the same values at $\mathbf{x}_1, \mathbf{x}_2$ and at $\mathbf{x}_0$; under these conditions $\mathbf{x} = \mathbf{x}_0$ is the result of the first differential correction for the joint problem.

The **identification penalty** K/m approximates the minimum of the penalty $\Delta Q(\mathbf{x})$, normalized by the number of observations: in the linear approximation $K/m = \Delta Q(\mathbf{x}_0)$. Since K is translation invariant

$$
\begin{aligned}
\mathbf{x}_0 &\to \mathbf{x}_0 + \mathbf{v}, \quad \mathbf{x}_1 \to \mathbf{x}_1 + \mathbf{v}, \quad \mathbf{x}_2 \to \mathbf{x}_2 + \mathbf{v} \\
K &\to K + 2\mathbf{v} \cdot (C_1\,\mathbf{x}_1 + C_2\,\mathbf{x}_2 - C_0\,\mathbf{x}_0) + \mathbf{v} \cdot (C_1 + C_2 - C_0)\mathbf{v} = K,
\end{aligned}
$$

we can compute K after a translation by $-\mathbf{x}_1$, that is assuming $\mathbf{x}_1 \to \mathbf{0}$,

$\mathbf{x}_2 \to \mathbf{x}_2 - \mathbf{x}_1 = \Delta\mathbf{x}$, and $\mathbf{x}_0 \to \Gamma_0 C_2 \, \Delta\mathbf{x}$:

$$K = \Delta\mathbf{x} \cdot C_2 \, \Delta\mathbf{x} - (\mathbf{x}_0 - \mathbf{x}_1) \cdot C_0 \, (\mathbf{x}_0 - \mathbf{x}_1) = \Delta\mathbf{x} \cdot C\Delta\mathbf{x}, \qquad (7.2)$$

with $C = C_2 - C_2 \, \Gamma_0 \, C_2$. Alternatively, translating by $-\mathbf{x}_2$, that is with $\mathbf{x}_2 \to \mathbf{0}$, $\mathbf{x}_1 \to -\Delta\mathbf{x}$ and $\mathbf{x}_0 \to \Gamma_0 \, C_1 \, (-\Delta\mathbf{x})$:

$$K = \Delta\mathbf{x} \cdot C_1 \, \Delta\mathbf{x} - (\mathbf{x}_0 - \mathbf{x}_2) \cdot C_0 \, (\mathbf{x}_0 - \mathbf{x}_2) = \Delta\mathbf{x} \cdot (C_1 - C_1 \, \Gamma_0 \, C_1) \, \Delta\mathbf{x}$$

and the same matrix C can be defined by the alternative expression $C = C_1 - C_1 \, \Gamma_0 \, C_1$. Both these formulae only assume that $\Gamma_0 = C_0^{-1}$ exists, then

$$C = C_2 - C_2 \, \Gamma_0 \, C_2 = C_1 - C_1 \, \Gamma_0 \, C_1. \qquad (7.3)$$

The above equality is true in exact arithmetic, but might be violated in a numerical computation if the matrix C_0 is badly conditioned. We can summarize the conclusions by the formula

$$Q(\mathbf{x}) \simeq Q_0 + \frac{1}{m} \Delta\mathbf{x} \cdot C \, \Delta\mathbf{x} + \frac{1}{m} (\mathbf{x} - \mathbf{x}_0) \cdot C_0 \, (\mathbf{x} - \mathbf{x}_0)$$

which allows also to assess the uncertainty of the identified solution, by defining confidence ellipsoids with matrix C_0.

This algorithm has a geometrical interpretation in terms of intersections of the two families of confidence ellipsoids. To result in a low penalty, say $m\Delta Q < \epsilon$, a **compromise solution** $\mathbf{x}_0$ has to belong to the intersection of the two confidence ellipsoids $m_1 \, \Delta Q_1 < \epsilon$ and $m_2 \, \Delta Q_2 < \epsilon$.

Probabilistic interpretation

If $\mathbf{x}_i^*$ is the nominal solution of the differential corrections with normal matrix C_i and covariance matrix $\Gamma_i = C_i^{-1}$, the probability density of the initial conditions $\mathbf{x}_i$ according to the Gaussian model (see Section 5.7) is

$$\mathsf{p}_{\mathbf{X}_i}(\mathbf{x}_i) = N(\mathbf{x}_i, \Gamma_i) = \frac{\sqrt{\det C_i}}{(2\pi)^{N/2}} \exp\left(-\frac{1}{2}(\mathbf{x}_i - \mathbf{x}_i^*) \cdot C_i(\mathbf{x}_i - \mathbf{x}_i^*)\right).$$

Let us assume $\mathbf{X}_1$ and $\mathbf{X}_2$ are independent random variables, that is their joint probability density function is $\mathsf{p}_{\mathbf{X}_1\mathbf{X}_2}(\mathbf{x}_1, \mathbf{x}_2) = \mathsf{p}_{\mathbf{X}_1}(\mathbf{x}_1) \cdot \mathsf{p}_{\mathbf{X}_2}(\mathbf{x}_2)$. This hypothesis is justified because the set of observations of the two independent discoveries is disjoint. Then the probability of the identification $P_I = \mathcal{P}(\mathbf{X}_1 = \mathbf{X}_2)$ is obtained as

$$P_I = \int_{\mathbb{R}^6} \mathsf{p}_{\mathbf{X}_1, \mathbf{X}_2}(\mathbf{x}, \mathbf{x}) \, d\mathbf{x} = \int_{\mathbb{R}^6} \mathsf{p}_{\mathbf{X}_1}(\mathbf{x}) \cdot \mathsf{p}_{\mathbf{X}_2}(\mathbf{x}) \, d\mathbf{x}.$$

The product $p_{\mathbf{x}_1}(\mathbf{x}) \cdot p_{\mathbf{x}_2}(\mathbf{x})$ is not the probability density of the identification orbit, because the integral over the entire initial conditions space is not equal to 1. Indeed, to obtain the conditional density function of the identification orbit (under the hypothesis $\mathbf{x}_1 = \mathbf{x}_2$), the product has to be renormalized by dividing by the probability of the identification P_I.

Then both the probability P_I and the conditional density function of the identified orbit can be computed starting from the product

$$p_{\mathbf{x}_1}(\mathbf{x}) \cdot p_{\mathbf{x}_2}(\mathbf{x}) \;=\; \frac{\sqrt{\det(C_1 C_2)}}{(2\pi)^N} \exp\Big\{-\frac{1}{2}\big[(\mathbf{x} - \mathbf{x}_1) \cdot C_1(\mathbf{x} - \mathbf{x}_1)$$
$$+\, (\mathbf{x} - \mathbf{x}_2) \cdot C_2(\mathbf{x} - \mathbf{x}_2)\big]\Big\}$$

and replacing the sum of two quadratic forms in the exponent with the single quadratic form centered in $\mathbf{x}_0$, with normal matrix C_0:

$$p_{\mathbf{x}_1}(\mathbf{x}) \cdot p_{\mathbf{x}_2}(\mathbf{x}) \;=\; \frac{\sqrt{\det(C_1 C_2)}}{(2\pi)^N} \exp\Big\{-\frac{1}{2}\big[(\mathbf{x} - \mathbf{x}_0) \cdot C_0\,(\mathbf{x} - \mathbf{x}_0) + K\big]\Big\}$$
$$=\; N(\mathbf{x}_0, \Gamma_0)(\mathbf{x}) \cdot \frac{\sqrt{\det(C_1 C_2)}}{(2\pi)^{N/2}\sqrt{\det C_0}} \exp\left(-\frac{K}{2}\right).$$

To simplify this expression, let us assume both C_1 and C_2 are positive definite. Then they can be diagonalized simultaneously, that is, there is an orthogonal matrix S such that

$$SC_1 S^T = \mathrm{diag}[\lambda_{1j}], \quad SC_2 S^T = \mathrm{diag}[\lambda_{2j}], \quad SC_0 S^T = \mathrm{diag}[\lambda_{1j} + \lambda_{2j}],$$

$$SCS^T = SC_2 S^T - SC_2 S^T\, S\, C_0^{-1} S^T\, SC_2 S^T = \mathrm{diag}\left[\frac{\lambda_{1j}\lambda_{2j}}{\lambda_{1j} + \lambda_{2j}}\right].$$

From this we can compute the determinants

$$\frac{\det(C_1 C_2)}{\det(C_1 + C_2)} = \det(S)^{-2} \frac{\prod_{j=1}^{N} \lambda_{1j} \prod_{j=1}^{N} \lambda_{2j}}{\prod_{j=1}^{N}(\lambda_{1j} + \lambda_{2j})} = \det(C)$$

and find that the factor multiplying $N(\mathbf{x}_0, \Gamma_0)$ has a simple interpretation

$$p_{\mathbf{x}_1}(\mathbf{x}) \cdot p_{\mathbf{x}_2}(\mathbf{x}) \;=\; N(\mathbf{x}_0, \Gamma_0)(\mathbf{x}) \cdot \frac{\sqrt{\det C}}{(2\pi)^{N/2}} \exp\left[-\frac{1}{2} K\right]$$
$$=\; N(\mathbf{x}_0, \Gamma_0)(\mathbf{x}) \cdot N(\mathbf{0}, C^{-1})(\Delta\mathbf{x}).$$

The probabilistic interpretation of the above formula is

$$\frac{p_{\mathbf{x}_1}(\mathbf{x}) \cdot p_{\mathbf{x}_2}(\mathbf{x})}{P_I} = N(\mathbf{x}_0, \Gamma_0).$$

with the identification probability estimated as $P_I = N(\mathbf{0}, C^{-1})(\Delta\mathbf{x})$.

In conclusion, the probability that the identification is true is the value of the Gaussian $N(\mathbf{x}_1, C^{-1})(\mathbf{x}_2) = N(\mathbf{x}_2, C^{-1})(\mathbf{x}_1)$, computed by using the normal matrix C (the same used to compute the identification penalty K). Assuming the identification is true, the identification orbit has the normal distribution $N(\mathbf{x}_0, \Gamma_0)$. The correspondence between the probabilistic and the optimization interpretations is maintained in the linear identification theory, as in the linear orbit determination theory.

7.3 Semilinear orbit identification

The applicability of the linear identification algorithm depends upon the nonlinearity, that is upon the difference between the confidence regions and the confidence ellipsoids of the two separate solutions $\mathbf{x}_1$ and $\mathbf{x}_2$.

Nonlinearity

As discussed in Section 5.5, there are two main sources of nonlinearity in the confidence regions. First, each of the two separate confidence regions for the solutions $\mathbf{x}_1$ and $\mathbf{x}_2$ could be already strongly nonlinear. Second, even assuming that the two separate confidence regions are well approximated by the confidence ellipsoids, the initial conditions have to be determined at times t_1, close to the times of the first m_1 observations, and t_2, close to the last m_2 observations, respectively.[2] When the orbits, determined at t_1 and t_2, respectively, are propagated to a common time, say $t_0 = (t_1 + t_2)/2$, the conditioning number of the propagated normal matrices increases at least quadratically with the time spans $|t_0 - t_i|$. Thus, as the time span between the two arcs increases, the confidence ellipsoids at the common time t_0 become more and more elongated, they are worse and worse approximations of the confidence regions, and the intersection of the confidence regions may have nothing to do with the intersection of the confidence ellipsoids.

There is a third source of nonlinearity which is specific to the coordinates used to represent the initial conditions of the orbit (see Section 10.3 for a discussion of different coordinates). In the Keplerian orbital elements $(a, e, I, \Omega, \omega, \ell)$ the values $e = 0$ and $I = 0$ correspond to singularities, in which some of the angle variables (Ω, ω, ℓ) are not uniquely defined. If the values $e = 0$ and/or $I = 0$ are within the confidence ellipsoid, then

---

[2] The model problem of Sections 5.1 and 5.6 already shows that the best time for obtaining well-conditioned normal and covariance matrices is the average of the observation times.

linearity fails even when the confidence region is small. For this reason non-singular elements like the **equinoctial elements** (a, h, k, p, q, λ) are used, see (Broucke and Cefola 1972), with

$$h = e \sin(\Omega + \omega), \quad k = e \cos(\Omega + \omega)$$

$$p = \tan(I/2) \sin \Omega, \quad q = \tan(I/2) \cos \Omega, \quad \lambda = \ell + \Omega + \omega.$$

The variables (h, k, p, q, λ) are defined for $e = 0$ and/or $I = 0$ (by $h = k = 0$ for $e = 0$ and by $p = q = 0$ for $I = 0$) and they are smooth as functions of Cartesian initial conditions.

Restricted orbit identification

To find the orbit identifications among a large catalog of orbits it is necessary to start from step (i) of the procedure outlined in Section 7.1: to select a small subset of the couples of orbit, and with a simple algorithm, not including any orbit propagation. Thus we would like to compare orbital elements which are constant in the two-body approximation, e.g., excluding λ in equinoctial elements. This also removes the effect of the nonlinearity in the propagation of λ, which occurs even in the two-body problem, as already appears from our model problem, see Section 5.6. We may also take advantage of the fact that some elements are typically better determined than others, even with a short observed arc: this is the case for the orbital plane variables, either (I, Ω) in Keplerian or (p, q) in equinoctial ones.

Thus we need to perform a **restricted identification**, computing a penalty K_2 in a two-dimensional space of elements (p, q) and/or a penalty K_5 in a five-dimensional space (a, h, k, p, q). In general, we split the vector $\mathbf{x}$ of estimated parameters into two components $\mathbf{h}$ and $\mathbf{g}$, and let $\mathbf{g}$ contain the elements to which the comparison is restricted.

The normal and covariance matrices C and Γ are decomposed as in Section 5.4. Then the marginal uncertainty of $\mathbf{g}$ (for arbitrary $\mathbf{h}$) can be described by the penalty, with respect to the minimum point $\mathbf{g}^*$

$$m\Delta Q \simeq (\mathbf{g} - \mathbf{g}^*) \cdot C^{\mathbf{gg}} (\mathbf{g} - \mathbf{g}^*), \quad C^{\mathbf{gg}} = C_{\mathbf{gg}} - C_{\mathbf{gh}} C_{\mathbf{hh}}^{-1} C_{\mathbf{hg}}$$

and by the marginal covariance matrix $\Gamma_{\mathbf{gg}} = (C^{\mathbf{gg}})^{-1}$. Note that this penalty as a function of $\mathbf{g}$ has been obtained by changing the value of $\mathbf{h}$ from the nominal $\mathbf{h}^*$ to a suitable point of the regression subspace.

We use this restricted penalty formula for the restricted identification problem: let $\mathbf{x}_1 = (\mathbf{h}_1, \mathbf{g}_1)$ and $\mathbf{x}_2 = (\mathbf{h}_2, \mathbf{g}_2)$ be the separate nominal solutions for the two arcs, and $C^{\mathbf{gg}}(\mathbf{x}_1)$ and $C^{\mathbf{gg}}(\mathbf{x}_2)$ the corresponding marginal normal matrices. The variables $\mathbf{h}$ are given as a function of $\mathbf{g}$ by:

$$\begin{cases} \mathbf{h}_1(\mathbf{g}) &= \mathbf{h}_1 - C_{\mathbf{h}}^{-1}(\mathbf{x}_1)C_{\mathbf{hg}}(\mathbf{x}_1)\,(\mathbf{g} - \mathbf{g}_1) \\ \mathbf{h}_2(\mathbf{g}) &= \mathbf{h}_2 - C_{\mathbf{h}}^{-1}(\mathbf{x}_2)C_{\mathbf{hg}}(\mathbf{x}_2)\,(\mathbf{g} - \mathbf{g}_2). \end{cases} \tag{7.4}$$

By the same formalism of the previous section:

$$\begin{aligned} \frac{m}{2}\Delta Q &\simeq (\mathbf{g} - \mathbf{g}_0) \cdot C_0^{\mathbf{gg}}\,(\mathbf{g} - \mathbf{g}_0) + K_{\mathbf{g}} \\ C_0^{\mathbf{gg}} &= C_1^{\mathbf{gg}}(\mathbf{x}_1) + C_2^{\mathbf{gg}}(\mathbf{x}_2) \\ \mathbf{g}_0 &= \left(C_0^{\mathbf{gg}}\right)^{-1}\left(C_1^{\mathbf{gg}}(\mathbf{x}_1)\,\mathbf{g}_1 + C_2^{\mathbf{gg}}(\mathbf{x}_2)\,\mathbf{g}_2\right) \\ C^{\mathbf{g}} &= C_2^{\mathbf{gg}}(\mathbf{x}_2) - C_2^{\mathbf{gg}}(\mathbf{x}_2)\left(C_0^{\mathbf{gg}}\right)^{-1}C_2^{\mathbf{gg}}(\mathbf{x}_2) \\ &= C_1^{\mathbf{gg}}(\mathbf{x}_1) - C_1^{\mathbf{gg}}(\mathbf{x}_1)\left(C_0^{\mathbf{gg}}\right)^{-1}C_1^{\mathbf{gg}}(\mathbf{x}_1) \\ K_{\mathbf{g}} &= (\mathbf{g}_2 - \mathbf{g}_1) \cdot C^{\mathbf{g}}\,(\mathbf{g}_2 - \mathbf{g}_1). \end{aligned}$$

$K_{\mathbf{g}}$ is not the same as the complete minimum penalty K of the previous section, but it is obtained by assuming that $\mathbf{x}_1 = (\mathbf{h}_1(\mathbf{g}_0), \mathbf{g}_0)$ in the computation of ΔQ_1, $\mathbf{x}_2 = (\mathbf{h}_2(\mathbf{g}_0), \mathbf{g}_0)$ in the computation of ΔQ_2, $\mathbf{g}_0$ being the proposed restricted identification. Thus $K_{\mathbf{g}} \leq K$: $K_{\mathbf{g}}$ is the minimum of the penalty over the space of variables $(\mathbf{g}, \mathbf{h}_1, \mathbf{h}_2)$, while K is the minimum under the additional constraint $\mathbf{h}_1 = \mathbf{h}_2$, and the minimum of a function can only increase when constraints are added.

The penalty $K_{\mathbf{g}}$ can be used as a preliminary control, that is, if $K_{\mathbf{g}} > \Sigma$, for a positive parameter Σ, then also $K > \Sigma$ and many couples can be discarded without doing a computation with the larger matrices. This allows us to select a subset of couples candidates for identification with the linear identification algorithm.

Multistage identification procedure

An effective procedure for proposing orbit identifications can be obtained by a sequence of filtering stages:

(i) restricted identification comparing only $\mathbf{g} = (p, q)$, selecting the couples with two-dimensional penalty K_2 below a control $\Sigma_2 > 0$;

(ii) restricted identification comparing only $\mathbf{g} = (a, h, k, p, q)$, selecting the couples with five-dimensional penalty K_5 below a control $\Sigma_5 > 0$;

(iii) full identification between the orbits $\mathbf{x}_1$ and $\mathbf{x}_2$ propagated to a common time t_0, selecting the couples with full penalty K below some control $\Sigma > 0$.

The three filters are applied in series, that is each one is applied only to the couples passing the previous one. After passing all three filter stages, the proposed identification has to be confirmed by differential corrections, starting from the first guess $\mathbf{x}_0$ of eq. (7.1), and quality control.

To control computational complexity, the most critical is filter 1, because it has to be applied to all the $\simeq N^2/2$ couples in a catalog containing N orbits. If N is very large, it may be necessary to use an algorithm of computational complexity $\mathcal{O}(N \log N)$ which could be very similar to the one discussed in Section 11.3. Because the orbital plane variables are indeed better determined, the control Σ_2 can be quite tight (Milani *et al.* 2000a, Figure 2). This helps in decreasing the number of couples passed to filter 2: in a test reported by Milani *et al.* (2000a), with $\Sigma_2 = 30$ only a fraction $\simeq 0.006$ of the couples passed filter 1. The second filter with $\Sigma_5 = 5\,000$ passed a fraction 0.07 of those passed by filter 1. The choice of the values for Σ_5 and Σ is not easy; a method to optimally select them for best performance is in (Milani *et al.* 2005c, Section 5).

The most tricky stage is filter 3, for two reasons. First, propagation of both $\mathbf{x}_1$ and $\mathbf{x}_2$, with covariance and normal matrices, to a common time $t_0 = (t_1 + t_2)/2$ specifically for each couple passing filter 2 would be computationally expensive. A possible solution is to prepare in advance a number of propagated orbit catalogs at suitably selected times, in such a way that filter 3 can be applied to the catalog corresponding to the epoch most suitable for the given couple. Although the propagation of an orbit with a full N-body model is computationally expensive, the complexity for propagating the entire catalog is $\mathcal{O}(N)$.

Second, the propagation including the variable λ is anyway nonlinear. If the time spans $|t_0 - t_i|$ are long, the shape of the confidence regions are different from the ellipsoids computed with the propagated normal matrices (see Figure 5.2), and the intersections of the confidence ellipsoids can be very different from those of the confidence regions. If the time interval between the two separate observed arcs is not too long, and the nonlinear effects are not too pronounced, this can be compensated by selecting a value of Σ much larger than the linear identification algorithm would suggest. For example, by using the standard χ^2 tables the probability of such an identification would be ridiculously small. Because of the exponential decrease of the normal probability density, the probabilistic interpretation based on the linear Gaussian formulae is incompatible with nonlinear effects, even if they are moderate; see Chapter 12. We call such an algorithm **semilinear identification**. In the test of Milani *et al.* (2000a) the value $\Sigma = 1\,000\,000$ was used, with a very large increase of the computational load because only a fraction 0.01 of the differential corrections attempted were convergent (low accuracy). The number of successful identifications was very encouraging, although this number was a small fraction of the single apparition orbits cataloged, leaving the suspicion that there might be many more to be discovered. For an even more intensive effort to find large numbers of identifications, see Section 10.2.

7.4 Nonlinear orbit identification

We would like to find algorithms allowing us to cope with fully nonlinear identification problems, e.g., with the case of two short observed arcs, with poorly determined orbits, separated in time by years. This is a very difficult problem, which has not been fully solved yet. To find such an algorithm we need a better understanding of the nonlinearity arising in the identification problem, and for this we restart from our model problem.

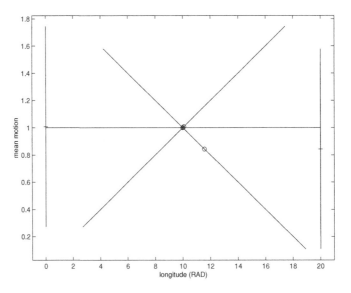

Fig. 7.1. The identification in the coordinates (n, λ), the latter being considered a real number. The two vertical segments at $\lambda \simeq 0$ and at $\lambda \simeq 20$ are in fact very thin confidence ellipses, representing the uncertainty (with confidence parameter $\sigma = 3$) at the central times of the two arcs. The slanted lines are also thin ellipses, the propagated confidence ellipses at the central time $t_0 \simeq 10$. Their intersection contains the true identification orbit, which is easily found with the linear identification formula starting from the two nominal solutions (marked with small circles).

Model identification problem

The main effect of nonlinearity can be illustrated in our model problem. In all the figures of this subsection we use an example with two arcs that are 20 orbital periods apart, each with four observations spanning $\simeq 0.005$ periods. The RMS observation error is 0.001 radians. The true orbit has $a = 1$ (in units such that $n = 1$).

If we attempt the identification of the two arcs in the space of the (n, λ) coordinates, considering λ as a real number (as if we could observe the number of revolutions), the problem is exactly linear (see Figure 7.1) and the linear formulae discussed above provide a very good first guess for the identification orbit. Indeed the first guess $\mathbf{x}_0$ and the final nominal solution, obtained by differential corrections, are very close.

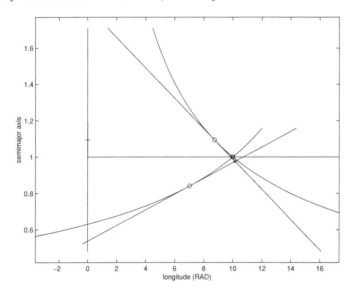

Fig. 7.2. The identification in the coordinates (a, λ), the latter being considered a real number. The slanted lines are in fact thin confidence ellipses for the two orbits propagated to the time t_0, while the two curves (tangent at the nominal solutions) are thin semilinear confidence boundaries.

If we use the same algorithm in the space of the (a, λ) coordinates (still considering $\lambda \in \mathbb{R}$), the problem becomes nonlinear. In Figure 7.2 we show the confidence ellipses of the two orbits, as obtained by propagating the normal matrices to the common time t_0 (by the formula of Section 5.5) and also the curves obtained by propagation, point by point, of the confidence ellipses at the times t_1 and t_2, respectively, to the common time t_0 (the so-called semilinear approximation, further discussed in Section 7.5). The nonlinear confidence regions have a connected intersection, although it is disjoint from the intersection of the linear confidence ellipses. Nevertheless, the first guess $\mathbf{x}_0$ computed with the linear identification formula, which belongs to the intersection of the linear confidence ellipses, is good enough to allow convergence of differential corrections to the true identification orbit, which belongs to the intersection of the nonlinear confidence regions.

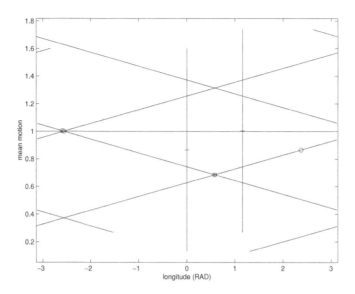

Fig. 7.3. The identification in the coordinates (n, λ), the latter being considered as an angle. The spurious solution with $n \simeq 0.69$ is near one of the four intersections of the wrapped confidence ellipses, the one which would be suggested by the linear identification formula.

The problem becomes more difficult if we take into account the fact that λ is an angle: when, after many years, an asteroid is independently rediscovered there is no way to know a priori how many revolutions have been completed between the two discoveries. Figure 7.3 shows that the problem is no longer linear, not even in the (n, λ) coordinates; indeed the confidence ellipses, linearly propagated to the common time t_0 and then wrapped on the cylinder obtained by identification of $\lambda = -\pi$ with $\lambda = +\pi$, have (in this example) an intersection with four connected components. The first guess for the identification orbit $\mathbf{x}_0$ obtained by the linear identification formulae turns out, in this case, to belong to a different connected component from the one containing the true solution. Thus the differential corrections starting from $\mathbf{x}_0$ converge to the spurious solution closest to $\mathbf{x}_0$.

Of course in the coordinates (a, λ) the two nonlinear effects (from the nonlinearity of the integral flow and from the wrapping on a cylinder) combine and result in a geometrically complicated situation. As shown in Figure 7.4, the intersection of both the linear and the nonlinear confidence regions can have a dozen connected components; the number of connected components does not even need to be the same.

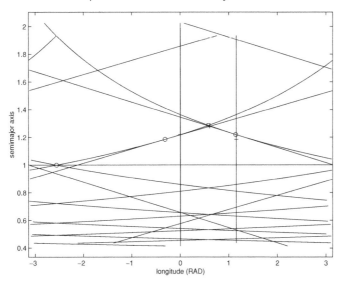

Fig. 7.4. The identification in the coordinates (n, λ), the latter as an angle. The spurious solution for $a \simeq 1.29$ corresponds to the one found in the other coordinates. The nonlinear confidence regions have an intersection with more connected components than the linear ones.

Cycle slip estimation

We propose a solution to the nonlinearity problem as shown above in the model problem. Let us split the orbital elements, in equinoctial coordinates with a replaced by the mean motion n, into the part corresponding to our model problem and the part containing the variables connected with eccentricity and inclination:

$$\mathbf{x} = \begin{bmatrix} \mathbf{g} \\ \mathbf{h} \end{bmatrix}, \quad \mathbf{g} = \begin{bmatrix} e \sin \varpi \\ e \cos \varpi \\ \tan(I/2) \sin \Omega \\ \tan(I/2) \cos \Omega \end{bmatrix}, \quad \mathbf{h} = \begin{bmatrix} n \\ \lambda \end{bmatrix}.$$

Let the central times of the two separate arcs of observations be t_1, t_2 with $t_1 < t_2$, and $\mathbf{x}_i, C_i, \Gamma_i$ for $i = 1, 2$ be the nominal orbital elements, the normal and the covariance matrices corresponding to each arc. We want to find a partial identification based only upon $\mathbf{h}_1 = (n_1, \lambda_1)$ and $\mathbf{h}_2 = (n_2, \lambda_2)$. From each arc we have a marginal confidence interval for the mean motion

$$n_i^- = n_i - \sigma \cdot \text{RMS}(n_i) \leq n \leq n_i^+ = n_i + \sigma \cdot \text{RMS}(n_i);$$

then the common range of values for n is

$$n^- = \max(n_1^-, n_2^-, 0) \leq n \leq n^+ = \min(n_1^+, n_2^+).$$

If the interval $[n^-, n^+]$ is not empty, let us select a time t_0 such that $t_1 < t_0 < t_2$. The two-body predictions for $\lambda(t_0)$ are

$$\lambda_1 + n^- (t_0 - t_1) \leq \lambda_{10} \leq \lambda_1 + n^+ (t_0 - t_1)$$
$$\lambda_2 + n^+ (t_0 - t_2) \leq \lambda_{20} \leq \lambda_2 + n^- (t_0 - t_2)$$

and by subtracting the inequalities

$$n^- \Delta t - \Delta\lambda \leq \lambda_{10} - \lambda_{20} \leq n^+ \Delta t - \Delta\lambda$$

where $\Delta\lambda = \lambda_2 - \lambda_1$ and $\Delta t = t_2 - t_1 > 0$. To obtain an intersection of the two lines of possible predictions for the time t_0, that is a common possible orbit, the predictions λ_{10} and λ_{20} need to be equal as angle variables: $\lambda_{10} - \lambda_{20} = 2\pi k$, with k an arbitrary integer (in fact $k \geq -1$). From this equation it is possible to find the finite number of possible values k for the number of cycles slipped

$$\frac{n^- \Delta t - \Delta\lambda}{2\pi} \leq k \leq \frac{n^+ \Delta t - \Delta\lambda}{2\pi}$$

implying that $k^- \leq k \leq k^+$ with

$$k^- = \text{Ceiling}\left(\frac{n^- \Delta t - \Delta\lambda}{2\pi}\right), \quad k^+ = \text{Floor}\left(\frac{n^+ \Delta t - \Delta\lambda}{2\pi}\right).$$

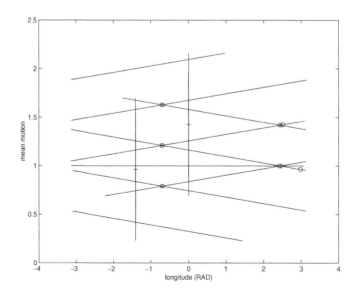

Fig. 7.5. The identification in the coordinates (n, λ), the latter being considered as an angle. For each intersection of the confidence ellipse, the cycle slip algorithm computes a separate first guess; all five of them lead to convergent differential corrections, with different quality of fit.

Selecting a value of the **cycle slip** number k in this range implies selecting the coordinates $\mathbf{h}_0$ for the preliminary orbit at epoch t_0

$$n_k = \frac{2\pi k + \Delta\lambda}{\Delta t}$$

$$\lambda_{0k} \equiv n_k(t_0 - t_1) + \lambda_1 \ (\mathrm{mod}\ 2\pi) \equiv n_k(t_0 - t_2) + \lambda_2 \ (\mathrm{mod}\ 2\pi).$$

This algorithm can be illustrated with our model problem, containing only the variables $\mathbf{h}$. Figure 7.5 shows a case with five intersections of the confidence ellipses. By testing five different first guess orbits, corresponding to $k = 4, 5, 6, 7, 8$, the differential corrections converge to five different solutions. The true solution corresponds to $k = 5$ and has a normalized residuals RMS $\simeq 1.01$. Note that the linear algorithm, without cycle slip, would provide a $\mathbf{h}_0$ close to the intersection with $n \simeq 1.21$, corresponding to $k = 6$; from this, the differential corrections converge to a spurious solution with normalized RMS $\simeq 0.91$. This example shows that to choose among alternative solutions for the identification orbit is far from easy, especially when the data are of poor quality, as in this example.

Constrained orbit identification

If the portion $\mathbf{h}_0$ of the identification orbit has been selected already, then the other part $\mathbf{g}_0$ should be selected in a consistent way. In the two-body approximation, the value of $\mathbf{g}$ is independent of time, thus $\mathbf{g}_i$ are the predictions for epoch t_0 from t_i, for $i = 1, 2$. The uncertainty of these predictions can be computed, given the fixed value of $\mathbf{h} = \mathbf{h}_0$, by the formula for the conditional case of Section 5.4: the nominal conditional value is

$$\mathbf{g}_{0i} = \mathbf{g}_i - C_{\mathbf{gg}}^{-1}(\mathbf{x}_i)\, C_{\mathbf{gh}}(\mathbf{x}_i)\, (\mathbf{h}_0 - \mathbf{h}_i) \quad \text{for } i = 1, 2$$

and the normal matrices are the same $C_{\mathbf{gg}}(\mathbf{x}_i)$. The **conditional identification penalty** K_C contains three terms: one from forcing $\mathbf{h}_1(t_0)$ to $\mathbf{h}_0$, one from forcing $\mathbf{h}_2(t_0)$ to $\mathbf{h}_0$, the third by the compromise between $\mathbf{g}_{01}$ and $\mathbf{g}_{02}$:

$$
\begin{aligned}
K_C &= K_{\mathbf{h}}^1 + K_{\mathbf{h}}^2 + K_{\mathbf{g}} \\
K_{\mathbf{h}}^1 &= (\mathbf{h}_0 - \mathbf{h}_1(t_0)) \cdot C^{\mathbf{hh}}(\mathbf{x}_1)\, (\mathbf{h}_0 - \mathbf{h}_1(t_0)) \\
K_{\mathbf{h}}^2 &= (\mathbf{h}_0 - \mathbf{h}_2(t_0)) \cdot C^{\mathbf{hh}}(\mathbf{x}_2)\, (\mathbf{h}_0 - \mathbf{h}_2(t_0)) \\
K_{\mathbf{g}} &= (\mathbf{g}_{02} - \mathbf{g}_{01}) \cdot C^{\mathbf{g}}\, (\mathbf{g}_{02} - \mathbf{g}_{01}) \\
C^{\mathbf{g}} &= C_{\mathbf{gg}}(\mathbf{x}_1) - C_{\mathbf{gg}}(\mathbf{x}_1)\, [C_{\mathbf{gg}}(\mathbf{x}_1) + C_{\mathbf{gg}}(\mathbf{x}_2)]^{-1}\, C_{\mathbf{gg}}(\mathbf{x}_1)
\end{aligned}
$$

and the point of minimum penalty is at

$$\mathbf{g}_0 = [C_{\mathbf{gg}}(\mathbf{x}_1) + C_{\mathbf{gg}}(\mathbf{x}_2)]^{-1}\, (C_{\mathbf{gg}}(\mathbf{x}_1)\, \mathbf{g}_{01} + C_{\mathbf{gg}}(\mathbf{x}_2)\, \mathbf{g}_{02}).$$

7.5 Recovery and precovery

The problem is how to describe the uncertainty of the position of the asteroid on the celestial sphere, taking into account the nonlinearity of the relationship between orbital elements at some epoch and the observations to be predicted at some other time.

The confidence ellipse

Let a least squares orbit be available with initial conditions $\mathbf{x}$ at some epoch t_0, with normal and covariance matrices C, Γ. At some later time t_1 an observation is either performed or planned. An **astrometric observation** is a map G from the elements space to the celestial sphere, parameterized by two coordinates (usually right ascension and declination) $\mathbf{y} = (\alpha, \delta)$:

$$\mathbf{y}(t_1) = G(\mathbf{x}(t_1)), \quad G : W \longrightarrow \mathbb{R}^2, \quad W \subset \mathbb{R}^6$$

with W an open set.[3] $\mathbf{x}(t_1)$ is the state vector at time t_1, a function of the initial conditions $\mathbf{x} = \mathbf{x}(t_0)$ through the integral flow $\mathbf{x}(t_1) = \Phi_{t_0}^{t_1}(\mathbf{x}(t_0))$. The composition of the observation function with the integral flow

$$\mathbf{y} = F(\mathbf{x}) = G\left(\Phi_{t_0}^{t_1}(\mathbf{x})\right), \quad F : W \longrightarrow \mathbb{R}^2$$

is the *astrometric prediction function*; its Jacobian matrix can be computed by means of the state transition matrix $D\Phi_{t_0}^{t_1}$, by $DF = DG\, D\Phi_{t_0}^{t_1}$.

The astrometric prediction function F maps the orbital elements space onto the observation space, and the confidence region $\Delta Q \leq \sigma^2$ into a **confidence prediction region** in the observation space. The linearized function DF maps the displacement (from the least squares solution $\mathbf{x}^*$) in the orbital elements space $\Delta \mathbf{x} = \mathbf{x} - \mathbf{x}^*$, into linearized deviations from the prediction $\mathbf{y}^* = F(\mathbf{x}^*)$:

$$\Delta \mathbf{y} = \mathbf{y} - \mathbf{y}^* = DF(\mathbf{x}^*)\, \Delta \mathbf{x}$$

and therefore maps the confidence ellipsoid $\Delta \mathbf{x} \cdot C\, \Delta \mathbf{x} \leq \sigma^2$ onto a confidence ellipse in the observation coordinate plane: $\Delta \mathbf{y} \cdot C_{\mathbf{y}}\, \Delta \mathbf{y} \leq \sigma^2$. The matrix $C_{\mathbf{y}}$ is the normal matrix for the observations $\mathbf{y}$ (at a given time t_1), and the inverse $\Gamma_{\mathbf{y}} = C_{\mathbf{y}}^{-1}$ is the corresponding covariance matrix. In the probabilistic interpretation, by using standard result from the theory of multivariate Gaussian distribution (see Section 3.3), the covariance matrix is transformed by $\Gamma_{\mathbf{y}} = DF\, \Gamma\, DF^T$, then the normal matrix is computed as $C_{\mathbf{y}} = \Gamma_{\mathbf{y}}^{-1}$.

---

[3] For example, W can be the *Poincaré domain* of the orbits with negative energy.

To obtain the same result within the optimization interpretation, and also to gain some geometrical insight, let us consider in the $\mathbf{x}$ space the rows of DF, that is the gradients of the observable angles. If the two angular variables observed are independent[4] the subspace spanned by the rows of DF has dimension 2. Thus $\mathbf{x} \in \mathbb{R}^6$ can be decomposed into a component $\mathbf{g}$ in this subspace, and a component $\mathbf{h}$ in the four-dimensional orthogonal subspace. That is, there is a rotation matrix R in $\mathbb{R}^6$ such that

$$R\,\mathbf{x} = \begin{bmatrix} \mathbf{g} \\ \mathbf{h} \end{bmatrix}, \quad \mathbf{g} \in \mathbb{R}^2, \ \mathbf{h} \in \mathbb{R}^4$$

and then the map DF is an isomorphism between $\mathbf{g}$ and $\mathbf{y}$:

$$DF = A \circ \Pi_{\mathbf{g}} \circ R \tag{7.5}$$

where $\Pi_{\mathbf{g}}$ is the 2×6 matrix of the projection on the two-dimensional subspace and A is an invertible 2×2 matrix. The normal and covariance matrix can be transformed into the new coordinate system by R, R^T; then we can use the formulae of Section 5.4 and compute the marginal covariance for $\mathbf{g}$:

$$\Gamma_{\mathbf{gg}}^{-1} = C^{\mathbf{gg}} = C_{\mathbf{gg}} - C_{\mathbf{gh}} \, C_{\mathbf{hh}}^{-1} \, C_{\mathbf{hg}}.$$

Since A is invertible, the same formulae for covariance propagation of Section 5.5 apply, that is

$$C_{\mathbf{y}} = \left(A^{-1}\right)^T C_{\mathbf{gg}} \, A^{-1}, \quad \Gamma_{\mathbf{y}} = A \, \Gamma_{\mathbf{gg}} \, A^T$$

and by combining all the transformations of the covariance with eq. (7.5)

$$\Gamma_{\mathbf{y}} = A \, \Pi_{\mathbf{g}} \, R \, \Gamma \, R^T \, \Pi_{\mathbf{g}}^T \, A^T = DF \, \Gamma \, DF^T,$$

that is, the same formula of the probabilistic interpretation.

This linear prediction formalism is used as a matter of routine in the collaborative case, and it has been proposed to use it systematically for asteroid astrometry. However, the astrometric prediction function F is nonlinear, and there is no guarantee that the confidence ellipse is a good approximation of the confidence prediction region: this is indeed not the case when a poorly determined orbit is used to predict the observations at a time t_1 very far from the last observation used in the orbit determination.

---

[4] This condition is violated only where (α, δ) are singular coordinates for the celestial sphere, that is for $\delta = \pm \pi/2$.

Semilinear predictions

We would like to have an algorithm to compute an approximation of the fully nonlinear confidence prediction region which is a better approximation than the linear one, and nevertheless can be computed explicitly. The astrometric prediction function contains the integral flow, thus in realistic cases to compute it accurately we can only numerically propagate a finite number of orbits from time t_0 to time t_1.

The geometric idea of the *semilinear confidence boundary* comes from the regression subspace of $\mathbf{h}$ given $\mathbf{g}$, that is the dimension 2 linear subspace

$$\mathbf{h} - \mathbf{h}^* = -C_{\mathbf{hh}}^{-1} \, C_{\mathbf{hg}} \, (\mathbf{g} - \mathbf{g}^*)$$

where $\mathbf{h}^*, \mathbf{g}^*$ are the nominal values

$$\begin{bmatrix} \mathbf{g}^* \\ \mathbf{h}^* \end{bmatrix} = R \, \mathbf{x}^*, \quad \mathbf{g}^* \in \mathbb{R}^2, \ \mathbf{h}^* \in \mathbb{R}^4.$$

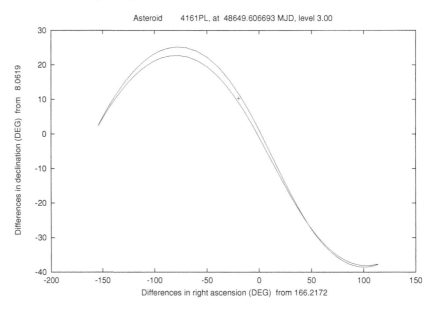

Fig. 7.6. Simulated recovery of the asteroid 4161 PLS, lost in September 1960, at the epoch of the discovery of the asteroid 1992 BU, that is 31 year later. The recovery observation (cross) is well inside the semilinear confidence boundary corresponding to the $\sigma = 3$ level. Reprinted from (Milani 1999) with permission from Elsevier.

The values of $\mathbf{h}$ from the above formula have the property that, for a given $\mathbf{g}$, thus for a given linearized prediction $\Delta\mathbf{y} = DF \, \Delta\mathbf{x}$, the quadratic penalty $\Delta_{\mathbf{h}}Q = \Delta\mathbf{x}^T \, C \, \Delta_{\mathbf{x}}/m$ is minimum. In particular, the intersection of the regression subspace of $\mathbf{h}$ given $\mathbf{g}$ with the outer surface of the confidence ellipsoid $\Delta\mathbf{x}^T \, C \, \Delta\mathbf{x} = \sigma^2$ is an ellipse γ in the $\mathbf{x}$ space, which projects (by

$\Pi_g \circ R$) onto the marginal confidence ellipse in the **g** space, thus also (by DF) onto the confidence ellipse in the **y** space.

We define the **semilinear confidence boundary** as the image, by the nonlinear astrometric prediction function F, of the ellipse γ defined above into the **y** space of astrometric observations. In practical cases, the semilinear boundary is very different from the linear ellipse when the size of both figures is large, e.g., several degrees. This happens when the orbit has been determined by using only a short observed arc and/or when the prediction is for a time remote from the observations. Figure 7.6 shows a rather extreme case of an asteroid lost after being observed only during a survey in 1960 and recovered 31 years later. In such a case the mean longitude at the prediction time has an enormous uncertainty, thus the semilinear boundary follows the curvature of the image of a long segment of the λ coordinate axis in the space of equinoctial elements.

7.6 Attribution

In the attribution case the problem is how to define the observation space in which the comparison between prediction and available data has to be performed. We assume an orbit $\mathbf{x}_1$ has been fit to the first set of m_1 observations, with epoch time t_1, and the uncertainty is described by the covariance and normal matrices Γ_1, C_1. The second arc includes m_2 scalar observations.

It is possible to compute a prediction for each of the m_2 observations, with its uncertainty, and to apply a test on the size of each of the normalized residuals, but this is inefficient for two reasons. First, the predictions for observations close in time are correlated, thus the marginal uncertainty of each one gives a less stringent control than performing a single test for all, with a full normal matrix for the vector prediction in $\mathbb{R}^{2\,m_2}$. Second, to compute an accurate prediction for a sequence of times requires us to propagate the orbit to each one of the (distinct) times t_i of the second arc. The number m_2 can be comparatively large, and still we can have a too short arc to fit an orbit, when many observations are taken over a short time span.

Thus it is useful to synthesize the information contained in the second arc into a vector observation at a single time t_2: this is the *attributable*.

Attributables

Let $(\rho, \alpha, \delta) \in \mathbb{R}^+ \times [-\pi, \pi) \times (-\pi/2, \pi/2)$ be spherical coordinates for the topocentric position of a celestial body. The angular coordinates (α, δ) are defined by a topocentric reference system that can be arbitrarily selected.

Usually, in the applications, α is the right ascension and δ the declination with respect to an equatorial reference system (e.g., J2000).

We shall call **attributable** a vector

$$A = (\alpha, \delta, \dot{\alpha}, \dot{\delta}) \in [-\pi, \pi) \times (-\pi/2, \pi/2) \times \mathbb{R}^2,$$

representing the angular position and velocity of the body at a time $\bar{t}$ in the selected reference frame. A natural operation is trying to *attribute* the data used for A to an already existing orbit, hence the name.

A detection of a moving object today is done by comparing two or more images of the same field, taken at short intervals of time. Thus an attributable can be computed from a short arc of astrometric observations of a celestial body. Given the observed values $(t_i, \alpha_i, \delta_i)$ for $i = 1, m$ with $m \geq 2$, we can compute an attributable with its uncertainty. We can fit both angular coordinates with linear functions of time, that is with the same fit of the model problem of Section 5.1. More precisely, let $\bar{t}$ be the mean of the t_i and let the fit solution at time $\bar{t}$ be $(\alpha, \dot{\alpha}, \delta, \dot{\delta})$; this solution is obtained with the regression line formulae, together with the two 2×2 normal matrices $C_{(\alpha, \dot{\alpha})}, C_{(\delta, \dot{\delta})}$ and covariance matrices $\Gamma_{(\alpha, \dot{\alpha})}, \Gamma_{(\delta, \dot{\delta})}$. The normal matrix C_A of A is composed just by joining the two normal matrices, and is not singular provided the observations refer to ≥ 2 distinct times; its inverse Γ_A is also composed by joining the two 2×2 covariance matrices.

On the other hand, if there are $m \geq 3$ observations and the time span is not too short, a more accurate estimate of the attributable A is obtained by fitting both angular coordinates as a function of time with a quadratic model. Then the solution $(\alpha, \dot{\alpha}, \ddot{\alpha}, \delta, \dot{\delta}, \ddot{\delta})$ is obtained with the standard formulae of the least squares problem, together with the two 3×3 covariance matrices $\Gamma_{(\alpha, \dot{\alpha}, \ddot{\alpha})}, \Gamma_{(\delta, \dot{\delta}, \ddot{\delta})}$. The marginal covariance matrix Γ_A of A, whatever the values of $(\ddot{\alpha}, \ddot{\delta})$, is obtained by extracting the relevant 4×4 submatrix, and the normal matrix is computed by $C_A = \Gamma_A^{-1}$.

Note that the observations can be weighted. If there are only two observations with equal weight $1/\sigma^2$ and difference in times $2\Delta t$, then the correlations $\mathrm{Corr}(\alpha, \dot{\alpha}), \mathrm{Corr}(\delta, \dot{\delta})$ are zero and C_A, Γ_A are diagonal:[5] the standard deviation of both angles is $\sigma/\sqrt{2}$ and the standard deviation of the angular rates is $\sqrt{2}\,\sigma/\Delta t$.

---

[5] We are assuming that the α and δ error components of an astrometric observation are not correlated, otherwise the 4×4 normal and covariance matrix of all the variables could be full. This assumption would fail if the timing was a significant source of error.

Prediction for an attributable

Predictions of an attributable $A(t)$ are a straightforward generalization of the standard ephemerides $(\alpha(t), \delta(t))$ discussed in Section 7.5. Let us assume the prediction function G maps an open set of the initial conditions space into a four-dimensional space, that is the vector of observables is

$$\mathbf{y}(\bar{t}) = (\alpha(\bar{t}), \delta(\bar{t}), \dot{\alpha}(\bar{t}), \dot{\delta}(\bar{t})) = G(\mathbf{x}(\bar{t})).$$

Given initial conditions $\mathbf{x}$ at time t_0 with covariance Γ, the prediction function $F = G \circ \Phi_{t_0}$ is also four-dimensional and its partial derivatives form the matrix DF of dimension 4×6: the covariance and normal matrix, by the same argument of Section 7.5, are the 4×4 matrices obtained from Γ by

$$\Gamma_{\mathbf{y}} = (DF)\,\Gamma\,(DF)^T, \qquad C_{\mathbf{y}} = \Gamma_{\mathbf{y}}^{-1}.$$

The matrix $\Gamma_{\mathbf{y}}$ can be used to assess the uncertainty of all the components of the attributable, e.g., the RMS uncertainty of the angles (α, δ) (see Figure 8.4) as well as that of the angular rates $(\dot{\alpha}, \dot{\delta})$. The normal matrix $C_{\mathbf{y}}$ can be used to define the metric used in the attribution algorithm.

Attribution penalty

Let $\mathbf{x}_1$ be the attributable, that is the four-dimensional vector representing the set of observations to be attributed, and C_1 be the 4×4 normal matrix of the fit used to compute it. Let $\mathbf{x}_2$ be the predicted attributable, computed from the known least squares orbit, and Γ_2 be the covariance matrix of such a four-dimensional prediction, obtained by propagation of the covariance of the orbital elements (as discussed above). Then $C_2 = \Gamma_2^{-1}$ is the corresponding normal matrix. With this new interpretation for the symbols $\mathbf{x}_1, \mathbf{x}_2, C_1, C_2$, the algorithm for linear attribution uses the same formulae of Section 7.2 applied in the four-dimensional attributable space:

$$
\begin{aligned}
C_0 &= C_1 + C_2, \qquad \Gamma_0 = C_0^{-1} \\
K_4 &= (\mathbf{x}_2 - \mathbf{x}_1) \cdot [C_1 - C_1\,\Gamma_0\,C_1]\,(\mathbf{x}_2 - \mathbf{x}_1) \\
\mathbf{x}_0 &= \Gamma_0\,[C_1\,\mathbf{x}_1 + C_2\,\mathbf{x}_2].
\end{aligned}
\tag{7.6}
$$

In particular, the **attribution penalty** K_4/m ($m = 8$, the number of scalar components of the two attributables) is computed and used as a control to filter out the orbit-attributable pairs which cannot belong to the same object (unless the observations are exceptionally poor). For the orbit-attributable couples with K_4 below some control value, the next stages are to select a preliminary orbit and to perform differential corrections.

If, for an orbit-attributable couple, the orbit is good enough, it could be used as a preliminary orbit without modifications. That is, the orbit computed with the data of the first arc is used as a first guess for the differential correction iterations, fitting the observations of both arcs. In more difficult cases, e.g., when the orbit for the first arc has not been obtained by a least squares fit, but is itself a preliminary orbit, a better preliminary orbit can be estimated from the four-dimensional compromise attributable $\mathbf{x}_0$. This is discussed in Section 8.5.

Attribution procedure

As in the case of orbit identification, a procedure to try to attribute a large number of attributables to a large orbit catalog needs to use a sequence of filters. We have experimented with the following filters:

(i) comparison of the two-dimensional prediction (α, δ) from the orbit at the attributable time with the angles of the attributable;

(ii) computation of the attribution penalty K_4 from the attributable and the predicted attributable from the orbit;

(iii) confirmation by differential corrections and quality control.

The choice of the controls for selection in each filter is very delicate, and needs to be based on experience since there is no analytic estimate available. The extensive tests of Milani *et al.* (2001a) have resulted in a large number (thousands) of identifications. Additional and even more extensive tests have been performed in the context of the simulation of future Solar System surveys, see (Milani *et al.* 2005a, Milani *et al.* 2008) and Chapters 8 and 11.

The case which has not been fully studied yet is the one in which two very different data sets are used, e.g., attributables from a current survey and a catalog of orbits from historic data. If the cataloged orbits are based on much longer observed arcs, possibly with lower accuracy observations, an ad hoc quality control procedure needs to be used, based on the increase of the quality control parameters as a result of the attribution rather than the absolute value of the controls.

8

LINKAGE

In Chapter 7 we explained how to compute an attributable

$$A = (\alpha, \delta, \dot{\alpha}, \dot{\delta}) \in [-\pi, \pi) \times (-\pi/2, \pi/2) \times \mathbb{R}^2 \qquad (8.1)$$

at a certain time $\bar{t}$, given ≥ 2 observations of a celestial body. Throughout this chapter, we shall use only the information contained in the attributable to try to achieve identifications and therefore orbit determination. This chapter is based on (Milani *et al.* 2001a, Milani *et al.* 2004, Milani *et al.* 2005a, Tommei *et al.* 2007, Gronchi *et al.* 2008) and ongoing research.

8.1 Admissible region

Let A be an attributable at time $\bar{t}$ for a celestial body $\mathcal{B}$ (e.g., an asteroid). We denote by $\mathbf{r}$ and $\mathbf{q}$ the heliocentric position vectors of the body and the observer on the Earth at time $\bar{t}$. Let $r = \|\mathbf{r}\|$, $q = \|\mathbf{q}\|$ be the Euclidean norms of these vectors. We also write $(\rho, \alpha, \delta) \in \mathbb{R}^+ \times [-\pi, \pi) \times (-\pi/2, \pi/2)$ for the spherical coordinates of the topocentric position $\boldsymbol{\rho} = \mathbf{r} - \mathbf{q}$ of the body, with $\rho = \|\boldsymbol{\rho}\|$. The information contained in the attributable A leaves completely unknown the topocentric distance ρ and the radial velocity $\dot{\rho}$ of $\mathcal{B}$.[1] The purpose of this section is to constrain the possible values of $\rho, \dot{\rho}$ with the hypothesis that the observed object is a Solar System body.

Excluding interstellar orbits

We introduce the following notation: let

$$\mathcal{E}_\odot(\rho, \dot{\rho}) = \frac{1}{2}\|\dot{\mathbf{r}}(\rho, \dot{\rho})\|^2 - k^2 \frac{1}{r(\rho)}, \qquad (8.2)$$

---

[1] The same quantities are called range and range-rate in the context of spacecraft tracking.

with $k = 0.017\,202\,098\,95$ the Gauss constant, be the two-body energy of the heliocentric orbit of $\mathcal{B}$, in the approximation neglecting the mass of $\mathcal{B}$. Note that we are using 1 AU as the unit of length and 1 ephemeris day as the unit of time; we do not need to specify the unit of mass as $\mathcal{E}_\odot(\rho, \dot\rho)$ is the two-body energy per unit mass of $\mathcal{B}$. We describe the region excluding interstellar orbits, that is satisfying the condition

$$\mathcal{E}_\odot(\rho, \dot\rho) \le 0. \tag{8.3}$$

In particular we shall show that this region can have either one or two connected components. The heliocentric position of $\mathcal{B}$ is given by

$$\mathbf{r} = \mathbf{q} + \rho\,\hat{\boldsymbol{\rho}}, \tag{8.4}$$

where $\hat{\boldsymbol{\rho}}$ is the unit vector in the observation direction. Using the spherical coordinates (ρ, α, δ), the heliocentric velocity $\dot{\mathbf{r}}$ of $\mathcal{B}$ is

$$\dot{\mathbf{r}} = \dot{\mathbf{q}} + \dot\rho\,\hat{\boldsymbol{\rho}} + \rho\,\dot\alpha\,\hat{\boldsymbol{\rho}}_\alpha + \rho\,\dot\delta\,\hat{\boldsymbol{\rho}}_\delta, \tag{8.5}$$

where $\hat{\boldsymbol{\rho}}_\alpha = \partial\hat{\boldsymbol{\rho}}/\partial\alpha$, $\hat{\boldsymbol{\rho}}_\delta = \partial\hat{\boldsymbol{\rho}}/\partial\delta$, and $\dot{\mathbf{q}}$ is the heliocentric velocity of the observer. In coordinates

$$
\begin{aligned}
\hat{\boldsymbol{\rho}} &= (\cos\alpha\cos\delta, \sin\alpha\cos\delta, \sin\delta),\\
\hat{\boldsymbol{\rho}}_\alpha &= (-\sin\alpha\cos\delta, \cos\alpha\cos\delta, 0),\\
\hat{\boldsymbol{\rho}}_\delta &= (-\cos\alpha\sin\delta, -\sin\alpha\sin\delta, \cos\delta),
\end{aligned}
$$

$$\hat{\boldsymbol{\rho}} \cdot \hat{\boldsymbol{\rho}}_\alpha = \hat{\boldsymbol{\rho}} \cdot \hat{\boldsymbol{\rho}}_\delta = \hat{\boldsymbol{\rho}}_\alpha \cdot \hat{\boldsymbol{\rho}}_\delta = 0, \quad \|\hat{\boldsymbol{\rho}}\| = \|\hat{\boldsymbol{\rho}}_\delta\| = 1, \quad \|\hat{\boldsymbol{\rho}}_\alpha\| = \cos\delta.$$

Thus the squared norms of the heliocentric position and velocity are

$$r^2(\rho) = \rho^2 + 2\rho\,\mathbf{q}\cdot\hat{\boldsymbol{\rho}} + \|\mathbf{q}\|^2, \tag{8.6}$$

$$
\begin{aligned}
\|\dot{\mathbf{r}}(\rho, \dot\rho)\|^2 &= \dot\rho^2 + 2\dot\rho\dot{\mathbf{q}}\cdot\hat{\boldsymbol{\rho}} + \rho^2\left(\dot\alpha^2\cos^2\delta + \dot\delta^2\right)\\
&\quad + 2\rho\left(\dot\alpha\dot{\mathbf{q}}\cdot\hat{\boldsymbol{\rho}}_\alpha + \dot\delta\dot{\mathbf{q}}\cdot\hat{\boldsymbol{\rho}}_\delta\right) + \|\dot{\mathbf{q}}\|^2.
\end{aligned} \tag{8.7}
$$

We shall use the coefficients[2]

$$
\begin{array}{llll}
c_0 &= \|\mathbf{q}\|^2 & c_3 &= 2\dot\alpha\,\dot{\mathbf{q}}\cdot\hat{\boldsymbol{\rho}}_\alpha + 2\dot\delta\,\dot{\mathbf{q}}\cdot\hat{\boldsymbol{\rho}}_\delta\\
c_1 &= 2\dot{\mathbf{q}}\cdot\hat{\boldsymbol{\rho}} & c_4 &= \|\dot{\mathbf{q}}\|^2\\
c_2 &= \dot\alpha^2\cos^2\delta + \dot\delta^2 = \eta^2 & c_5 &= 2\mathbf{q}\cdot\hat{\boldsymbol{\rho}},
\end{array} \tag{8.8}
$$

[2] For more accurate results, the position $\mathbf{q}(\bar{t})$ and the velocity $\dot{\mathbf{q}}(\bar{t})$ should be computed consistently with the interpolation used for $\hat{\boldsymbol{\rho}}(\bar{t})$, by using the Poincaré observer interpolation method.

and the polynomial expressions

$$
\begin{aligned}
\|\dot{\mathbf{r}}(\rho,\dot\rho)\|^2 &= 2\mathcal{T}_\odot(\rho,\dot\rho) = \dot\rho^2 + c_1\dot\rho + c_2\rho^2 + c_3\rho + c_4, \\
r^2 = S(\rho) &= \rho^2 + c_5\rho + c_0, \qquad W(\rho) = c_2\rho^2 + c_3\rho + c_4.
\end{aligned}
\tag{8.9}
$$

By substituting the last expressions in (8.2), condition (8.3) reads

$$
2\mathcal{E}_\odot(\rho,\dot\rho) = \dot\rho^2 + c_1\dot\rho + W(\rho) - 2k^2/\sqrt{S(\rho)} \le 0.
$$

To have real solutions for $\dot\rho$, the discriminant of $\mathcal{E}_\odot$, as a polynomial of degree 2 in $\dot\rho$, must be non-negative, i.e.,

$$
c_1^2/4 - W(\rho) + 2k^2/\sqrt{S(\rho)} \ge 0.
$$

Let us set $\gamma = c_4 - c_1^2/4$ (note that $\gamma \ge 0$), and define $P(\rho) = c_2\rho^2 + c_3\rho + \gamma$; then condition (8.3) implies

$$
2k^2/\sqrt{S(\rho)} \ge P(\rho).
\tag{8.10}
$$

The polynomial $P(\rho)$ is non-negative for each ρ: it is the opposite of the discriminant of $\mathcal{T}_\odot(\rho,\dot\rho)$, regarded as a polynomial in the variable $\dot\rho$. $\mathcal{T}_\odot$ is a kinetic energy and is non-negative, thus its discriminant is non-positive. Also $S(\rho)$ is non-negative, thus we can square both sides of (8.10) and obtain an inequality involving a polynomial of degree 6

$$
4k^4 \ge V(\rho) = P^2(\rho)S(\rho) = \sum_{i=0}^{6} A_i \rho^i,
\tag{8.11}
$$

with coefficients

$$
\begin{aligned}
A_0 &= c_0\gamma^2, \quad A_1 = c_5\gamma^2 + 2c_0c_3\gamma, \quad A_2 = \gamma^2 + 2c_3c_5\gamma + c_0(c_3^2 + 2c_2\gamma), \\
A_3 &= 2c_3\gamma + c_5(c_3^2 + 2c_2\gamma) + 2c_0c_2c_3, \\
A_4 &= c_3^2 + 2c_2\gamma + 2c_2c_3c_5 + c_0c_2^2, \quad A_5 = c_2(2c_3 + c_2c_5), \quad A_6 = c_2^2.
\end{aligned}
$$

The most important property of the region defined by (8.3) is that *it has at most two connected components*. For a proof, see (Milani et al. 2004).

If the motion of the observer is approximated by a circular heliocentric orbit, it is possible to show that the region of Solar System orbits is connected when the direction of observation is at quadrature (orthogonal to the direction to the Sun). At opposition (direction of observation opposite to the Sun) there can be two connected regions only if the path of the body $\mathcal{B}$ on the celestial sphere is retrograde, as in the following example (see Figure 8.1): we have used the attributable $(\alpha, \delta, \dot\alpha, \dot\delta) = (0, 0, -0.09, 0.01)$, with

$\dot{\alpha}, \dot{\delta}$ in degrees per day (assuming the Earth is at $(x, y, z) = (1, 0, 0)$ in equatorial coordinates, lengths in AU). We have also plotted the level curves for small positive and negative values of $\mathcal{E}_\odot$, showing the qualitative change.

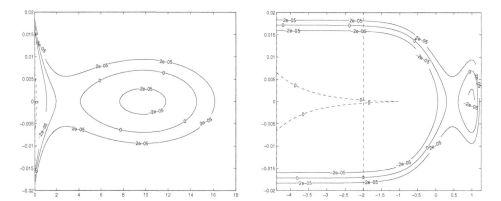

Fig. 8.1. An example with two connected components. On the left three level curves of $\mathcal{E}_\odot$, including the zero level curve, and $\mathcal{E}_\oplus = 0$ (dashed curve) in the plane $(\rho, \dot{\rho})$; on the right the same plot in the plane $(\log_{10}(\rho), \dot{\rho})$. Reproduced with permission of Springer from Milani *et al.* (2004).

To know the number of components of the region defined by (8.3) we have to compute the roots of the degree six polynomial $V(\rho) - 4k^4$. There are fast and reliable algorithms in the numerical analysis literature providing the **roots of a polynomial** (as a complex vector), with rigorous upper bounds for the errors including rounding off. We use the algorithm by Bini (1997) and the corresponding public domain software.[3]

The inner boundary

A difficulty in the practical usage of the region defined by condition (8.3) as a tool for the identification problem is that it is not a compact set, that is, the observed object could be at an arbitrarily small distance from the observer. This makes it impossible to sample it with a finite number of points in such a way that the corresponding orbits are representative of the range of ephemeris uncertainty (see the discussion in Section 8.4). There are several ways to assign an inner boundary to the region where $\mathcal{B}$ could be, based on different practical considerations:

- an inner boundary can be assigned requiring that $\mathcal{B}$ is not a satellite of the Earth, i.e., by imposing a condition on the geocentric energy $\mathcal{E}_\oplus(\rho, \dot{\rho})$;

---

[3] For the Fortran 77 version see `http://www.netlib.org/numeralgo/na10`;
for Fortran 90 see `http://users.bigpond.net.au/amiller/pzeros.f90`.

- a minimal distance can be dictated by physical limitations, such as the Earth's atmosphere or the Earth's radius $R_\oplus$ in the geocentric approximation;
- a minimal distance can be assigned by requiring that $\mathcal{B}$ is not too small, if photometric measurements are supplied together with the astrometry used to compute the attributable.

Excluding satellites of the Earth

We look for a simple description of the region satisfying the condition $\mathcal{E}_\oplus(\rho, \dot{\rho}) \geq 0$. A simplifying approximation is obtained by assuming that the observations are geocentric: with $\mathbf{q}_\oplus$ the heliocentric position of the Earth's center, assuming $\mathbf{r} = \boldsymbol{\rho} + \mathbf{q}_\oplus$, the geocentric energy is

$$\mathcal{E}_\oplus(\rho, \dot{\rho}) = \frac{1}{2}\|\dot{\boldsymbol{\rho}}\|^2 - k^2\mu_\oplus\frac{1}{\rho} \geq 0, \tag{8.12}$$

where $\mu_\oplus$ is the ratio between the mass of the Earth and the mass of the Sun. By using $\|\dot{\boldsymbol{\rho}}(\rho, \dot{\rho})\|^2 = \dot{\rho}^2 + \rho^2\,\eta^2$, where $\eta = \sqrt{\dot{\alpha}^2\cos^2\delta + \dot{\delta}^2}$ is the **proper motion**, (8.12) becomes

$$\dot{\rho}^2 + \rho^2\,\eta^2 - 2k^2\mu_\oplus\frac{1}{\rho} \geq 0,$$

that is

$$\dot{\rho}^2 \geq G(\rho), \qquad \text{with} \quad G(\rho) = \frac{2k^2\mu_\oplus}{\rho} - \eta^2\rho^2, \tag{8.13}$$

where $G(\rho) > 0$ for $0 < \rho < \rho_0 = \sqrt[3]{(2k^2\mu_\oplus)/\eta^2}$.

However, condition (8.12) is meaningful only inside the sphere of influence of the Earth, otherwise the dynamics of $\mathcal{B}$ is dominated by the Sun, not by the Earth. Thus we need to introduce the condition

$$\rho \geq R_{SI} = a_\oplus\sqrt[3]{\mu_\oplus/3} \tag{8.14}$$

where R_{SI} is the radius of the sphere of influence, and $a_\oplus$ is the semimajor axis of the Earth. To exclude the satellites of the Earth we have to assume that either (8.12) or (8.14) apply. If $\rho_0 \leq R_{SI}$ the region of the satellites to be excluded is defined simply by eq. (8.13); this occurs for

$$\rho_0^3 = 2k^2\mu_\oplus/\eta^2 \leq R_{SI}^3 = a_\oplus^3\,\mu_\oplus/3$$

thus, taking into account Kepler's third law $a_\oplus^3\,n_\oplus^2 \simeq k^2$ with $n_\oplus$ the mean motion of the Earth, we have $\rho_0 \leq R_{SI}$ if and only if $\eta \geq \sqrt{6}\,n_\oplus$. Otherwise, if $\rho_0 > R_{SI}$, the boundary of the region containing satellites of the Earth is formed by a segment of the straight line $\rho = R_{SI}$ and the two arcs of the $\dot{\rho}^2 = G(\rho)$ curve with $0 < \rho < R_{SI}$, as in Figures 8.1 and 8.2.

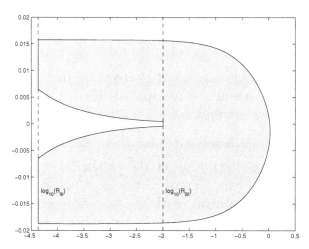

Fig. 8.2. The qualitative features of the region of heliocentric orbits in the $(\rho, \dot{\rho})$ plane: by combining conditions (8.3), (8.12), (8.14) and $\rho \geq R_\oplus$, we are left with the domain sketched. This figure refers to a case with only one connected component and the plot is in the plane $(\log_{10}(\rho), \dot{\rho})$.

The shape of the inner boundary

To understand the shape of the boundary of the Earth satellites region we need to find possible intersections between the curves $\mathcal{E}_\oplus = 0$ and $\mathcal{E}_\odot = 0$. However, if $\mathcal{E}_\oplus$ is computed in a geocentric approximation, these intersections are physically meaningful only if they occur for $R_\oplus < \rho < R_{SI}$, that is, during a close approach to the Earth, but above its physical surface. In (Milani *et al.* 2004) we prove that for $R_\oplus \leq \rho \leq R_{SI}$ the condition $\mathcal{E}_\oplus(\rho, \dot{\rho}) \leq 0$ implies $\mathcal{E}_\odot(\rho, \dot{\rho}) \leq 0$.

This result shows that the intersections of the two zero-energy curves occur only where they do not matter; it also implies that the region of Solar System orbits excluding the satellites of the Earth does not have more connected components than the region satisfying condition (8.3) only. This applies only for particular values of the mass, radius, and orbital elements of the planet on which the observer is located. It is a physical property of the Earth, not a general property of whatever planet. A larger planet, such as Jupiter, can have satellites whose velocity would lead to a hyperbolic orbit with respect to the Sun, if Jupiter was not controlling the orbit; the Earth cannot have satellites with this behavior.

The tiny object boundary

An alternative method to assign a lower limit to the distance is to impose the condition that the object is not a "shooting star" (very small and very close to the Earth). We assume that the size is controlled by setting a maximum for the **absolute magnitude** H

$$H(\rho) \leq H_{max}. \tag{8.15}$$

If some value of the apparent magnitude is available, then the absolute magnitude H can be computed from h, the average of the measured apparent magnitudes, using the relation

$$H = h - 5\log_{10}\rho - x(\rho), \tag{8.16}$$

where the correction $x(\rho)$ accounts for the distance from the Sun and the phase effect.[4] For small ρ (e.g., $\rho < 0.01$ AU) the correction $x(\rho)$ has a negligible dependence upon ρ because the distance from the Sun is $\simeq 1$ AU and the phase is close to the angle between $\hat{\rho}$ and the opposition direction. Thus we can approximate $x(\rho)$ with a quantity x_0 independent of ρ. Also for larger values of ρ this is an acceptable approximation. Moreover, we are using ρ, the distance at the reference time $\bar{t}$, for all the epochs of the observations including photometry: this is a fair approximation unless the relative change of distance during the time span of the observed arc is relevant, which can happen only for very small distances. In this approximation, condition (8.15) becomes

$$H_{max} \geq H = h - 5\log_{10}\rho - x_0 \Longrightarrow \log_{10}\rho \geq \frac{h - H_{max} - x_0}{5} \stackrel{def}{=} \log_{10}\rho_H,$$

that is, given the apparent magnitude h, we have a minimum distance $\rho_H = \rho(H_{max})$ for the object to be of significant size. If we use $H_{max} = 30$ (a few meters diameter) and $x_0 = 0$ then, for example

$$h = 20 \Longrightarrow \rho \geq 0.01 \text{ AU}, \quad h = 15 \Longrightarrow \rho \geq 0.001 \text{ AU}.$$

In any case, the absolute magnitude of the object is not a function of $\dot{\rho}$ and the region satisfying condition (8.15) is just a half-plane $\rho \geq \rho_H$. We call the **tiny object boundary** the straight line $\rho = \rho_H$.

Provided $\rho_H \geq R_\oplus$ (for $H_{max} = 30$ this occurs for $h \geq 8.1$) it is possible to use the same arguments of the theorem on the intersection between the energy curves to show that condition (8.15) does not increase the number of connected components with respect to the region defined by excluding the satellites of the Earth. On the contrary it is quite possible that the geometry of the region becomes simpler. If $H_{max} = 30$ and $h > 20$ the entire sphere of influence of the Earth is excluded by condition (8.15), thus condition (8.14) is implied by (8.15) and condition (8.12) becomes irrelevant.

---

[4] See the IAU definition of absolute magnitude (Bowell *et al.* 1989).

Definition of admissible region

We wish to determine a region which is a good approximation of the subset in the $\rho > 0$ half-plane where the object $\mathcal{B}$ we are searching for has to be located. Thus we adapt the definition to the goal of the population orbit determination at hand, e.g., it has to be different in a search for objects in heliocentric limited orbit (asteroids, comets, trans-neptunian objects), for objects passing very close to the Earth (meteoroids), and for objects in geocentric orbits (artificial satellites, space debris).

We give as an example a definition appropriate in a search for objects in heliocentric orbit with significant size, thus we can assume that $\rho(H_{max}) > R_{SI}$. Given an attributable A and selecting a maximum absolute magnitude H_{max}, we define as the **admissible region** the set

$$\mathcal{D}(A) = \{(\rho, \dot\rho) : \rho \geq \rho_H, \ \mathcal{E}_\odot(\rho, \dot\rho) \leq 0\}. \tag{8.17}$$

This definition does not use any geocentric approximation, avoiding the problems discussed in Sections 8.7 and 9.4. For smaller objects the portions of the conditions (8.12) and (8.14) should be taken into account.

8.2 Sampling of the admissible region

The admissible region consists of at most two compact connected components. Its boundary has an outer part, given by arcs of the curve $\mathcal{E}_\odot(\rho, \dot\rho) = 0$, symmetric with respect to the line $\dot\rho = -c_1/2$. The boundary also has an inner part consisting, in the simplest case, of a segment of the line $\rho = \rho(H_{max})$; for smaller objects, with $\rho(H_{max}) < R_{SI}$, the inner boundary has a more complicated shape like the one shown in Figure 8.2.

To sample the admissible region we start by sampling its boundary. We would like to select points that are equispaced on the boundary, that is, if the boundary is parameterized by its arc length s, then the distance of each couple of consecutive points corresponds to a fixed increment of s. To avoid the computation of the arc length parameter we use the following idea: we choose a large number of points, equispaced in one of the two coordinates, and then we use an elimination rule to be iterated until we are left with the desired number of points. It can be shown (Milani *et al.* 2004) that the remaining points are close to the ideal distribution, equispaced in arc length.

Delaunay triangulations

Consider the polygonal domain $\tilde{\mathcal{D}}$ defined by connecting with edges the sample of boundary points of the admissible region $\mathcal{D}$; we shall define a method to triangulate $\tilde{\mathcal{D}}$. A **triangulation** of $\tilde{\mathcal{D}}$ is a pair (Π, τ), where $\Pi = \{P_1, \ldots, P_N\}$ is a set of points (the *nodes*) of the domain, and $\tau - \{T_1, \ldots, T_k\}$ is a set of triangles with vertices in Π such that:

(i) $\bigcup_{i=1,k} T_i = \tilde{\mathcal{D}}$;
(ii) for each $i \neq j$ the set $T_i \cap T_j$ is either empty or a *vertex* or an *edge* of a triangle.

To each triangulation (Π, τ) we can associate the *minimum angle*, that is the minimum among the angles of all the triangles T_i. Among all possible triangulations of a convex domain a **Delaunay triangulation** is characterized by these properties (Bern and Eppstein 1992):

(i) it maximizes the minimum angle;
(ii) it minimizes the maximum circumcircle;
(iii) for each triangle T_i, the interior part of its circumcircle does not contain any nodes of the triangulation (Risler 1991).

These properties are all equivalent for convex domains.

If the domain is a convex quadrangle whose vertexes Π are not on the same circle, then there exist two possible triangulations $(\Pi, \tau_1), (\Pi, \tau_2)$: by property (iii), only one of these is Delaunay (see Figure 8.3). In this case the Delaunay triangulation can be obtained from the other one by an *edge-flipping* technique, which consists of substituting the diagonal $P_1 P_3$ (*non-Delaunay edge*) of the quadrangle, corresponding to the common edge, with the diagonal $P_2 P_4$ (*Delaunay edge*). The edge-flipping also results in an increase of the minimum angle.

If, in addition to the set of points Π, we give as input also some edges $P_i P_j$, for example the boundary edges as we do for $\tilde{\mathcal{D}}$, we refer to the corresponding triangulation containing the prescribed edges as a **constrained triangulation**.

The domain $\tilde{\mathcal{D}}$ is in general not convex; in this case we need to give as input the edges along the boundary. Then there still exists a constrained triangulation that maximizes the minimum angle (also minimizes the maximum circumcircles, i.e., (i), (ii) hold), called *constrained Delaunay triangulation* (Bern and Eppstein 1992), but property (iii) is not guaranteed. Figure 8.3 suggests how to transform any triangulation of $\tilde{\mathcal{D}}$ into a constrained Delaunay: for each triangle T_i, we iterate a procedure over the adjacent triangles; if the common edge with an adjacent triangle is not Delaunay, we apply the edge-flipping technique. Repeating this procedure until all edges of the triangulation are Delaunay or edges of the boundary of $\tilde{\mathcal{D}}$, at each step the

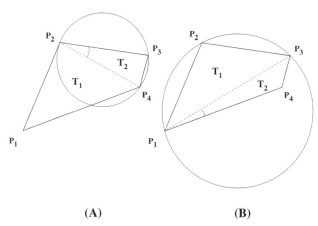

(A)   **(B)**

Fig. 8.3. Possible triangulations of the quadrangle $P_1 P_2 P_3 P_4$: the one in (A) is a Delaunay triangulation. We mark in both cases the minimum angle and we draw the circumcircles corresponding to triangle $P_2 P_3 P_4$ (left plot) and to triangle $P_1 P_2 P_3$ (right plot). Reproduced with permission of Springer from Milani *et al.* (2004).

minimum angle increases and at the end we obtain the triangulation that maximizes the minimum angle (Delaunay 1934).

The procedure adopted to triangulate our domain uses as input the sampling of the boundary described above and the polygon formed by these boundary points. The first phase is to generate a constrained Delaunay triangulation (Π_0, τ_0) with these boundary points and boundary edges. Once the initial triangulation is obtained, we refine it by adding new points internal to the domain, keeping at each insertion the Delaunay property. At each step we add a point extending to the internal part of the domain the discrete density defined on the boundary points by the quantities $\rho(P_j) = \min_{l \neq j} d(P_l - P_j)$ where d is some distance.[5] Let G_i be the barycenters of the triangles T_i; we define the corresponding densities

$$\tilde{\rho}(G_i) = \frac{1}{3} \sum_{m=1}^{3} \rho(P_{i_m})$$

$(P_{i_m}, m = 1, 2, 3,$ belongs to the same triangle T_i) and we add as a new point the barycenter $G_{\bar{k}}$ that maximizes the minimum distance (weighted with its density $\tilde{\rho}(G_{\bar{k}})$) from the nodes of the triangulation. Then we eliminate the corresponding triangle $T_{\bar{k}}$ and we add to τ the triangles obtained joining the edges of $T_{\bar{k}}$ with the new point (keeping at each triangle insertion the Delaunay optimal property by means of the edge-flipping technique). We

---

[5] $\rho(P_j)$ is indeed an approximation of the inverse of a density function.

iterate this insertion procedure until

$$\max_{G_i} \left(\min_j \left\{ \frac{d(G_i, P_j)}{\tilde{\rho}(G_i)} \right\} \right) > \sigma, \tag{8.18}$$

where σ is a fixed small parameter. In (de' Michieli Vitturi 2004) it is shown that the algorithm converges and the final number of triangles is $< \mu(\partial \tilde{\mathcal{D}}) \, n_0 / \sqrt{3} \, \sigma$, where n_0 is the number of points on the boundary of length $\mu(\partial \tilde{\mathcal{D}})$.

When no new point needs to be added, either because some maximum number has been reached, or because the convergence criterion (8.18) is fulfilled, we apply to the triangulation obtained as above a mesh improvement technique, generalizing **Laplacian smoothing** (Winslow 1964, Field 1988). We move every internal point P_j of the triangulation to the center of mass (weighted with the density defined above) of the polygon formed by all its neighboring points (i.e., the ones connected to P_j by an edge), if it lies inside the polygon. This technique improves the quality of the triangulation, but it can produce a triangulation that is not Delaunay, so that we apply again the edge-flipping technique at the end of the smoothing algorithm. The final result is a triangulation optimal from the point of view of property (i), that is avoiding as much as possible "flattened" triangles.

The definition of Delaunay triangulation uses distances and angles, thus it depends on the metric selected for the space $(\rho, \dot{\rho})$, in fact its own definition is based on computations of distances and angles. In particular we can select a strictly increasing function $f(\rho)$ and perform the triangulation of the admissible region with the metric $ds^2 = df(\rho)^2 + d\dot{\rho}^2$, i.e., we can work in the plane $(f(\rho), \dot{\rho})$ endowed with the Euclidean metric. In our work we have selected an *adaptive metric*, defined by the function

$$f(\rho) = 1 - \exp\left(-\frac{\rho^2}{2\, s^2}\right). \tag{8.19}$$

Since $f'(\rho)$ is maximum at $\rho = s$, by choosing the parameter s we select which part of the admissible region should be more densely sampled. For example, we can use $s = \rho_{max}$, the largest root of the polynomial equation obtained by selecting the equal sign in (8.11); with this choice we enhance the portion of the space $(\rho, \dot{\rho})$ farthest from the observer. If our purpose is to search for objects in a particular portion of the $(\rho, \dot{\rho})$ space, then we can use a metric selected ad hoc. For example, to enhance the region near the Earth we can use either a smaller s, or possibly $f(\rho) = \log_{10}(\rho)$ as in Figure 8.1.

8.3 Attributable orbital elements

Given a short arc of observations, after computing the attributable we are left with a totally undetermined point in the $(\rho, \dot{\rho})$ plane. Following Section 8.1, we can assume that this point belongs to an admissible region of Solar System orbits, and we can sample this compact region by a finite Delaunay triangulation. Each node defines a **virtual asteroid**, that is a possible, but by no means determined, set of six quantities[6]

$$X = [\alpha, \delta, \dot{\alpha}, \dot{\delta}, \rho, \dot{\rho}]$$

which are topocentric spherical polar coordinates, in a different order. A set of six initial conditions uniquely determines the orbit of an asteroid, thus it is a set of orbital elements, belonging to a new type (different from the classical coordinates, such as Keplerian, equinoctial, cometary, Cartesian, etc.). We shall call such data a set of **attributable orbital elements**.

Distance-dependent corrections

We need to refer a set of orbital elements to an epoch time t_0; we can also obtain a value for the absolute magnitude H if there are photometric measurements together with the astrometric ones. The values of these quantities are not equal to the mean observation time $\bar{t}$ and the mean apparent magnitude h computed with the attributable, but require distance dependent corrections. An observation at time $\bar{t}$ of an asteroid needs to be corrected for **aberration**. The light spends a significant time $\delta t = \rho/c$, with c the speed of light, to reach the observer from $\mathcal{B}$; this means that the asteroid is observed at time $\bar{t}$ for its position at the time $\bar{t} - \delta t = t_0$, the epoch time of the orbital elements, which is a function of ρ.

Equation (8.16) describing the relationship between apparent magnitude h and the absolute magnitude H has the form $H = h + Z(\rho)$, thus also $H = H(\rho)$. In conclusion, both t_0 and H change for different virtual asteroids with the same attributable.

Two approximations are used in the above definitions. The values of the angles (α, δ) are corrected for aberration with an approximation to some order[7] in δt. A single $\delta t(\rho)$ and a single $Z(\rho)$ are used for all observations, while the value of ρ is not constant in time: this approximation may fail if the distance changes significantly during the arc time span, that is if $\dot{\rho}\,\delta t$ is of the order of ρ, in practice this can happen only for very small ρ.

---

[6] Five of these are measured by real numbers, while α is an angle, defined mod 2π; this is important whenever we compute a difference of two such vectors.

[7] For higher order aberration corrections see Chapter 17.

To convert from attributable elements to Cartesian ones we use eqs. (8.4), and (8.5), thus $(\mathbf{r}, \dot{\mathbf{r}})$ are functions of $(\rho, \dot{\rho})$, of $\hat{\boldsymbol{\rho}} = \hat{\boldsymbol{\rho}}(\alpha, \delta)$, and other quantities depending upon the attributable. However, they also contain $\mathbf{q}$ and $\dot{\mathbf{q}}$, which represent the position and velocity of the observer at time $\bar{t} = t_0 + \delta t$. The observer position is close to the geocenter position at the same time: $\mathbf{P} = \mathbf{q} - \mathbf{q}_\oplus$ is small, $|\mathbf{P}| \simeq R_\oplus \simeq 4 \times 10^{-5}|\mathbf{q}_\oplus|$, but a significant contribution to the heliocentric velocity comes from the observer geocentric velocity, which has size $|\dot{\mathbf{P}}| = \Omega_\oplus R_\oplus \cos\theta \simeq 0.5\cos\theta$ km/s $\leq |\dot{\mathbf{q}}_\oplus|/60$, where $\Omega_\oplus$ is the angular velocity of the Earth's rotation and θ is the latitude of the observer.

The problem is that the main frequency of the geocentric motion is $\Omega_\oplus$, more than two orders of magnitude faster than the mean motion of both the Earth and the asteroid. Thus the quadratic interpolation used to compute the attributable fails to represent correctly the motion of the observer (see Figure 9.1), unless the time span of the observations used for the attributable is much shorter than one day. In (Poincaré 1906), to solve a similar problem (see Section 9.4), it is suggested to make a quadratic interpolation for the geocentric position vectors at the individual observations times to estimate $\mathbf{P}(\bar{t})$ and $\dot{\mathbf{P}}(\bar{t})$. These values must be used in eqs. (8.4) and (8.5) when the time span of the observations used to form the attributable is of the order of one day or more, and can improve the results even for shorter arcs.

Structure of the confidence regions

The problem is how to represent the uncertainty of a set of attributable orbital elements, obtained from a given attributable. This case is quite different from the customary one, in which the uncertainty of a set of orbital elements is described by a positive definite 6×6 covariance matrix, computed in the differential corrections, by a fit to ≥ 3 observations well separated in time and in direction, see Section 10.5.

Among the attributable orbital elements, the first four coordinates are the attributable A, computed by a least squares fit, thus with a positive definite 4×4 covariance matrix Γ_A. The last two coordinates are the point $B = (\rho, \dot{\rho})$, to be selected in the admissible region. To describe the uncertainty of the attributable orbital elements $X = [A, B]$ we need to translate into a mathematical formalism the intuitive statement that the attributable A is measured, the point B is just conjectured.

The inverse of the covariance matrix Γ_A, which is used in the least squares fit to compute A, is the 4×4 conditional normal matrix C_A, appearing in a probabilistic interpretation in the Gaussian probability density for the variables A *assuming* that B has a given value, that is assuming the selected virtual asteroid. C_A can be built with the design matrix, giving the partials of the observations (α_i, δ_i) with respect to the four coordinates of A. Thus also Γ_A is the conditional covariance matrix[8] of the attributable. We can formally define the conditional covariance matrix for the attributable elements X as the 6×6 symmetric matrix

$$\Gamma_X = \begin{bmatrix} \Gamma_A & \mathbf{0} \\ \mathbf{0} & \mathbf{0} \end{bmatrix},$$

with $\mathbf{0}$ suitable matrices with null coefficients. Γ_X is obviously not positive definite: it has the B subspace as kernel (null space). The 2×2 submatrix in the lower right-hand corner is $\Gamma_B = \mathbf{0}$ because the value of B has been assumed at some exact value, no uncertainty. The **companion matrix**

$$C_X = \begin{bmatrix} C_A & \mathbf{0} \\ \mathbf{0} & \mathbf{0} \end{bmatrix}$$

is the conditional normal matrix in 6-space. C_X and Γ_X are not inverse of each other, but pseudo-inverse, that is Γ_X is indeed the matrix providing the least squares differential correction for X when B is constrained to a fixed value, see eq. (10.9).

A covariance matrix which is not positive definite, such as Γ_X, can be used in the same way (with some caution) as a conventional covariance matrix to compute the uncertainty of predictions, such as future observations. The covariance Γ_X can be propagated and/or transformed to a covariance matrix in some other coordinate system, e.g., Cartesian coordinates Y. Then, given the Jacobian matrix $\partial Y / \partial X$,

$$\Gamma_Y = \frac{\partial Y}{\partial X} \, \Gamma_X \, \frac{\partial Y}{\partial X}^T \qquad (8.20)$$

is also not positive definite, with a two-dimensional null space, containing the radial direction in both position and velocity. It is possible to propagate also the normal matrix, by using the inverse Jacobian matrix $\partial X / \partial Y$

$$C_Y = \frac{\partial X}{\partial Y}^T \, C_X \, \frac{\partial X}{\partial Y}.$$

C_Y also has a null space of dimension 2, C_Y and Γ_Y are pseudo-inverse.

In the formulae of this section we have used so far a rather standard notation; from now on we will face the following ambiguity: a normal matrix and a covariance matrix are functions of the values of the variables for which they are computed. The matrices resulting from the differential corrections

---

[8] The conditional covariance matrix is the inverse of the conditional normal matrix, see Sections 5.4 and 3.3.

process are the ones at convergence, e.g., if the vector A has to be determined, and the nominal least squares solution is A_1, the normal matrix C_A must be computed by using the design matrix computed in A_1; then the notation should stress this:

$$C_A \big|_{A=A_1}, \quad \Gamma_A \big|_{A=A_1}.$$

However, we shall also use the abbreviated version C_{A_1}, Γ_{A_1}. A similar problem occurs for partial derivatives of a function $F(A)$: confusion is possible between the variable, with respect to which derivation is performed, and the value assumed by the corresponding argument. We shall use the short notation:

$$\frac{\partial F}{\partial A}\bigg|_{A_1} = \frac{\partial F}{\partial A}\bigg|_{A=A_1}.$$

Quasi-product structure

As discussed in Section 8.1, for each value A of the attributable we can define an admissible region $\mathcal{D}(A)$ in the plane of $B = (\rho, \dot{\rho})$, such that for $B \in \mathcal{D}(A)$ the attributable orbital elements $X = [A, B]$ belong to a Solar System body. The set $\mathcal{D}(A)$ is compact with at most two connected components, and its boundary can be explicitly computed.

Even if we cannot determine the value of B from the observations (no significant curvature information, see Section 9.1), we can assume that, if the exact value of the attributable is A, the value of B is contained in $\mathcal{D}(A)$. The existence of an observable real body with B outside $\mathcal{D}(A)$ is not impossible, but is either very unlikely (observable hyperbolic comets are rare) or outside the scope of this investigation (artificial satellites of the Earth do exist, but they have to be handled separately, see Section 8.7).

Thus the confidence region, describing the uncertainty of the attributable orbital elements $X = [A, B]$, is defined by

$$Z_X(\sigma) = \left\{ [A, B] \,\big|\, (A - A_1)^T \, C_{A_1} \, (A - A_1) \leq \sigma^2 \text{ and } B \in \mathcal{D}(A) \right\} \quad (8.21)$$

where $\sigma > 0$ is a parameter, A_1 is the nominal (least squares) value of the four attributable coordinates at time $\bar{t}_1$, and C_{A_1} is the corresponding normal matrix. This set is not a Cartesian product, although in many cases it can be approximated by the product of a confidence ellipsoid in the space of A times the admissible region computed with the nominal attributable A_1:

$$Z_X^1(\sigma) = \left\{ A \,\big|\, (A - A_1)^T \, C_{A_1} \, (A - A_1) \leq \sigma^2 \right\} \times \mathcal{D}(A_1). \quad (8.22)$$

The *quasi-product structure* of eq. (8.21) and its approximation with the product of eq. (8.22) will play an important role in the following.

Sampling the confidence region

The practical problem is how to sample the confidence region $Z_X(\sigma)$ with a finite number of virtual asteroids. Our approach is to use the nodes of a Delaunay triangulation of the admissible region $\mathcal{D}(A_1)$, the points $\{B_1^i = (\rho_i, \dot{\rho}_i)\}_{i=1,k}$ in $\mathcal{D}(A_1)$; then the orbits of the virtual asteroids are defined by the attributable orbital elements

$$\{X^i = [A_1, B_1^i]\}\ i = 1, k$$

with epoch times $t_1^i = \bar{t}_1 - \rho_i/c$. The sampling is adequate for prediction if

(i) the sampling of $\mathcal{D}(A_1)$ by the nodes $\{B_1^i\}$ is dense enough;
(ii) the uncertainty in the A subspace is not too large, and anyway is appropriately accounted for by the covariance matrix Γ_{A_1};
(iii) $\mathcal{D}(A)$ is not too different from $\mathcal{D}(A_1)$ for values of A far from the nominal, but still inside the confidence ellipsoid for A.

All the above are hypotheses to be verified in concrete cases. Some parameters, such as the number of points in the Delaunay triangulation, can be adjusted to meet the requirements of condition (i). Condition (ii) refers to the reliability of the astrometric measurement error model (see Section 5.8); condition (iii) remains to be investigated.

8.4 Predictions from an attributable

We now discuss how to compute a prediction, starting from a set of virtual asteroids, that is from a set of attributable orbital elements with uncertainty

$$X^i, t_1^i, H;\quad \Gamma_{X^i}$$

with $X^i = [A_1, B_1^i]$, obtained as described in the previous section. The process of prediction consists of two steps. The first is the *orbit propagation* Φ from X^i at the epoch time t_1^i to the prediction time $\bar{t}_2$: this gives a set of orbital elements with uncertainty

$$Y^i, \bar{t}_2, H;\quad \Gamma_{Y^i}$$

where the new covariance matrix Γ_{Y^i} is given by an equation analogous to (8.20). As already mentioned, the elements Y^i can be given in a different coordinate system, e.g., Cartesian coordinates. It follows again from formula

(8.20) that the conditional covariance matrix Γ_{Y^i} has rank 4, that is, it is not positive definite with a two-dimensional null space.

The second step is to compute the observation function $\mathcal{A} : Y^i \mapsto A^i$ with A^i the attributable predicted at the new observation epoch $\bar{t}_2$. Since the light leaves the observed body at a time earlier than $\bar{t}_2$ an aberration correction has to be applied again. The Jacobian matrix of partial derivatives of the prediction function F is the 4×6 matrix

$$\frac{\partial \mathcal{A}}{\partial Y}\bigg|_{Y^i} .$$

Generically[9] this matrix has rank 4. A formula similar to (8.20) for covariance propagation holds also for mappings between spaces of different dimensions, provided the rank of the Jacobian matrix is maximum, see (3.10):

$$\Gamma_{A^i} = \frac{\partial \mathcal{A}}{\partial Y}\bigg|_{Y^i} \Gamma_{Y^i} \left[\frac{\partial \mathcal{A}}{\partial Y}\bigg|_{Y^i} \right]^T .$$

By (8.20), taking into account the zeros of Γ_{X^i}, this formula implies

$$\Gamma_{A^i} = \frac{\partial \mathcal{A}'}{\partial X}\bigg|_{X^i} \Gamma_{X^i} \left[\frac{\partial \mathcal{A}'}{\partial X}\bigg|_{X^i} \right]^T = \frac{\partial \mathcal{A}'}{\partial A}\bigg|_{X^i} \Gamma_{A_1} \left[\frac{\partial \mathcal{A}'}{\partial A}\bigg|_{X^i} \right]^T \qquad (8.23)$$

where $\mathcal{A}' = \mathcal{A} \circ \Phi$ and the derivatives are with respect to the attributable A at time $\bar{t}_1$. What is the rank of the 4×4 matrix Γ_{A^i}? The following two statements give a partial answer. First, for $\bar{t}_2 \to \bar{t}_1$, A^i has A_1 as limit, the transformation between the two attributables approaches the identity and $\Gamma_{A^i} \to \Gamma_{A_1}$, which has rank 4. Thus for $\bar{t}_2 - \bar{t}_1$ small enough the rank of Γ_{A^i} is 4. However, we do not know how small $\bar{t}_2 - \bar{t}_1$ has to be for this to be guaranteed.

Second, generically the rows of $\partial \mathcal{A}'/\partial X$ are linearly independent, and they do not belong to the null space of Γ_{X^i}. Thus generically Γ_{A^i} has rank 4. However, a matrix can be of maximum rank but numerically degenerate if its conditioning number is larger than the inverse of the machine accuracy. In this case, the matrix has an inverse in exact arithmetic, but the computation of the inverse is numerically unstable and requires the utmost caution.

Thus we expect, in almost all cases, the matrix Γ_{A^i} to be invertible. We can think of Γ_{A^i} as the marginal covariance matrix associated to the portion A^i of the attributable orbital elements $Y^i = [A^i, B_2^i]$. Indeed, the uncertainty of the attributable A^i is computed without making any assumption on the non-measured quantities $B_2^i = (\rho_2^i, \dot{\rho}_2^i)$. By the rule dual to the one used

---

[9] The precise mathematical definition of a generic property is not simple; we can describe it by saying that this occurs almost always.

for the conditional matrices, the **marginal normal matrix** $C_{A^i} = \Gamma_{A^i}^{-1}$ generically exists, but it may be difficult to compute.[10] If the inverse matrix

$$M = \left[\frac{\partial \mathcal{A}'}{\partial A} \bigg|_{X^i} \right]^{-1} \qquad (8.24)$$

exists, then C_{A^i} can be computed by the formula derived from (8.23)

$$C_{A^i} = M^T \, C_{A_1} \, M. \qquad (8.25)$$

Thus it is possible in most (maybe not all) cases, to define a confidence ellipsoid for the prediction A^i, corresponding to the assumption B_1^i, in 4-space of the attributables A' at time $\bar{t}_2$:

$$Z_{A^i}(\sigma) = \left\{ A' \, \big| (A' - A^i)^T \, C_{A^i} \, (A' - A^i) \le \sigma^2 \right\}. \qquad (8.26)$$

This is actually the interior part of a three-dimensional ellipsoid in the four-dimensional space of the attributables, where the second attributable is predicted to be, within a confidence level described by the parameter σ. However, this confidence parameter σ cannot be interpreted as a χ. In fact, it is not possible to provide a probabilistic prediction model, unless there is a way to assign probabilities to the points in the admissible region.[11]

Triangulated ephemerides

We can draw the conclusions from the discussion in this section and give a definition of the confidence region for the prediction A^i even in the case we are discussing, that is when the only information available is an attributable.

The confidence region for the attributable orbital elements derived from the attributable A_1 is $Z_X(\sigma)$ defined by (8.21); we assume it can be approximated by the product $Z_X^1(\sigma)$, defined by (8.22). The image on the attributables space at time $\bar{t}_1$ of the admissible region $\mathcal{D}(A_1)$ is a two-dimensional compact manifold with boundary $V = \mathcal{A}(\Phi(\mathcal{D}(A_1)))$. We have no way to explicitly compute this manifold as a function of $B = (\rho, \dot{\rho})$, because the map $X \to \mathcal{A}'(X)$ does not have an analytic expression ($\mathcal{A}'(X)$ is the predicted attributable at the second time). We can compute a triangulation of this manifold by using the image of the already computed triangulation $\{B_1^i\}, i = 1, k$, of $\mathcal{D}(A_1)$. The nodes of the triangulation $A^i = \mathcal{A}(\Phi(X^i))$ in the four-dimensional observations space at $\bar{t}_2$ are the predictions from the VAs X^i, in turn defined by the nodes B_1^i.

---

[10] The marginal normal matrix is the inverse of the marginal covariance matrix, see Sections 5.4 and 3.3.
[11] A population model could provide this informative a priori probability density in the $(\rho, \dot{\rho})$ plane. We have not yet tested this possibility.

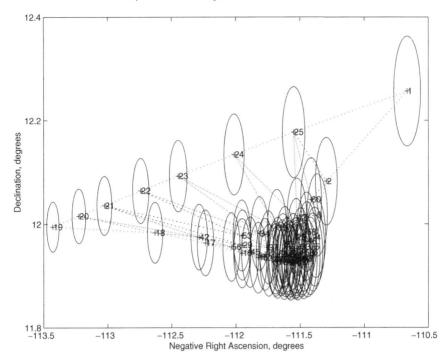

Fig. 8.4. For the asteroid 2003 BH$_{84}$, the observations 12 days after the discovery have been predicted in the triangulated form by using only the attributable computed with the observations of the discovery night. The ellipses indicate the projected uncertainty coming from the fit of the attributable. The recovery attributable, computed with the actually observed data of the later night, is at $-\alpha = -111°.8$, $\delta = 11°.9$, well within the bundle of ellipses. Reprinted from (Milani *et al.* 2005a) with permission from Elsevier.

The **triangulated ephemerides** have to be computed in the four-dimensional predictions space, although some two-dimensional projections can be used to have a good perception of the uncertainty of the attributable and assess the difficulty of a planned recovery, see Figure 8.4.

We associate to each node of the triangulated ephemerides its covariance. We have to think of each node surrounded by its own confidence ellipsoid Z_A^i, defined by eq. (8.26); thus the projections, such as in Figure 8.4, are surrounded by a confidence ellipse. This is an approximation to the *tubular neighborhood* $T(V)$ of the two-manifold V which would be obtained by the union of confidence ellipsoids centered in every point of V.

This tubular neighborhood, so difficult to be computed, plays an essential role in different kinds of identification: recovery, precovery, and linkage. For recovery/precovery, given a candidate attributable found either in the sky or in the archives, to decide if the objects are the same we need to assess not only how much each observation is close to the prediction(s), but also whether this discrepancy can be accounted for by the prediction uncertainty.

When planning what area (either in the sky or in the archive images) has to be scanned, the answer is simply that we need the covered area to include the projection on the celestial sphere of $T(V)$. This can be approximated by the union of the ellipses, the projection of each Z_A^i on the celestial sphere. It does not matter how many ellipses overlap, because we are not computing a probability density. Figure 8.4 suggests that, with some care to take into account the lower density of the predicted observations along the tiny object boundary, this approximation can give a good idea of the region to be scanned for a recovery/precovery of a certain body.

8.5 Linkage by sampling the admissible region

We assume that for a given object $\mathcal{B}$ the only observational information available is contained in two attributables, A_1 at time $\bar{t}_1$ and A_2 at time $\bar{t}_2$. Neither from the first nor from the second can we compute an orbit, thus we have a linkage problem.

The idea is to generate a swarm of virtual asteroids X^i, sampling as described in the previous section the confidence region of one of the two attributables, let us say A_1. Then we compute, from each of the X^i, a prediction A^i for the epoch $\bar{t}_2$, each with its covariance matrix Γ_{A^i}. Generically these covariance matrices will be invertible, and the corresponding normal matrices C_{A^i} can be computed from eq. (8.25). We also know the normal matrix C_2 of the attributable A_2. Thus for each virtual asteroid X^i we can compute an attribution penalty according to eq. (7.6)

$$K_4^i = (A_2 - A^i) \cdot [C_2 - C_2 \, \Gamma_0 \, C_2] \; (A_2 - A^i), \quad \Gamma_0 = [C_2 + C_{A^i}]^{-1}$$

and use the values as a criterion to select some of the virtual asteroids to proceed to the orbit computation. Note that the identification penalty K_4^i, computed for a given node B_1^i of the triangulation of $\mathcal{D}(A_1)$, does not need to be small. First, we cannot know a priori whether the two objects observed at times $\bar{t}_1$ and $\bar{t}_2$ are indeed the same. Second, even if they were the same, the value of B_1^i could be totally wrong with respect to the true values of the distance and its derivative at time $\bar{t}_1$. In both cases the two attributables cannot fit, and this will be revealed by a large value of K_4^i.

Thus the procedure might be as follows. If for all nodes i the value of the penalty is large, say $K_4^i > K_{max}$, then we discard the couple (A_1, A_2) as not likely to belong to the same body. If there are some nodes B_1^i such that $K_4^i \leq K_{max}$, then we proceed to the next step.

The value of the control K_{max} to be used is difficult to establish a priori, because we lack an analytical theory. We cannot use χ^2-tables for dimension 8, because we are sampling the confidence region with a finite number of points B_1^i, thus we cannot assume that the minimum among the K_4^i is the absolute minimum we could get by trying all values of $B_1 \in \mathcal{D}(A_1)$, that is

$$\min_{i=1,k} K_4^i \geq \min_{B_1 \in \mathcal{D}(A_1)} K_4(A_1, B_1) \tag{8.27}$$

and we cannot compute analytically the safety margin to be left to take into account this difference. We conclude that the value of K_{max} to be used in large-scale production of linkages can only be dictated by the analysis of the results of large-scale tests, such as the ones cited in Chapter 11.

The procedure described above provides us also with a number of best fitting corrected attributables $A_2^i = \Gamma_0 \left[C_{A^i} A^i + C_2 A_2 \right]$, according to the third equation of (7.6). Each A_2^i comes with its penalty value K_4^i, which is not too large, that is, an orbit with B_1^i as distance and radial velocity at time t_1^i and giving the attributable A_2^i as observation at time $\bar{t}_2$, can fit both A_1 and A_2 with not too large residuals; the fit is performed in the eight-dimensional space of the residuals of both attributables.

To start differential corrections we need a set of orbital elements to be used as first guess, with a consistent set of six coordinates at the same epoch; such a set is called a **preliminary orbit**. There is no requirement that such an orbit is accurate: it is only hoped that it belongs to the convergence domain of the differential corrections. To achieve this, we have a number of options, the simpler one is to use the attributable A_2^i and the value $B_2^i = (\rho_2^i, \dot{\rho}_2^i)$ as computed from the orbit $X^i = [A_1, B_1^i]$ at t_1^i. The epoch of this set of initial conditions is $\bar{t}_2 - \rho_2^i/c$. Another possibility is to use the attributable back-propagated (linearly) to time $\bar{t}_1$, starting from A_2^i

$$A_1^i = A_1 + M \left(A_2^i - A^i \right),$$

with M from eq. (8.24), and the value B_1^i of the node, at the epoch t_1^i. In both cases, we find an orbit which could fit both attributables if the quadratic approximation of (7.6) is good enough.

In Figure 8.5 we show the linkage procedure for the asteroid 2003 BH$_{84}$. We use the attributable of the discovery night and another attributable computed with the observations made five days later. The nodes of the triangulated admissible region with identification penalty $K_4^i < (0.6)^2$ are encircled and the edges joining them are enhanced with solid lines. From each of the encircled nodes a constrained differential corrections procedure, explained in Chapter 10, allows us to obtain some orbits that fit better to the observations.

The latter are represented here by the points with the same labels as the encircled nodes; a linear fit shows that they are quite well aligned.

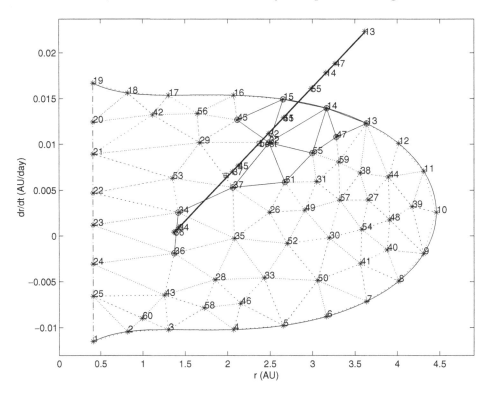

Fig. 8.5. Attributable from the discovery night of 2003 BH_{84} identified with the attributable from five days later. The solid lines join the nodes with penalty $K_4^i < (0.6)^2$, used to compute preliminary orbits: from each of them we have started a constrained differential correction procedure, finding solutions on the line of variations (see Chapter 10). Reprinted from (Milani *et al.* 2005a) with permission from Elsevier.

8.6 Linkage by the two-body integrals

We shall describe a method (Gronchi *et al.* 2008) to produce preliminary orbits starting from two attributables A_1, A_2 of the same Solar System body at two epochs $\bar{t}_1$, $\bar{t}_2$. It is based on the use of the two-body integrals, which had been proposed also in (Taff and Hall 1977).

Angular momentum and energy

For a given attributable A the angular momentum vector per unit mass can be written as a polynomial function of the radial distance and velocity $\rho, \dot{\rho}$

$$\mathbf{c}(\rho, \dot{\rho}) = \mathbf{r} \times \dot{\mathbf{r}} = \mathbf{D}\dot{\rho} + \mathbf{E}\rho^2 + \mathbf{F}\rho + \mathbf{G}, \qquad (8.28)$$

$$\text{where} \quad \begin{cases} \mathbf{D} = \mathbf{q} \times \hat{\boldsymbol{\rho}} \\ \mathbf{E} = \dot{\alpha}\hat{\boldsymbol{\rho}} \times \hat{\boldsymbol{\rho}}_\alpha \mid \dot{\delta}\hat{\boldsymbol{\rho}} \times \hat{\boldsymbol{\rho}}_\delta = \eta\,\hat{\mathbf{n}} \\ \mathbf{F} = \dot{\alpha}\mathbf{q} \times \hat{\boldsymbol{\rho}}_\alpha + \dot{\delta}\mathbf{q} \times \hat{\boldsymbol{\rho}}_\delta + \hat{\boldsymbol{\rho}} \times \dot{\mathbf{q}} \\ \mathbf{G} = \mathbf{q} \times \dot{\mathbf{q}} \end{cases} \qquad (8.29)$$

depend only on the attributable A and the motion of the observer $\mathbf{q}, \dot{\mathbf{q}}$ at the attributable time $\bar{t}$. The vectors $\hat{\boldsymbol{\rho}}, \hat{\boldsymbol{\rho}}_\alpha, \hat{\boldsymbol{\rho}}_\delta$ have been defined in Section 8.1.

For the given A the two-body energy as a function of $\rho, \dot{\rho}$

$$2\mathcal{E}(\rho, \dot{\rho}) = \dot{\rho}^2 + c_1\dot{\rho} + c_2\rho^2 + c_3\rho + c_4 - \frac{2k^2}{\sqrt{\rho^2 + c_5\rho + c_0}} \qquad (8.30)$$

depends on $A, \mathbf{q}, \dot{\mathbf{q}}$ only through the coefficients c_j of (8.8).

Equating the integrals

Now take two attributables $A_1 = (\alpha_1, \delta_1, \dot{\alpha}_1, \dot{\delta}_1)$, $A_2 = (\alpha_2, \delta_2, \dot{\alpha}_2, \dot{\delta}_2)$ at epochs $\bar{t}_1, \bar{t}_2$; we shall use subscripts 1 and 2 referring to the different epochs. If A_1, A_2 correspond to the same physical object, then the angular momentum vectors at the two epochs must coincide: this gives

$$\mathbf{D}_1\dot{\rho}_1 - \mathbf{D}_2\dot{\rho}_2 = \mathbf{J}(\rho_1, \rho_2) \qquad (8.31)$$

where

$$\mathbf{J}(\rho_1, \rho_2) = \mathbf{E}_2\rho_2^2 - \mathbf{E}_1\rho_1^2 + \mathbf{F}_2\rho_2 - \mathbf{F}_1\rho_1 + \mathbf{G}_2 - \mathbf{G}_1.$$

Relation (8.31) is a system of three equations in the four unknowns $\rho_1, \dot{\rho}_1, \rho_2, \dot{\rho}_2$, with constraints

$$\rho_1 > 0, \quad \rho_2 > 0.$$

By scalar multiplication of (8.31) with $\mathbf{D}_1 \times \mathbf{D}_2$ we eliminate the variables $\dot{\rho}_1, \dot{\rho}_2$ and obtain the scalar equation

$$\mathbf{D}_1 \times \mathbf{D}_2 \cdot \mathbf{J}(\rho_1, \rho_2) = 0. \qquad (8.32)$$

The left-hand side of (8.32) is a quadratic form in the variables ρ_1, ρ_2: we write it as

$$q(\rho_1, \rho_2) = q_{20}\rho_1^2 + q_{10}\rho_1 + q_{02}\rho_2^2 + q_{01}\rho_2 + q_{00} \qquad (8.33)$$

with

$$\begin{aligned} q_{20} &= -\mathbf{E}_1 \cdot \mathbf{D}_1 \times \mathbf{D}_2, & q_{02} &= \mathbf{E}_2 \cdot \mathbf{D}_1 \times \mathbf{D}_2, \\ q_{10} &= -\mathbf{F}_1 \cdot \mathbf{D}_1 \times \mathbf{D}_2, & q_{01} &= \mathbf{F}_2 \cdot \mathbf{D}_1 \times \mathbf{D}_2, \\ q_{00} &= (\mathbf{G}_2 - \mathbf{G}_1) \cdot \mathbf{D}_1 \times \mathbf{D}_2. \end{aligned}$$

In the geocentric approximation for the observations we note that $\mathbf{G}_i$ is the angular momentum of the Earth at epoch $\bar{t}_i$, $i = 1, 2$, thus $\mathbf{G}_1 = \mathbf{G}_2$ and $q_{00} = 0$. In this case there is a spurious solution $\rho_1 = \rho_2 = 0$.

The use of the angular momentum integral to determine an orbit of a Solar System body is already present in (Mossotti 1816). More recently, Kristensen (1995) proposed a method to compute a preliminary orbit from two short arcs of observations in which the basic idea is to equate the angular momentum vectors at the two mean epochs of observations.

For the given A_1, A_2 we can also equate the corresponding two-body energies $\mathcal{E}_1, \mathcal{E}_2$. By vector multiplication of (8.31) with $\mathbf{D}_1$ and $\mathbf{D}_2$, projecting on the direction of $\mathbf{D}_1 \times \mathbf{D}_2$, we obtain

$$\dot{\rho}_1(\rho_1, \rho_2) = \frac{(\mathbf{J} \times \mathbf{D}_2) \cdot (\mathbf{D}_1 \times \mathbf{D}_2)}{|\mathbf{D}_1 \times \mathbf{D}_2|^2}, \quad \dot{\rho}_2(\rho_1, \rho_2) = \frac{(\mathbf{J} \times \mathbf{D}_1) \cdot (\mathbf{D}_1 \times \mathbf{D}_2)}{|\mathbf{D}_1 \times \mathbf{D}_2|^2}$$

(8.34)

and, substituting into $\mathcal{E}_1 = \mathcal{E}_2$,

$$\mathcal{F}_1(\rho_1, \rho_2) - \frac{2k^2}{\sqrt{\mathcal{G}_1(\rho_1)}} = \mathcal{F}_2(\rho_1, \rho_2) - \frac{2k^2}{\sqrt{\mathcal{G}_2(\rho_2)}},$$

(8.35)

for some polynomial functions $\mathcal{F}_1(\rho_1, \rho_2)$, $\mathcal{F}_2(\rho_1, \rho_2)$, $\mathcal{G}_1(\rho_1)$, $\mathcal{G}_2(\rho_2)$ with degrees $\deg(\mathcal{F}_1) = \deg(\mathcal{F}_2) = 4$ and $\deg(\mathcal{G}_1) = \deg(\mathcal{G}_2) = 2$. By squaring twice we obtain the polynomial equation

$$p(\rho_1, \rho_2) = \left[(\mathcal{F}_1 - \mathcal{F}_2)^2 \mathcal{G}_1 \mathcal{G}_2 - 4k^4 (\mathcal{G}_1 + \mathcal{G}_2) \right]^2 - 64k^8 \mathcal{G}_1 \mathcal{G}_2 = 0 \quad (8.36)$$

with total degree 24. Some spurious solutions may have been added.

Intersections between the curves

We study the semialgebraic intersection problem

$$\begin{cases} p(\rho_1, \rho_2) = 0 \\ q(\rho_1, \rho_2) = 0, \end{cases} \qquad \rho_1, \rho_2 > 0 \qquad (8.37)$$

with classical algebraic geometry methods, see (Cox *et al.* 1996). Let us write

$$p(\rho_1, \rho_2) = \sum_{j=0}^{20} a_j(\rho_2) \, \rho_1^j,$$

(8.38)

where $\qquad \deg(a_j) = \begin{cases} 20 & \text{for } j = 0 \ldots 4 \\ 24 - (j+1) & \text{for } j = 2k - 1 \quad \text{with } k \geq 3 \\ 24 - j & \text{for } j = 2k \qquad \text{with } k \geq 3 \end{cases}$

and

$$q(\rho_1, \rho_2) = b_2\, \rho_1^2 + b_1\, \rho_1 + b_0(\rho_2) \tag{8.39}$$

for some univariate polynomial coefficients a_i, b_j, depending on ρ_2.

We consider the resultant $\mathrm{Res}(\rho_2)$ of p, q with respect to ρ_1: it is a polynomial with degree ≤ 48, defined as the determinant of the Sylvester matrix

$$\mathtt{Sylv}(\rho_2) = \begin{pmatrix} a_{20} & 0 & b_2 & 0 & \cdots & \cdots & 0 \\ a_{19} & a_{20} & b_1 & b_2 & 0 & \cdots & 0 \\ \vdots & \vdots & b_0 & b_1 & b_2 & \cdots & \vdots \\ \vdots & \vdots & 0 & b_0 & b_1 & \cdots & \vdots \\ a_0 & a_1 & \vdots & \vdots & \vdots & b_0 & b_1 \\ 0 & a_0 & 0 & 0 & 0 & 0 & b_0 \end{pmatrix}. \tag{8.40}$$

The positive real roots of $\mathrm{Res}(\rho_2)$ are the only possible values of ρ_2 for a solution (ρ_1, ρ_2) of (8.37). Thus we can use the following scheme to compute the solutions of (8.37):

(1) find the positive roots $\rho_2(k)$ of $\mathrm{Res}(\rho_2)$ by using a global solution method such as (Bini 1997);

(2) for each k solve $q(\rho_1, \rho_2(k)) = 0$ and compute the two possible values for $\rho_1(k, 1), \rho_1(k, 2)$, discarding negative solutions;

(3) compute $p(\rho_1(k, 1), \rho_2(k)), p(\rho_1(k, 2), \rho_2(k))$ and select $\rho_1(k)$ from the pair that gives zero (at least in exact arithmetic: in practice we select the pair giving the smaller absolute value);

(4) discard spurious solutions, resulting by squaring to obtain (8.36) from (8.35);

(5) for the obtained values of ρ_1, ρ_2 compute the values of $\dot{\rho}_1(k), \dot{\rho}_2(k)$ by (8.34);

(6) change coordinates to Cartesian heliocentric elements at times $t_1(k), t_2(k)$, corrected by the aberration due to the finite velocity of the light, and by using the Poincaré observer interpolation method (see Section 8.3);

(7) change coordinates to Keplerian elements at times $t_1(k), t_2(k)$.

Note that the values of the angular momentum vector and of the energy at a given time fix the values of the Keplerian elements a, e, I, Ω. The two attributables A_1, A_2 at epochs $\bar{t}_1, \bar{t}_2$ give eight scalar data, thus the problem is over-determined. From a non-spurious pair $(\tilde{\rho}_1, \tilde{\rho}_2)$, the solution of (8.37), we obtain the same values of a, e, I, Ω at both times $\tilde{t}_i = \bar{t}_i - \tilde{\rho}_i/c, i = 1, 2$, with the aberration correction, but we must check the compatibility conditions

$$\omega_1 = \omega_2, \qquad \ell_1 = \ell_2 + n(\tilde{t}_1 - \tilde{t}_2), \tag{8.41}$$

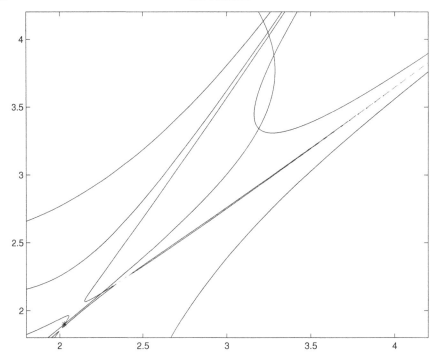

Fig. 8.6. Intersections of the curves $p = 0, q = 0$ in the plane ρ_1, ρ_2. This example is for asteroid (243) Ida and the intersection corresponding to the true object is marked with an asterisk.

where ω_1, ω_2 and ℓ_1, ℓ_2 are the arguments of perihelion and the mean anomalies of the body at times $\tilde{t}_1, \tilde{t}_2$ and $n = ka^{-3/2}$ is the mean motion of the celestial body. The first of conditions (8.41) corresponds to the fifth integral of the Kepler problem, from the Laplace–Lenz vector integral (4.11), the second involves a two-body propagation (e.g., by the Kepler equation).

In Figure 8.6 we show the intersections of the algebraic curve $p(\rho_1, \rho_2) = 0$ with the conic $q(\rho_1, \rho_2) = 0$ from two attributables obtained from the orbit of the asteroid (243) Ida, assuming geocentric observations at epochs > 38 days apart. There are eight intersections with $\rho_1, \rho_2 > 0$: seven are shown in the figure and one is very near the origin. Removing spurious solutions only three are left, one of which gives a hyperbolic orbit and one grossly fails the conditions (8.41); the only solution left gives a very good preliminary orbit.

This method has been developed recently (Gronchi *et al.* 2008) and has not yet been submitted to a large-scale test. From a few examples it appears to have a good potential in providing preliminary orbits for difficult linkages between very short observed arcs, providing little information beyond the respective attributables, and separated by a time interval much larger than the arc time spans.

8.7 The space debris problem

Near-Earth space is filled by more than $300\,000$ artificial debris particles
with diameter larger than 1 cm (Rossi, 2005). This population is similar
to the asteroidal one because its long-term evolution is affected by high-
velocity mutual collisions. Another analogy is that there is an impact risk,
that is space assets (e.g., the International Space Station) could be seriously
damaged by a collision with some debris (see Chapter 12). The space where
the debris is placed can be divided into three main regions: low-Earth orbit
(LEO), below about 2000 km, medium-Earth orbit (MEO), above 2000 km
and below $36\,000$ km and geosynchronous Earth orbit (GEO) at about $36\,000$
km of altitude.

In this section we outline the theoretical basis of the orbit determination
algorithms for space debris. The main problem to compute the orbits of the
observed space debris is the identification[12] of two or more sets of observational data. Exactly as for asteroids, the data contained in the observations
during a single pass over an observing station are not enough to obtain a
least squares orbit solution. As an example, if the image moves together
with the fixed stars, the debris produces a trail the two extremes of which
are measured: this gives us a tracklet. The information contained in such
data is just the mean angular position and the first time derivatives, that is
the attributable, defined by the same formula (8.1) as the asteroid case.

Admissible regions for Earth satellites

Following Tommei *et al.* (2007), we use a new interpretation of eq. (8.4)
with the geocentric position $\mathbf{r}$ of the debris, the geocentric position $\mathbf{q}$ of the
observer, and the topocentric position $\boldsymbol{\rho}$ of the debris: $\mathbf{r} = \boldsymbol{\rho}+\mathbf{q}$ still applies,
and eq. (8.2) is replaced by

$$\mathcal{E}_{\oplus}(\rho,\dot{\rho}) = \frac{1}{2}\|\dot{\mathbf{r}}(\rho,\dot{\rho})\|^2 - \frac{G\,m_{\oplus}}{r(\rho)}. \qquad (8.42)$$

Then a definition of admissible region such that only satellites of the Earth
are allowed includes the condition

$$\mathcal{E}_{\oplus}(\rho,\dot{\rho}) \leq 0. \qquad (8.43)$$

Given the attributable A at time $\bar{t}$ obtained from the observations, eqs.
(8.5), (8.6), (8.7), and (8.8) are the same, as well as the derivation leading
to the degree 6 inequality (8.11). Thus the same conclusions apply, namely

---

[12] Identification is also called, in this context, correlation; we do not use this terminology to avoid
confusion with the other well-established meaning of the word, see Section 3.1.

the region in the $(\rho, \dot{\rho})$ half-plane $\rho > 0$ fulfilling (8.43) has at most two connected components. One component has an open inner boundary $\rho > 0$, if a second component exists, it is compact.

The admissible region needs to be compact for the reasons explained in the asteroid case, thus a condition defining an inner boundary needs to be added. The choice for the inner boundary depends upon the specific orbit determination task: a simple method is to add constraints $\rho_{min} \le \rho \le \rho_{max}$ allowing us, e.g., to focus the search of identifications on one of the three classes LEO, MEO, and GEO, as in (Tommei *et al.* 2007). Another natural choice for the inner boundary is to take $\rho \ge h_{atm}$ where h_{atm} is the thickness of a portion of the Earth's atmosphere in which a satellite cannot remain in orbit for a significant time span.[13] As an alternative, it is possible to constrain the semimajor axis of the satellite to be larger that $R_{\oplus} + h_{atm} = \bar{R}$, and this leads to an equation

$$\mathcal{E}_{\oplus}(\rho, \dot{\rho}) \ge -\frac{G\,m_{\oplus}}{2\,\bar{R}} \tag{8.44}$$

which defines another degree 6 inequality with the same coefficients as eq. (8.11) but for a different constant term. Figure 8.7 shows the interplay of different definitions of the inner boundary.

Another possible way to find an inner boundary is to exclude trajectories impacting the Earth in less than one revolution, that is to use an inequality on the perigee q (Farnocchia 2008)

$$q = a(1 - e) \ge \bar{R}. \tag{8.45}$$

By substituting into the two-body formulae from Section 4.2 we obtain

$$\sqrt{1 + \frac{2\mathcal{E}_{\oplus}||\mathbf{c}||^2}{G^2\,m_{\oplus}^2}} \le 1 + \frac{2\mathcal{E}_{\oplus}\bar{R}}{G\,m_{\oplus}}. \tag{8.46}$$

Since the left-hand side is $e \ge 0$, we need to impose the condition

$$1 + \frac{2\mathcal{E}_{\oplus}\bar{R}}{G\,m_{\oplus}} \ge 0$$

on the right-hand side; this is again $a \ge \bar{R}$. By squaring (8.46) we obtain

$$||\mathbf{c}||^2 \ge 2\,\bar{R}\,(G\,m_{\oplus} + \mathcal{E}_{\oplus}\bar{R}). \tag{8.47}$$

Given the expressions (8.28) for $\mathbf{c}$ and (8.42) for $\mathcal{E}_{\oplus}$, the above condition is an algebraic inequality in the variables $(\rho, \dot{\rho})$; by another squaring it is possible to convert it into a polynomial equation of degree 10 in ρ and

---

[13] In practice, h_{atm} is just few times the scale height $\mathcal{H}$ of the atmosphere, see Section 14.4.

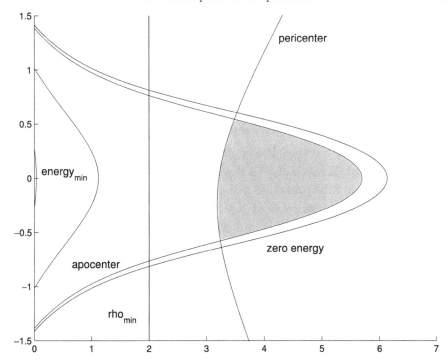

Fig. 8.7. The admissible region for an Earth satellite must be a subset of the region with negative geocentric energy. Additional constraints may be added by using the physical boundary $\bar{R} = R_\oplus + h_{atm}$ defined by the atmosphere and the dynamical boundary defined by the sphere of influence $r \leq R_{SI}$: $a \geq \bar{R}, q \geq \bar{R}$ and $Q \leq R_{SI}$ can be used.

degree 4 in $\dot{\rho}$. Figure 8.7 also shows this inner boundary, as well as an alternative outer boundary constraining the apocenter Q at some large value (the equations are analogous). The main limitation of this approach is that we do not have a rigorous proof that the region defined by eqs. (8.45) and (8.43) has at most two connected components.

Sampling

The admissible region for space debris can be used in the same way as in the asteroid case, to be sampled by a swarm of **virtual debris**, which is analogous to the virtual asteroids. In this way the linkage problem is transformed into a multiple hypothesis attribution problem, for which the theory is the same as in Sections 8.4 and 8.5.

If the Delaunay triangulation method of Section 8.2 is used, the starting point is a sampling of the boundary of the admissible region. For the outer boundary this is the same as in the asteroid case, for the inner one we can use either a minimum ρ or a minimum perigee q, that is (8.45). The first choice is simpler and leads more easily to a reliable algorithm, because with condition (8.45) we cannot be sure of the number of connected components. A reasonable approach would be to triangulate a region with the simplest inner boundary, then discard the nodes which turn out to have a ballistic orbit with $q < \bar{R}$.

The two-body integrals method for optical observations

Linkage between attributables belonging to space debris can also be obtained by the prime integrals method. All the formulae of Section 8.6 are the same, with the geocentric interpretation of $\mathbf{r}$ and $\mathbf{q}$ and with $Gm_\oplus$ in place of k^2. The main difference is a simplification: the geocentric motion of the observer $\mathbf{q}$ has the frequency $\Omega_\oplus$, which cannot be much faster than the mean motion of the satellites, in most cases it is slower. Thus step (6) of the algorithm of the previous section is simplified, without any need for the Poincaré observer interpolation method.

As an example, this method could be suitable for GEO debris: with two tracklets belonging to consecutive nights, that is spaced by about one orbital period, it could provide accurate preliminary orbits. There is just one pitfall to be avoided: two attributables spaced in time by one day, taken from the same station, result in an approximate rank deficiency in eq. (8.32), because $\mathbf{D}_1$ and $\mathbf{D}_2$ are nearly the same (Gronchi *et al.* 2008). Thus images of the same portion of the GEO belt have to be taken at different hours.

This method has not yet been submitted to large-scale tests; however, we have a program under way to test and compare the prime integrals and the recursive attribution methods specifically for debris in the GEO region.

Radar attributable and admissible regions

Active artificial satellites and space debris can also be observed by radar: because of the $1/\rho^4$ dependence of the signal-to-noise for radar observations, range and range-rate are currently measured only for debris in LEO. When a return signal is acquired, the antenna pointing angles are also available. Given the capability of modern radars to scan very rapidly the entire visible sky,[14] radar can be used to discover all the debris above a minimum size while visible from an antenna, or a system of antennas (Mehrholz *et al.* 2002).

When a radar observation is performed we assume that the measured quantities (all with their own uncertainty) are the range, the range-rate,

---

[14] The pointing of the radar can also be achieved by phased array technology, without physical motion of the antenna.

and also the antenna pointing direction, that is the debris apparent position on the celestial sphere, expressed by two angular coordinates such as right ascension α and declination δ. The time derivatives of these angular coordinates, $\dot{\alpha}$ and $\dot{\delta}$, are not measured: therefore the concept of attributable must be modified and an admissible region defined in the $(\dot{\alpha}, \dot{\delta})$ plane.

We define as **radar attributable** a vector

$$A_{rad} = (\alpha, \delta, \rho, \dot{\rho}) \in [-\pi, \pi) \times (-\pi/2, \pi/2) \times \mathbb{R}^+ \times \mathbb{R}, \qquad (8.48)$$

containing the information from a radar observation, at time t; note that, by analogy with other cases, we assume t is the receive time. Given a radar attributable A_{rad}, we define as *radar admissible region* for a space debris the set of values of $(\dot{\alpha}, \dot{\delta})$ such that, for the given radar attributable A_{rad}

$$\mathcal{E}_\oplus(\dot{\alpha}, \dot{\delta}) \le 0. \qquad (8.49)$$

In order to compute the admissible region we use the geocentric energy, given by the formula

$$2\mathcal{E}_\oplus = \dot{\rho}^2 + c_1 \dot{\rho} + c_2 \rho^2 + c_3 \rho + c_4 - \frac{2Gm_\oplus}{\sqrt{\rho^2 + c_5\rho + c_0}} \qquad (8.50)$$

analogous to eq. (8.30), as a function of the unknown quantities $\dot{\alpha}$ and $\dot{\delta}$. Among the coefficients (8.8), only $c_2 = \eta^2$ and c_3 depend on $\dot{\alpha}$ and $\dot{\delta}$, thus from eq. (8.30) we have a quadratic polynomial in $\dot{\alpha}$ and $\dot{\delta}$

$$2\mathcal{E}_\oplus = z_{11}\, \dot{\alpha}^2 + 2\, z_{12}\, \dot{\alpha}\, \dot{\delta} + z_{22}\, \dot{\delta}^2 + 2\, z_{13}\, \dot{\alpha} + 2\, z_{23}\, \dot{\delta} + z_{33}, \qquad (8.51)$$

with

$$
\begin{array}{llll}
z_{11} & = & \rho^2 \cos^2\delta & \quad z_{13} & = & \rho\, \dot{\mathbf{q}} \cdot \boldsymbol{\rho}_\alpha \\
z_{12} & = & 0 & \quad z_{23} & = & \rho\, \dot{\mathbf{q}} \cdot \boldsymbol{\rho}_\delta \\
z_{22} & = & \rho^2 & \quad z_{33} & = & \dot{\rho}^2 + c_1\, \dot{\rho} + c_4 - 2\, G\, m_\oplus/\sqrt{S(\rho)},
\end{array}
$$

where $S(\rho)$ is defined as in (8.9). The boundary of the admissible region in the $(\dot{\alpha}, \dot{\delta})$ plane is then given by

$$\mathcal{E}_\oplus(\dot{\alpha}, \dot{\delta}) = 0. \qquad (8.52)$$

For each value of A_{rad}, this equation represents a conic section in the $(\dot{\alpha}, \dot{\delta})$ plane; more precisely, since $z_{11}, z_{22} > 0$ and $z_{12} = 0$, it is an ellipse with its axes aligned with the coordinate axes. Actually, in a plane $(\dot{\alpha}\cos\delta, \dot{\delta})$, with the axes rescaled according to the metric of the tangent plane to the celestial sphere, the curves $\mathcal{E}_\oplus(\dot{\alpha}, \dot{\delta}) = 0$ are circles.

The region defined by negative geocentric energy $\mathcal{E}_\oplus(\dot{\alpha}, \dot{\delta}) \le 0$ is the inside of one ellipse (or circle in the rescaled coordinates). Thus it is a compact set, and the problem of defining an inner boundary of the admissible region is less important than in the optical attributable case. Anyway, it is possible to define an inner boundary by constraining the semimajor axis $a > \bar{R}$, that is by eq. (8.44), resulting in a concentric inner ellipse (circle), thus in an admissible region forming an elliptic (circular) annulus.

It is also possible to exclude the ballistic trajectories by imposing $q > \bar{R}$, that is by using inequality (8.47), in which $\dot{\alpha}, \dot{\delta}$ are to be considered as variables. Then the geocentric energy is given by (8.51), and the angular momentum by

$$\mathbf{c} = \mathbf{r} \times \dot{\mathbf{r}} = \mathbf{E} + \mathbf{F}\,\dot{\alpha} + \mathbf{G}\,\dot{\delta} \tag{8.53}$$

where
$$\begin{cases} \mathbf{E} = \mathbf{r} \times \dot{\mathbf{q}} + \dot{\rho}\,\mathbf{q} \times \hat{\boldsymbol{\rho}}, \\ \mathbf{F} = \mathbf{r} \times \rho\,\hat{\boldsymbol{\rho}}_{\alpha}, \\ \mathbf{G} = \mathbf{r} \times \rho\,\hat{\boldsymbol{\rho}}_{\delta}. \end{cases} \tag{8.54}$$

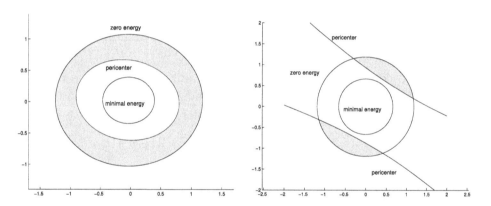

Fig. 8.8. Possible shapes of a radar admissible region, including the negative energy condition (outer ellipse) and the pericenter condition (intermediate ellipse on the left, hyperbola on the right); the lower bound to the semimajor axis (inner ellipse) is implied by the pericenter condition.

Note that the geocentric vector $\mathbf{r}$ is fully determined by the radar attributable as $\mathbf{r} = \mathbf{q} + \rho\,\hat{\boldsymbol{\rho}}$ and contains no unknown. By a sequence of simple algebraic passages, similar to those of the optical case, we can conclude that the admissible region defined by $\mathcal{E}_{\oplus} \leq 0$ and $q \geq \bar{R}$ is the set of solutions of the system of three inequalities

$$\mathcal{E}_{\oplus} \leq 0, \quad \mathcal{E}_{\oplus} \geq -G\,m_{\oplus}/2\bar{R}, \quad \|\mathbf{c}\|^2 \geq 2\,\bar{R}\,(G\,m_{\oplus} + \mathcal{E}_{\oplus}\bar{R}),$$

all three quadratic in the $(\dot{\alpha}, \dot{\delta})$ variables. Thus the admissible region can be geometrically described as a region bounded by three conics: the first two are concentric ellipses (circles in the rescaled coordinates), the third one can be either an ellipse or a hyperbola, with a different center and different symmetry axes. Figure 8.8 shows the possible qualitatively different cases.

The sampling of the radar admissible region could be obtained by a Delaunay triangulation, but the algorithm would be cumbersome because of the need to distinguish geometrically distinct cases for the boundary. A simpler method would be to use a *spider web* sampling for the annulus, that is, in the plane of the rescaled coordinates $(\dot\alpha \cos\delta, \dot\delta)$, to use a rectangular grid in the polar coordinates centered in the common center of the two circles. The virtual debris obtained in this way could then be tested for the condition $q > \bar{R}$ and the ones failing this test discarded, if ballistic trajectories (such as very recent launches) are not among the target of the survey.

The two-body integrals method for radar observations

The two-body integrals method for linkage of the previous section can also be applied to the case of two radar attributables A_1, A_2 with receive times t_1, t_2. The formulae for geocentric energy and angular momentum are given by (8.51) and (8.53), polynomials of degree 2 and 1 in the unknowns $(\dot\alpha, \dot\delta)$, respectively. Thus the system of four scalar equations obtained by equating the two energies and the two angular momentum vectors contains three linear equations and a quadratic one, and has overall algebraic degree 2. It follows that such a system can be solved by elementary algebra.

We use subscripts 1 and 2 referring to the different epochs: the angular momentum equations are

$$\mathbf{E}_1 + \mathbf{F}_1 \dot\alpha_1 + \mathbf{G}_1 \dot\delta_1 = \mathbf{E}_2 + \mathbf{F}_2 \dot\alpha_2 + \mathbf{G}_2 \dot\delta_2, \tag{8.55}$$

where $\mathbf{E}_i, \mathbf{F}_i, \mathbf{G}_i$ for $i = 1, 2$ are the quantities defined in (8.54) for each of the two radar attributables. The equation above is a system of three linear equations in four unknowns $(\dot\alpha_1, \dot\delta_1, \dot\alpha_2, \dot\delta_2)$ and can be solved for three unknowns as a function of one of the four. By scalar multiplication with $\mathbf{G}_1 \times \mathbf{F}_2$ we can obtain $\dot\alpha_1$ as a function of $\dot\delta_2$, and so on. This procedure fails only if the four vectors $\mathbf{F}_1, \mathbf{F}_2, \mathbf{G}_1, \mathbf{G}_2$ do not generate a linear space of dimension 3. Apart from coordinate singularities, this can happen only if $\mathbf{r}_1$ is parallel to $\mathbf{r}_2$, e.g., when the time interval is equal to the orbital period.

When the equations for, say, $(\dot\alpha_1, \dot\delta_1, \dot\alpha_2)$ as a function of $\dot\delta_2$ are substituted in the equation for the energies

$$\mathcal{E}_{\oplus,1}(\dot\alpha_1, \dot\delta_1) = \mathcal{E}_{\oplus,2}(\dot\alpha_2, \dot\delta_2)$$

we obtain a quadratic equation in $\dot\delta_2$, which can be solved by elementary algebra, giving at most two real solutions. Geometrically, the angular momentum equation (8.55) can be described by a straight line in a plane, e.g., in $(\dot{\alpha}_2, \dot{\delta}_2)$, where the energy equation defines a conic section.

The practical application of this method to a real case of one or more radar stations surveying for space debris requires some non-trivial additional steps. In particular, it is necessary to check that the assumption that a two-body model is a good approximation for the orbit over the time span from t_1 to t_2. Because the radar observations are limited to LEO, the debris performs between 12 and 16 orbits per day; this implies that in just one day the orbital perturbations due to the non-spherical shape of the Earth's orbit are significant, especially on the longitude of the node Ω. The largest changes to Ω are due to the secular perturbation

$$\Omega(t_2) - \Omega(t_1) \simeq -(t_2 - t_1)\, \frac{3}{2} n \left(\frac{R_\oplus}{a} \right)^2 \frac{J_2}{(1 - e^2)^2} \cos I$$

where $J_2 = -C_{20}$ is the coefficient of the second zonal spherical harmonic of the Earth's gravity field (see Section 13.2). It is possible to account for the precession of the node in a modified version of the two-body integrals method (Farnocchia 2008), obtaining a system of four algebraic equations, but the overall degree is 112. Such a system could be solved with the same methods described in the previous section and in (Gronchi *et al.* 2008), but it is not yet clear whether this would result in an efficient algorithm.

The simpler method, with explicit solution by elementary algebra, could be applicable if the radar system had the capability of acquiring two radar attributables from most debris within the same pass, or at most at the next pass (< 2 hours later). If this was possible, it would provide a very good example of how to transform a difficult problem of orbit determination, such as linkage, into one with an elementary solution.

9

METHODS BY LAPLACE AND GAUSS

In this chapter we discuss the classical methods by Laplace and Gauss to obtain some preliminary orbit, a solution of the two-body problem, given at least three observations of two angular coordinates (α, δ) on the celestial sphere. We will show that these procedures are controlled by the presence of curvature, that is information beyond that contained in an attributable. We also discuss the possibility of non-unique solutions. This chapter is based on our papers (Milani *et al.* 2008, Gronchi 2009).

9.1 Attributables and curvature

If there are $m \geq 3$ observations (α_i, δ_i) of a Solar System body $\mathcal{B}$ (e.g., an asteroid) at different times t_i, $i = 1, m$, we can compute an attributable A by fitting both angular coordinates as a function of time with a polynomial model. In most cases a degree 2 model, centered at the mean $\bar{t}$ of the times t_i, is satisfactory:

$$
\begin{aligned}
\alpha(t) &= \alpha(\bar{t}) + \dot{\alpha}(\bar{t})\,(t - \bar{t}) + \frac{1}{2}\ddot{\alpha}(\bar{t})\,(t - \bar{t})^2, \\
\delta(t) &= \delta(\bar{t}) + \dot{\delta}(\bar{t})\,(t - \bar{t}) + \frac{1}{2}\ddot{\delta}(\bar{t})\,(t - \bar{t})^2;
\end{aligned}
$$

the vector $(\alpha, \dot{\alpha}, \ddot{\alpha}, \delta, \dot{\delta}, \ddot{\delta})$ is obtained as a solution of the problem of Section 5.1, together with the two 3×3 covariance matrices $\Gamma_\alpha, \Gamma_\delta$. We are assuming that the α and δ error components are not correlated, otherwise the 6×6 covariance matrix of all the variables could be full.

Computation of curvature

The heliocentric position of the observed body $\mathcal{B}$ is the vector $\mathbf{r} \in \mathbb{R}^3$ and the topocentric position is

$$\boldsymbol{\rho} = \rho\,\hat{\boldsymbol{\rho}} = \mathbf{r} - \mathbf{q},$$

where $\mathbf{q}$ is the heliocentric position of the observer, $\hat{\boldsymbol{\rho}}$ is the unit vector defining the observation direction, ρ the topocentric distance of $\mathcal{B}$.

We shall use an orthonormal basis adapted to the apparent path of $\mathcal{B}$ on the celestial sphere, that is the image of $\hat{\boldsymbol{\rho}}(t)$. Following Danby (1988) we note that

$$\mathbf{v} = \frac{d\hat{\boldsymbol{\rho}}}{dt} = \eta\,\hat{\mathbf{v}}, \qquad \hat{\mathbf{v}} \cdot \hat{\boldsymbol{\rho}} = 0,$$

where $\eta = \|\mathbf{v}\|$ is the proper motion. By using the **arc length parameter** s, defined by $ds/dt = \eta$, we have $d\hat{\boldsymbol{\rho}}/ds = \hat{\mathbf{v}}$. Denoting with a prime the derivative with respect to s, the derivative $\hat{\mathbf{v}}'$ has the properties

$$\hat{\mathbf{v}}' \cdot \hat{\boldsymbol{\rho}} = \frac{d}{ds}[\hat{\mathbf{v}} \cdot \hat{\boldsymbol{\rho}}] - \hat{\mathbf{v}} \cdot \hat{\boldsymbol{\rho}}' = -1,$$

$$\hat{\mathbf{v}}' \cdot \hat{\mathbf{v}} = \frac{1}{2}\frac{d}{ds}\|\hat{\mathbf{v}}\|^2 = 0;$$

in the orthonormal basis $\{\hat{\boldsymbol{\rho}}, \hat{\mathbf{v}}, \hat{\mathbf{n}}\}$, with $\hat{\mathbf{n}} = \hat{\boldsymbol{\rho}} \times \hat{\mathbf{v}}$, we can express $\hat{\mathbf{v}}'$ as

$$\hat{\mathbf{v}}' = -\hat{\boldsymbol{\rho}} + \kappa\,\hat{\mathbf{n}}$$

for a scalar function κ called the **geodesic curvature** of the path. It measures the deviation of the path from a great circle (a geodesic on the sphere).

The second derivative of the path $\hat{\boldsymbol{\rho}}(t)$ with respect to t can be computed from $\hat{\mathbf{v}}'$ and is

$$\frac{d^2\hat{\boldsymbol{\rho}}}{dt^2} = -\eta^2\,\hat{\boldsymbol{\rho}} + \dot{\eta}\,\hat{\mathbf{v}} + \kappa\eta^2\,\hat{\mathbf{n}}. \tag{9.1}$$

The three components of the vector $d^2\hat{\boldsymbol{\rho}}/dt^2$ give us information on the *curvature* of the path: the component along $\hat{\mathbf{n}}$ is strictly related to the geodesic curvature, that along $\hat{\mathbf{v}}$ is called the **along-track acceleration**, and that along $\hat{\boldsymbol{\rho}}$ simply means that the path is on a sphere.

To compute the two components $\kappa\eta^2$, $\dot{\eta}$ of curvature starting from the values of $(\alpha, \delta, \dot{\alpha}, \dot{\delta}, \ddot{\alpha}, \ddot{\delta})$, obtained by a polynomial fit of the observations, we use the orthogonal basis $\{\hat{\boldsymbol{\rho}}, \hat{\boldsymbol{\rho}}_\alpha, \hat{\boldsymbol{\rho}}_\delta\}$, where

$$\hat{\boldsymbol{\rho}} = (\cos\delta\cos\alpha, \cos\delta\sin\alpha, \sin\delta),$$

$$\hat{\boldsymbol{\rho}}_\alpha = \frac{\partial\hat{\boldsymbol{\rho}}}{\partial\alpha} = (-\cos\delta\sin\alpha, \cos\delta\cos\alpha, 0),$$

$$\hat{\boldsymbol{\rho}}_\delta \;=\; \frac{\partial \hat{\boldsymbol{\rho}}}{\partial \delta} = (-\sin\delta\cos\alpha, -\sin\delta\sin\alpha, \cos\delta),$$

with $\|\hat{\boldsymbol{\rho}}\| = \|\hat{\boldsymbol{\rho}}_\delta\| = 1$, $\|\hat{\boldsymbol{\rho}}_\alpha\| = \cos\delta$. We have the following relations:

$$\hat{\mathbf{v}} \;=\; \hat{\boldsymbol{\rho}}' = \alpha'\,\hat{\boldsymbol{\rho}}_\alpha + \delta'\,\hat{\boldsymbol{\rho}}_\delta,$$

$$\hat{\mathbf{n}} \;=\; \hat{\boldsymbol{\rho}} \times (\alpha'\,\hat{\boldsymbol{\rho}}_\alpha + \delta'\,\hat{\boldsymbol{\rho}}_\delta) = -\frac{\delta'}{\cos\delta}\,\hat{\boldsymbol{\rho}}_\alpha + \alpha'\cos\delta\,\hat{\boldsymbol{\rho}}_\delta,$$

$$\hat{\mathbf{v}}' \;=\; (\alpha''\hat{\boldsymbol{\rho}}_\alpha + \delta''\hat{\boldsymbol{\rho}}_\delta) + (\alpha'^2\,\hat{\boldsymbol{\rho}}_{\alpha\alpha} + 2\alpha'\delta'\hat{\boldsymbol{\rho}}_{\alpha\delta} + \delta'^2\hat{\boldsymbol{\rho}}_{\delta\delta}),$$

and the second derivative vectors are

$$\hat{\boldsymbol{\rho}}_{\alpha\alpha} \;=\; \frac{\partial^2 \hat{\boldsymbol{\rho}}}{\partial \alpha^2} = (-\cos\delta\cos\alpha, -\cos\delta\sin\alpha, 0),$$

$$\hat{\boldsymbol{\rho}}_{\alpha\delta} \;=\; \frac{\partial^2 \hat{\boldsymbol{\rho}}}{\partial\alpha\,\partial\delta} = (\sin\delta\sin\alpha, -\sin\delta\cos\alpha, 0),$$

$$\hat{\boldsymbol{\rho}}_{\delta\delta} \;=\; \frac{\partial^2 \hat{\boldsymbol{\rho}}}{\partial \delta^2} = (-\cos\delta\cos\alpha, -\cos\delta\sin\alpha, -\sin\delta).$$

We also need the scalar products[1]

$$\begin{aligned}
\hat{\boldsymbol{\rho}}_{\alpha\alpha}\cdot\hat{\boldsymbol{\rho}}_\alpha &= 0 = \Gamma_{\alpha\alpha,\alpha} & \qquad \hat{\boldsymbol{\rho}}_{\alpha\alpha}\cdot\hat{\boldsymbol{\rho}}_\delta &= \sin\delta\cos\delta = \Gamma_{\alpha\alpha,\delta}\\
\hat{\boldsymbol{\rho}}_{\alpha\delta}\cdot\hat{\boldsymbol{\rho}}_\alpha &= -\sin\delta\cos\delta = \Gamma_{\alpha\delta,\alpha} & \qquad \hat{\boldsymbol{\rho}}_{\alpha\delta}\cdot\hat{\boldsymbol{\rho}}_\delta &= 0 = \Gamma_{\alpha\delta,\delta}\\
\hat{\boldsymbol{\rho}}_{\delta\delta}\cdot\hat{\boldsymbol{\rho}}_\alpha &= 0 = \Gamma_{\delta\delta,\alpha} & \qquad \hat{\boldsymbol{\rho}}_{\delta\delta}\cdot\hat{\boldsymbol{\rho}}_\delta &= 0 = \Gamma_{\delta\delta,\delta}
\end{aligned}$$

to compute the geodesic curvature

$$\kappa = \hat{\mathbf{v}}'\cdot\hat{\mathbf{n}} = (\delta''\alpha' - \alpha''\delta')\cos\delta + \alpha'\,(1+\delta'^2)\sin\delta$$

as a function of the derivatives with respect to the arc length. To obtain an expression containing the time derivatives we need to use

$$\alpha'' = \frac{1}{\eta}\frac{d}{dt}\left(\frac{\dot\alpha}{\eta}\right) = \frac{\eta\,\ddot\alpha - \dot\eta\,\dot\alpha}{\eta^3}$$

and the analog for δ''; with the terms containing $\dot\eta$ canceling out we obtain

$$\kappa\eta^2 = \frac{1}{\eta}\left[(\ddot\delta\dot\alpha - \ddot\alpha\dot\delta)\cos\delta + \dot\alpha(\eta^2 + \dot\delta^2)\sin\delta\right]. \tag{9.2}$$

To compute the along-track acceleration we consider the second derivative

$$\frac{d^2\hat{\boldsymbol{\rho}}}{dt^2} = \left(\ddot\alpha\,\hat{\boldsymbol{\rho}}_\alpha + \ddot\delta\,\hat{\boldsymbol{\rho}}_\delta\right) + \left(\dot\alpha^2\,\hat{\boldsymbol{\rho}}_{\alpha\alpha} + 2\,\dot\alpha\,\dot\delta\,\hat{\boldsymbol{\rho}}_{\alpha\delta} + \dot\delta^2\,\hat{\boldsymbol{\rho}}_{\delta\delta}\right)$$

so that

$$\dot\eta = \frac{d^2\hat{\boldsymbol{\rho}}}{dt^2}\cdot\hat{\mathbf{v}} = \frac{\ddot\alpha\,\dot\alpha\,\cos^2\delta + \ddot\delta\,\dot\delta - \dot\alpha^2\,\dot\delta\,\cos\delta\,\sin\delta}{\eta}. \tag{9.3}$$

---

[1] That is, the Riemannian connection of the sphere, as expressed by the Christoffel symbols.

9.2 The method of Laplace

In the orthonormal basis $\{\hat{\boldsymbol{\rho}}, \hat{\mathbf{v}}, \hat{\mathbf{n}}\}$ we can write the first and second time derivatives of the topocentric vector $\boldsymbol{\rho}$ as follows:

$$\dot{\boldsymbol{\rho}} = \dot{\rho}\,\hat{\boldsymbol{\rho}} + \rho\,\eta\,\hat{\mathbf{v}},$$

$$\ddot{\boldsymbol{\rho}} = (\rho\,\dot{\eta} + 2\,\dot{\rho}\,\eta)\hat{\mathbf{v}} + \rho\,\eta^2\,\kappa\hat{\mathbf{n}} + \ddot{\rho} - \rho\,\eta^2\hat{\boldsymbol{\rho}}.$$

The **Laplace method** uses the following approximations: $\mathbf{q}$ is the position of the center of the Earth (geocentric approximation) and the mass of all the planets is zero. Then the two-body formula can be used for the accelerations $\ddot{\boldsymbol{\rho}}$ and $\ddot{\mathbf{q}}$

$$\ddot{\boldsymbol{\rho}} = \frac{-\mu\,\mathbf{r}}{r^3} + \frac{\mu\,\mathbf{q}}{q^3}$$

where r is the heliocentric distance of the asteroid, q is the heliocentric distance of the Earth, μ the mass of the Sun times the gravitational constant. Note that the denominator of the first fraction is $r^3 = S(\rho)^{3/2}$ where

$$S(\rho) = \rho^2 + 2q\rho\cos\epsilon + q^2 \tag{9.4}$$

is the same polynomial appearing in Section 8.1, with $\mathbf{q} = q\,\hat{\mathbf{q}}$ and $\cos\epsilon = \hat{\mathbf{q}}\cdot\hat{\boldsymbol{\rho}}$. Equation (9.4) is a geometric relation among q, r, and ρ and it is often called the **geometric equation** in the orbit determination literature.

We compute the components of $\ddot{\boldsymbol{\rho}}$ along $\hat{\mathbf{n}}$ and $\hat{\mathbf{v}}$: using $\hat{\boldsymbol{\rho}}\cdot\hat{\mathbf{n}} = 0$ we have

$$\ddot{\boldsymbol{\rho}}\cdot\hat{\mathbf{n}} = \frac{-\mu\,\mathbf{q}\cdot\hat{\mathbf{n}}}{r^3} + \frac{\mu\,\mathbf{q}\cdot\hat{\mathbf{n}}}{q^3} = \rho\,\eta^2\,\kappa, \tag{9.5}$$

$$\ddot{\boldsymbol{\rho}}\cdot\hat{\mathbf{v}} = \frac{-\mu\,\mathbf{q}\cdot\hat{\mathbf{v}}}{r^3} + \frac{\mu\,\mathbf{q}\cdot\hat{\mathbf{v}}}{q^3} = \rho\,\dot{\eta} + 2\,\dot{\rho}\,\eta. \tag{9.6}$$

Let us define

$$C = \frac{\eta^2\,\kappa\,q^3}{\mu\,\hat{\mathbf{q}}\cdot\hat{\mathbf{n}}}; \tag{9.7}$$

then, in the two-body approximation, eq. (9.5) takes the form

$$1 - C\,\frac{\rho}{q} = \frac{q^3}{S(\rho)^{3/2}}, \tag{9.8}$$

which is often called the **dynamical equation**; in fact, it express only the $\hat{\mathbf{n}}$ component of the dynamics.

By substituting in (9.8) the possible values of ρ obtained by eq. (9.4) and squaring, we obtain a polynomial equation of degree eight in r

$$p(r) = C^2 r^8 - q^2\left(C^2 + 2C\cos\epsilon + 1\right)r^6 + 2q^5(C\cos\epsilon + 1)r^3 - q^8 = 0, \tag{9.9}$$

which is equivalent to eq. (9.8) if the left-hand side of (9.8) is positive, that is, only if $q/\rho > C$.

From relation (9.6) we can compute the value of $\dot\rho$ from a value of ρ that solves (9.5) and then define an orbit in attributable orbital elements.

9.3 The method of Gauss

For the times $t_i, i = 1, 2, 3$, let $\mathbf{r}_i, \boldsymbol{\rho}_i$ denote the heliocentric and topocentric position of the body, respectively; $\mathbf{q}_i$ is the heliocentric position of the observer. **The Gauss method** uses three observations corresponding to the positions

$$\mathbf{r}_i = \boldsymbol{\rho}_i + \mathbf{q}_i \quad i = 1, 2, 3 \tag{9.10}$$

at times $t_1 < t_2 < t_3$. We assume that $t_i - t_j, 1 \leq i, j \leq 3$, is much smaller than the period of the orbit and we write $\mathcal{O}(\Delta t)$ for the order of magnitude of the time differences. The coplanarity of the $\mathbf{r}_i$ implies

$$\lambda_1 \mathbf{r}_1 - \mathbf{r}_2 + \lambda_3 \mathbf{r}_3 = 0 \tag{9.11}$$

for some $\lambda_1, \lambda_3 \in \mathbb{R}$. The vector product of both sides of eq. (9.11) with $\mathbf{r}_1$ and $\mathbf{r}_3$ and the fact that the vectors $\mathbf{r}_i \times \mathbf{r}_j$ for $i \neq j$ have the same orientation as $\mathbf{c} = \mathbf{r}_h \times \dot{\mathbf{r}}_h$, for $h = 1, 2, 3$ (that is the angular momentum integral per unit mass at any of the three times) allows us to write

$$\lambda_1 = \frac{\mathbf{r}_2 \times \mathbf{r}_3 \cdot \hat{\mathbf{c}}}{\mathbf{r}_1 \times \mathbf{r}_3 \cdot \hat{\mathbf{c}}}, \qquad \lambda_3 = \frac{\mathbf{r}_1 \times \mathbf{r}_2 \cdot \hat{\mathbf{c}}}{\mathbf{r}_1 \times \mathbf{r}_3 \cdot \hat{\mathbf{c}}} \qquad \text{triangle area ratios.}$$

From (9.10) and the scalar product of $\hat{\boldsymbol{\rho}}_1 \times \hat{\boldsymbol{\rho}}_3$ with eq. (9.11) we obtain

$$\rho_2[\hat{\boldsymbol{\rho}}_1 \times \hat{\boldsymbol{\rho}}_3 \cdot \hat{\boldsymbol{\rho}}_2] = \hat{\boldsymbol{\rho}}_1 \times \hat{\boldsymbol{\rho}}_3 \cdot [\lambda_1 \mathbf{q}_1 - \mathbf{q}_2 + \lambda_3 \mathbf{q}_3]. \tag{9.12}$$

The differences $\mathbf{r}_i - \mathbf{r}_2, i = 1, 3$, can be expanded in powers of $t_{ij} = t_i - t_j = \mathcal{O}(\Delta t)$, e.g., by using the **f and g** series (Herrick 1971, Everhart and Pitkin 1983); thus $\mathbf{r}_i = f_i \mathbf{r}_2 + g_i \dot{\mathbf{r}}_2$, with

$$f_i = 1 - \frac{\mu}{2} \frac{t_{i2}^2}{r_2^3} + \mathcal{O}(\Delta t^3), \qquad g_i = t_{i2} \left(1 - \frac{\mu}{6} \frac{t_{i2}^2}{r_2^3} \right) + \mathcal{O}(\Delta t^4). \tag{9.13}$$

Then $\mathbf{r}_i \times \mathbf{r}_2 = -g_i \, \mathbf{c}$, $\mathbf{r}_1 \times \mathbf{r}_3 = (f_1 g_3 - f_3 g_1) \, \mathbf{c}$ and

$$\lambda_1 = \frac{g_3}{f_1 g_3 - f_3 g_1}, \qquad \lambda_3 = \frac{-g_1}{f_1 g_3 - f_3 g_1}, \tag{9.14}$$

$$f_1 g_3 - f_3 g_1 = t_{31} \left(1 - \frac{\mu}{6} \frac{t_{31}^2}{r_2^3} \right) + \mathcal{O}(\Delta t^4). \tag{9.15}$$

Using (9.13) and (9.15) in (9.14) we obtain

$$\lambda_1 = \frac{t_{32}}{t_{31}}\left[1 + \frac{\mu}{6r_2^3}(t_{31}^2 - t_{32}^2)\right] + \mathcal{O}(\Delta t^3), \tag{9.16}$$

$$\lambda_3 = \frac{t_{21}}{t_{31}}\left[1 + \frac{\mu}{6r_2^3}(t_{31}^2 - t_{21}^2)\right] + \mathcal{O}(\Delta t^3). \tag{9.17}$$

Let $V = \hat{\boldsymbol{\rho}}_1 \times \hat{\boldsymbol{\rho}}_2 \cdot \hat{\boldsymbol{\rho}}_3$. By substituting (9.16), (9.17) into (9.12), using relations $t_{31}^2 - t_{32}^2 = t_{21}(t_{31} + t_{32})$ and $t_{31}^2 - t_{21}^2 = t_{32}(t_{31} + t_{21})$, we can write

$$\begin{aligned} -V\rho_2 t_{31} = & \ \hat{\boldsymbol{\rho}}_1 \times \hat{\boldsymbol{\rho}}_3 \cdot (t_{32}\,\mathbf{q}_1 - t_{31}\,\mathbf{q}_2 + t_{21}\,\mathbf{q}_3) \tag{9.18}\\ & + \hat{\boldsymbol{\rho}}_1 \times \hat{\boldsymbol{\rho}}_3 \cdot \left[\frac{\mu}{6r_2^3}[t_{32}t_{21}(t_{31} + t_{32})\,\mathbf{q}_1 + t_{32}t_{21}(t_{31} + t_{21})\,\mathbf{q}_3]\right]\\ & + \mathcal{O}(\Delta t^4). \end{aligned}$$

If the $\mathcal{O}(\Delta t^4)$ terms are neglected, the coefficient of $1/r_2^3$ in (9.19) is

$$B(\mathbf{q}_1, \mathbf{q}_3) = \frac{\mu}{6}t_{32}t_{21}\hat{\boldsymbol{\rho}}_1 \times \hat{\boldsymbol{\rho}}_3 \cdot [(t_{31} + t_{32})\,\mathbf{q}_1 + (t_{31} + t_{21})\,\mathbf{q}_3]. \tag{9.19}$$

Then multiply (9.19) by $q_2^3/B(\mathbf{q}_1, \mathbf{q}_3)$ to obtain

$$-\frac{V\,\rho_2\,t_{31}}{B(\mathbf{q}_1, \mathbf{q}_3)}\,q_2^3 = \frac{q_2^3}{r_2^3} + \frac{A(\mathbf{q}_1, \mathbf{q}_2, \mathbf{q}_3)}{B(\mathbf{q}_1, \mathbf{q}_3)},$$

where

$$A(\mathbf{q}_1, \mathbf{q}_2, \mathbf{q}_3) = q_2^3\,\hat{\boldsymbol{\rho}}_1 \times \hat{\boldsymbol{\rho}}_3 \cdot [t_{32}\,\mathbf{q}_1 - t_{31}\,\mathbf{q}_2 + t_{21}\,\mathbf{q}_3].$$

Let

$$C_2 = \frac{V\,t_{31}\,q_2^4}{B(\mathbf{q}_1, \mathbf{q}_3)}, \qquad \gamma_2 = -\frac{A(\mathbf{q}_1, \mathbf{q}_2, \mathbf{q}_3)}{B(\mathbf{q}_1, \mathbf{q}_3)}; \tag{9.20}$$

then

$$C_2\frac{\rho_2}{q_2} = \gamma_2 - \frac{q_2^3}{r_2^3} \tag{9.21}$$

is the dynamical equation of the Gauss method.

After the possible values for r_2 have been found by (9.21) and the geometric equation $r_2^2 = \rho_2^2 + q_2^2 + 2\rho_2 q_2 \cos \epsilon_2$, the velocity vector $\dot{\mathbf{r}}_2$ can be computed by different methods, e.g., from the **Gibbs formula** (Herrick 1971, Chapter 8). Given the values of λ_1, λ_3, from the scalar product of eq. (9.11) with $\hat{\boldsymbol{\rho}}_1 \times \hat{\boldsymbol{\rho}}_2$ we obtain a linear equation for ρ_3, from the scalar product with $\hat{\boldsymbol{\rho}}_2 \times \hat{\boldsymbol{\rho}}_3$ a linear equation for ρ_1; from this we can compute $\mathbf{r}_1$ and $\mathbf{r}_3$. The Gibbs method provides $\dot{\mathbf{r}}_2$ in the form (Herrick 1971, Chapter 8)

$$\dot{\mathbf{r}}_2 = -d_1\,\mathbf{r}_1 + d_2\,\mathbf{r}_2 + d_3\,\mathbf{r}_3 \tag{9.22}$$

where

$$d_i = G_i + H_i \, r_i^{-3}, \quad i = 1, 2, 3,$$

$$G_1 = \frac{t_{32}^2}{t_{21} \, t_{32} \, t_{31}} \, , \; G_3 = \frac{t_{21}^2}{t_{21} \, t_{32} \, t_{31}} \, , \; G_2 = G_1 - G_3,$$

$$H_1 = \mu \, t_{32}/12 \, , \; H_3 = \mu \, t_{21}/12 \, , \; H_2 = H_1 - H_3.$$

When $\mathbf{r}_2$ and $\dot{\mathbf{r}}_2$ are available, they provide a set of initial conditions (at epoch $t_2 - \rho_2/c$), from which we can compute two-body solutions $\mathbf{r}_1, \mathbf{r}_3$ for the times $t_1 - \rho_1/c$, $t_3 - \rho_3/c$ (by using a two-body propagator, see Appendix A). Then the coefficients λ_1, λ_3 are available from eq. (9.11), and eq. (9.12) provides an improved value of ρ_2, from which a new iteration could be started. This is just one of the many iterative methods described in the literature to improve the preliminary orbit, with the goal of obtaining smaller residuals with respect to the three observations.

As shown by Celletti and Pinzari (2005), each step in these iterative procedures used to improve the preliminary orbits (which they call a **Gauss map**[2]) can be shown to increase the order in Δt in the approximation of the exact solutions to the two-body equation of motion. Celletti and Pinzari (2006) have also shown that the iteration of a Gauss map can diverge when the solution of the degree 8 equation is far from the fixed point, outside of the convergence domain. Thus the Gauss map should be used with some caution, e.g., with a recovery procedure in case of divergence.

9.4 Topocentric Gauss–Laplace methods

The critical difference between the methods of Gauss and Laplace is the following. Gauss uses a truncation (to order $\mathcal{O}(\Delta t^2)$) in the motion $\mathbf{r}(t)$ of the asteroid but the positions of the observer (be it coincident with the center of the Earth or not) are used in their exact values. Laplace uses a truncation to the same order of the relative motion $\boldsymbol{\rho}(t)$ (see eq. (9.2) in Section 9.1), thus implicitly approximating the motion of the observer. In this section we examine the consequences of the difference between the techniques.

Gauss–Laplace equivalence

To directly compare the two methods let us introduce in Gauss' method the same approximation to order $\mathcal{O}(\Delta t^2)$ in the motion of the center of the

---

[2] The classical treatises, such as (Crawford *et al.* 1930), use the term *differential corrections* for algorithms of the same class of the *Gauss map* in (Celletti and Pinzari 2005). We follow the terminology of the recent papers because, in modern usage, *differential corrections* refers to the iterative method to solve the least squares problem.

Earth, which we assumed to coincide with the observer. Using the f, g series for the Earth we obtain

$$\mathbf{q}_i = \left(1 - \frac{\mu}{2}\frac{t_{i2}^2}{q_2^3}\right)\mathbf{q}_2 + t_{i2}\,\dot{\mathbf{q}}_2 + \frac{\mu}{6}\frac{t_{i2}^3}{q_2^3}\left[\frac{3(\mathbf{q}_2 \cdot \dot{\mathbf{q}}_2)\,\mathbf{q}_2}{q_2^2} - \dot{\mathbf{q}}_2\right] + \mathcal{O}(\Delta t^4). \quad (9.23)$$

By substituting (9.23) in (9.19) we find

$$B(\mathbf{q}_1, \mathbf{q}_3) = \frac{\mu}{6}t_{32}t_{21}\,\hat{\boldsymbol{\rho}}_1 \times \hat{\boldsymbol{\rho}}_3 \cdot [3t_{31}\,\mathbf{q}_2 + t_{31}(t_{32} - t_{21})\,\dot{\mathbf{q}}_2 + \mathcal{O}(\Delta t^3)].$$

If $t_{32} - t_{21} = t_3 + t_1 - 2t_2 = 0$, that is, the interpolation for d^2/dt^2 is done at the central value t_2, then

$$B(\mathbf{q}_1, \mathbf{q}_3) = \frac{\mu}{2}t_{21}t_{32}t_{31}\,\hat{\boldsymbol{\rho}}_1 \times \hat{\boldsymbol{\rho}}_3 \cdot \mathbf{q}_2\,(1 + \mathcal{O}(\Delta t^2));$$

else, if $t_2 \neq (t_1 + t_3)/2$, the last factor is just $(1 + \mathcal{O}(\Delta t))$. Substituting (9.23) in (9.24) we have

$$\begin{aligned}
A(\mathbf{q}_1, \mathbf{q}_2, \mathbf{q}_3) &= -\frac{\mu}{2}t_{21}t_{32}t_{31}\,\hat{\boldsymbol{\rho}}_1 \times \hat{\boldsymbol{\rho}}_3 \cdot \left\{\mathbf{q}_2 \right. \\
&\quad \left. + \frac{1}{3}\,(t_{21} - t_{32})\left[\frac{3(\mathbf{q}_2 \cdot \dot{\mathbf{q}}_2)\,\mathbf{q}_2}{q_2^2} - \dot{\mathbf{q}}_2\right]\right\} + \mathcal{O}(\Delta t^5).
\end{aligned}$$

$$(9.24)$$

If, as above, $t_{32} - t_{21} = t_3 + t_1 - 2t_2 = 0$ then

$$A(\mathbf{q}_1, \mathbf{q}_2, \mathbf{q}_3) = -\frac{\mu}{2}t_{21}t_{32}t_{31}\,\hat{\boldsymbol{\rho}}_1 \times \hat{\boldsymbol{\rho}}_3 \cdot \mathbf{q}_2\,(1 + \mathcal{O}(\Delta t^2))$$

and we can conclude from (9.20) that

$$\gamma_2 = -\frac{A}{B} = 1 + \mathcal{O}(\Delta t^2);$$

else, if $t_2 \neq (t_1 + t_3)/2$, the last factor is just $(1 + \mathcal{O}(\Delta t))$. For V we use (9.1) to make a Taylor expansion of $\hat{\boldsymbol{\rho}}_i$ in t_2:

$$\hat{\boldsymbol{\rho}}_i = \hat{\boldsymbol{\rho}}_2 + t_{i2}\eta\hat{\mathbf{v}}_2 + \frac{t_{i2}^2}{2}(-\eta^2\hat{\boldsymbol{\rho}}_2 + \dot{\eta}\hat{\mathbf{v}}_2 + \kappa\eta^2\hat{\mathbf{n}}_2) + \mathcal{O}(\Delta t^3).$$

This implies that

$$\hat{\boldsymbol{\rho}}_1 \times \hat{\boldsymbol{\rho}}_3 \cdot \hat{\boldsymbol{\rho}}_2 = \frac{1}{2}\left[t_{12}\eta\hat{\mathbf{v}}_2 \times t_{32}^2\kappa\,\eta^2\hat{\mathbf{n}}_2 - t_{32}\eta\,\hat{\mathbf{v}}_2 \times t_{12}^2\kappa\,\eta^2\,\hat{\mathbf{n}}_2\right] \cdot \hat{\boldsymbol{\rho}}_2 + \mathcal{O}(\Delta t^4);$$

if $t_2 = (t_1 + t_3)/2$ then the term $\mathcal{O}(\Delta t^4)$ vanishes and the remainder is $\mathcal{O}(\Delta t^5)$. Thus

$$V = -\frac{\kappa\eta^3}{2}(t_{12}t_{32}^2 - t_{32}t_{12}^2)\,(1 + \mathcal{O}(\Delta t^2)) = \frac{\kappa\eta^3}{2}t_{21}t_{32}t_{31}\,(1 + \mathcal{O}(\Delta t^2)),$$

$$C_2 = \frac{V t_{31} q_2^4}{B} = \frac{\kappa \eta^3 t_{31} q_2^4 + \mathcal{O}(\Delta t^3)}{\mu \hat{\rho}_1 \times \hat{\rho}_3 \cdot \mathbf{q}_2 \left(1 + \mathcal{O}(\Delta t)\right)}.$$

In the denominator, $\hat{\rho}_1 \times \hat{\rho}_3$ computed to order Δt^2 is

$$\hat{\rho}_1 \times \hat{\rho}_3 = t_{31}\, \eta \, \hat{\mathbf{n}}_2 + \frac{t_{32}^2 - t_{12}^2}{2} \left(\dot{\eta} \, \hat{\mathbf{n}}_2 - \kappa \, \eta^2 \, \hat{\mathbf{v}}_2 \right) + \mathcal{O}(\Delta t^3). \qquad (9.25)$$

As a conclusion, if $t_{32} - t_{21} = t_3 + t_1 - 2t_2 = 0$ then

$$C_2 = \frac{\kappa \, \eta^3 \, t_{31} q_2^4 + \mathcal{O}(\Delta t^3)}{\mu \, t_{31} \, \eta \, q_2 \hat{\mathbf{q}}_2 \cdot \hat{\mathbf{n}}_2 + \mathcal{O}(\Delta t^3)} = \frac{\kappa \, \eta^2 \, q_2^3}{\mu \, \hat{\mathbf{q}}_2 \cdot \hat{\mathbf{n}}_2} \left(1 + \mathcal{O}(\Delta t^2)\right),$$

otherwise the last factor is $(1 + \mathcal{O}(\Delta t))$.

Thus, neglecting the difference between topocentric and geocentric observations, the coefficients of the two dynamical equations (9.8) and (9.21) are the same to order zero in Δt, also to order one if t_2 is the average time.

Topocentric Laplace method

Now let us remove the approximation that the observer lies at the center of the Earth and introduce *topocentric* observations into the Laplace method. The center of mass of the Earth is at $\mathbf{q}_\oplus$ but the observer is at $\mathbf{q} = \mathbf{q}_\oplus + \mathbf{P}$. We derive the dynamical equation by also taking into account the acceleration contained in the geocentric position of the observer $\mathbf{P}$, that is

$$\frac{d^2 \boldsymbol{\rho}}{dt^2} = -\frac{\mu \mathbf{r}}{r^3} + \frac{\mu \mathbf{q}_\oplus}{q_\oplus^3} - \ddot{\mathbf{P}}.$$

By scalar multiplication with $\hat{\mathbf{n}}$, using eq. (9.1), we obtain

$$\frac{d^2 \boldsymbol{\rho}}{dt^2} \cdot \hat{\mathbf{n}} = \rho \eta^2 \kappa = \mu \left[q_\oplus \frac{\hat{\mathbf{q}}_\oplus \cdot \hat{\mathbf{n}}}{q_\oplus^3} - q_\oplus \frac{\hat{\mathbf{q}}_\oplus \cdot \hat{\mathbf{n}}}{r^3} - P \frac{\hat{\mathbf{P}} \cdot \hat{\mathbf{n}}}{r^3} \right] - \ddot{\mathbf{P}} \cdot \hat{\mathbf{n}}.$$

The term $P \hat{\mathbf{P}} \cdot \hat{\mathbf{n}}/r^3$ can be neglected. This approximation is legitimate because $P/q_\oplus \le 4.3 \times 10^{-5}$ and the neglected term is smaller than the planetary perturbations. Thus we obtain the dynamical equation

$$C \frac{\rho}{q_\oplus} = (1 - \Lambda_n) - \frac{q_\oplus^3}{r^3} \qquad (9.26)$$

where

$$C = \frac{\eta^2 \kappa q_\oplus^3}{\mu \hat{\mathbf{q}}_\oplus \cdot \hat{\mathbf{n}}}, \qquad \Lambda_n = \frac{q_\oplus^2 \ddot{\mathbf{P}} \cdot \hat{\mathbf{n}}}{\mu \hat{\mathbf{q}}_\oplus \cdot \hat{\mathbf{n}}} = \frac{\ddot{\mathbf{P}} \cdot \hat{\mathbf{n}}}{(\mu/q_\oplus^2) \, \hat{\mathbf{q}}_\oplus \cdot \hat{\mathbf{n}}}. \qquad (9.27)$$

Note that Λ_n is singular only where C is also singular. The analog of eq. (9.6), again neglecting $\mathcal{O}(p/q_\oplus)$, is

$$\rho\ddot{\eta} + 2\dot{\rho}\dot{\eta} = \frac{\mu\,\hat{\mathbf{q}}_\oplus \cdot \hat{\mathbf{v}}}{q_\oplus^2}\left(1 - \Lambda_v - \frac{q_\oplus^3}{r^3}\right), \qquad \Lambda_v = \frac{q_\oplus^2\,\ddot{\mathbf{P}}\cdot\hat{\mathbf{v}}}{\mu\,\hat{\mathbf{q}}_\oplus \cdot \hat{\mathbf{v}}}. \qquad (9.28)$$

The important fact is that Λ_n and Λ_v are by no means small. The centripetal acceleration of the observer (towards the rotation axis of the Earth) has size $|\ddot{\mathbf{P}}| = \Omega_\oplus^2\,R_\oplus \cos\theta$ where $\Omega_\oplus$ is the angular velocity of the Earth rotation, $R_\oplus$ the radius of the Earth, and θ the latitude; the maximum of $|\ddot{\mathbf{P}}|$ is $\simeq 3.4$ $\mathrm{cm\,s^{-2}}$ and occurs at the equator. The term $\mu/q_\oplus^2$ in the denominator of Λ_n is the size of the heliocentric acceleration of the Earth, $\simeq 0.6$ $\mathrm{cm\,s^{-2}}$. Thus $|\Lambda_n|$ can be > 1, and the coefficient $1 - \Lambda_n$ can be very different from 1 (it may even be negative). Without taking into account the observer geocentric acceleration, the classical Laplace method is not a good approximation, except when the observations of different nights are taken from the same station at the same sidereal time, so that the observer acceleration cancels out.

The common procedure for the Laplace method is to go back to the geocentric observation case by applying a **topocentric correction**, simulating the observations as they would appear to an observer placed at the center of the Earth. Some value of ρ is assumed as a first guess, e.g., $\rho = 1$ AU (Leuschner 1913, page 15). If this value is approximately correct, by iterating the cycle (topocentric correction – Laplace determination of ρ) convergence is achieved. If the starting value is really wrong, e.g., if the object is undergoing a close approach to the Earth, the procedure may well diverge. These reliability problems discourage the use of the classical form of the Laplace method when processing a large data set, containing discoveries of different orbital classes and therefore spanning a wide range of distances.

The same argument applies to the algorithms used to improve Laplace preliminary orbits, e.g. (Leuschner 1913, Crawford *et al.* 1930). The difference with the Gauss map is that in the Laplace method the observations in the first approximation are treated as geocentric (or possibly corrected with an *assumed* distance), while in the Gauss method already the first approximation properly handles topocentric observations.

Topocentric, Gauss–Laplace equivalence

When taking into account the displacement $\mathbf{P}$, the Taylor expansion of $\mathbf{q}_i(t)$ of eq. (9.23) is not applicable. We need to use

$$\mathbf{q}_i = \mathbf{q}_2 + t_{i2}\dot{\mathbf{q}}_2 + \frac{t_{i2}^2}{2}\ddot{\mathbf{q}}_2 + \mathcal{O}(\Delta t^3)$$

where $\mathbf{q}_2(t)$ and its derivatives contain also $\mathbf{P}(t)$. By using eq. (9.25) and assuming $t_{21} = t_{32}$, (9.19) and (9.24) become

$$B(\mathbf{q}_1, \mathbf{q}_3) = \frac{\mu\,\eta}{2}\,t_{21}t_{32}t_{31}^2\,\hat{\mathbf{n}}_2 \cdot \mathbf{q}_2 + \mathcal{O}(\Delta t^6),$$

$$A(\mathbf{q}_1, \mathbf{q}_2, \mathbf{q}_3) = \frac{q_2^3\,\eta}{2}\,t_{21}t_{32}t_{31}^2\,\hat{\mathbf{n}}_2 \cdot \ddot{\mathbf{q}}_2 + \mathcal{O}(\Delta t^6).$$

Note that $\dot{\mathbf{q}}_2$ does not appear in A at this approximation level. Thus

$$h_0 = -\frac{A}{B} = -\frac{q_2^3\,\hat{\mathbf{n}}_2 \cdot \ddot{\mathbf{q}}_2 + \mathcal{O}(\Delta t^2)}{\mu\,\hat{\mathbf{n}}_2 \cdot \mathbf{q}_2 + \mathcal{O}(\Delta t^2)}$$

and, neglecting once again $P/q_\oplus$ terms,

$$h_0 = -\frac{q_2^3\,\hat{\mathbf{n}}_2 \cdot \ddot{\mathbf{q}}_{\oplus 2}}{\mu\,\hat{\mathbf{n}}_2 \cdot \mathbf{q}_2} - \frac{q_2^3\,\hat{\mathbf{n}}_2 \cdot \ddot{\mathbf{P}}_2}{\mu\,\hat{\mathbf{n}}_2 \cdot \mathbf{q}_2} + O(\Delta t^2)$$

$$= \frac{q_2^3}{q_{\oplus 2}^3} - \frac{q_2^3\,\hat{\mathbf{n}}_2 \cdot \ddot{\mathbf{P}}_2}{\mu\,\hat{\mathbf{n}}_2 \cdot \mathbf{q}_2} + \mathcal{O}(\Delta t^2).$$

Finally

$$\hat{\mathbf{n}}_2 \cdot \mathbf{q}_2 = q_2\,\hat{\mathbf{n}}_2 \cdot \left(\frac{\mathbf{q}_{\oplus 2}}{q_2} + \frac{\mathbf{P}_2}{q_2}\right) = q_2\left(\hat{\mathbf{n}}_2 \cdot \hat{\mathbf{q}}_{\oplus 2} + \mathcal{O}\left(\frac{P_2}{q_2}\right)\right),$$

then

$$\gamma_2 = 1 - \frac{q_{\oplus 2}^3\,\hat{\mathbf{n}}_2 \cdot \ddot{\mathbf{P}}_2}{\mu\,\hat{\mathbf{n}}_2 \cdot \mathbf{q}_2} + \mathcal{O}(\Delta t^2) + \mathcal{O}\left(\frac{P_2}{q_2}\right) = 1 - \Lambda_{n2} + \mathcal{O}(\Delta t^2) + \mathcal{O}\left(\frac{P_2}{q_2}\right)$$

where Λ_{n2} is the same quantity as Λ_n of eq. (9.27) computed at $t = t_2$.

The conclusion is that the Gauss method used with the heliocentric positions of the observer $\mathbf{q}_i = \mathbf{q}_{\oplus i} + \mathbf{P}_i$ is equivalent to the topocentric Laplace method to lowest order in Δt and neglecting the very small term $\mathcal{O}(P_2/q_2)$.

Problems in the topocentric Laplace method

The Laplace method, with the geocentric approximation, is not really equivalent to the Gauss method: by using the observer positions in (9.19) and (9.24), the Gauss method naturally accounts for topocentric observations. Can we account for topocentric observations in the Laplace method (without iterations) by adding the term Λ_n from eq. (9.27)? The answer is contained in (Poincaré 1906). To summarize the argument of Poincaré, we can use plots showing the shape of the topocentric correction as a function of time.

Figure 9.1 shows the simulated path of an approaching asteroid as seen

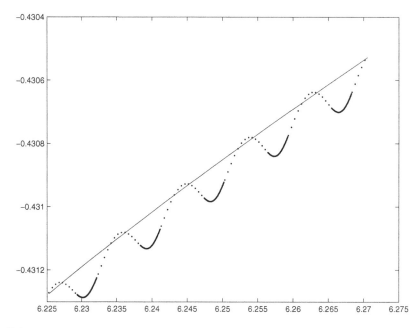

Fig. 9.1. The path in the sky of the near-Earth asteroid (101955) 1999 RQ$_{36}$ as it would have been seen in July 2005 if an observatory on Mauna Kea had been observing continuously. The solid portions of the curve are possible observations, the dotted ones are not possible (altitude $< 15°$). The continuous curve shows observations from the geocenter. Coordinates are RA and DEC in radians. Reprinted from (Milani *et al.* 2008) with permission from Elsevier.

from an observing station. The darker portions of the curve indicate possible observations, the dotted ones are practically impossible, with an altitude $< 15°$. The apparent motion of the asteroid from night to night cannot be approximated using parabolic motion segments fitted to a single night. For the geocentric path (continuous curve) the parabolic approximation to $\hat{\boldsymbol{\rho}}(t)$, used by Laplace, would be applicable. Topocentric observations contain more information beyond the attributable; to reduce the observations to the geocenter by using the topocentric correction is a bad strategy.

Poincaré suggests computing Λ_n, Λ_v by using, in place of $\ddot{\mathbf{P}}(\bar{t})$, a value obtained by interpolating[3] the positions $\mathbf{P}(t_i)$ at the times t_i of the observations (not limited to 3, one of the advantages of the Laplace method). Poincaré gives no examples, but we have implemented this procedure and found that it works (see Section 8.3 and 8.6, where the same method is used). This method has not undergone a large-scale test, thus its practical advantages have not yet been assessed.

When the observations are performed from an artificial satellite (such as

---

[3] Our translation of Poincaré: *It is necessary to avoid computing these quantities by starting from the law of rotation of the Earth.*

the Hubble Space Telescope or, in the future, from Gaia) the acceleration $\ddot{\mathbf{P}} \simeq 900$ cm s^{-2} and the Λ_n and Λ_v coefficients can be up to $\simeq 1500$. A few hours of observations extending to several orbits produce multiple kinks as in (Marchi *et al.* 2004, Figure 1), containing important orbital information.

9.5 Number of solutions

Charlier gave a geometric interpretation of the occurrence of alternative solutions for a Laplace method preliminary orbit (Charlier 1910, Charlier 1911). He realized that (neglecting the errors in the measurements and in the model) this depends only on the position of $\mathcal{B}$ in a reference plane defined by the Sun, the Earth, and the body at a given time, and he was able to divide this plane into four connected components by two algebraic curves, separating regions with a unique solution from those with two solutions.

In this section we have shown that the Gauss method allows us to take into account topocentric observations in a natural way, and also the Laplace method can be modified to consider this effect. In both cases from the two-body dynamics we obtain an equation like (9.30), with the same algebraic structure as eq. (9.8) of the geocentric Laplace method, but it depends on the additional parameter γ_2, and reduces to eq. (9.8) only for $\gamma_2 = 1$. Thus generically Charlier theory cannot be applied. We introduce a generalization of Charlier theory, providing a qualitative theory of alternative solutions also in the more realistic case of topocentric observations.

The intersection problem

Assume that we have three observations of a celestial body whose motion is dominated by the gravitational attraction of the Sun.

We write r, ρ, q, ϵ for the values of the quantities corresponding to $r_i, \rho_i, q_i, \epsilon_i$ at the average time $\bar{t}$. Note that q and $\mathbf{q}_i, i = 1, 2, 3$, can be obtained from planetary ephemerides and Earth rotation models, ϵ can be computed by interpolating the values of ϵ_i (computed in turn from $\alpha_i, \delta_i, \mathbf{q}_i$), while r, ρ are unknown because r_i, ρ_i are also.

Actually the results we shall present do not depend on the value of q. By choosing a different unit of length we could set $q = 1$ without loss of generality; we prefer to leave q in all the formulae, since different units may be used in the applications of the theory to specific problems.

The geometry of the three bodies immediately gives the relation

$$r^2 = q^2 + \rho^2 + 2q\rho \cos \epsilon \qquad \text{geometric equation.} \qquad (9.29)$$

Using the two-body dynamics we can deduce the following relation:

$$\mathcal{C}\,\frac{\rho}{q} = \gamma - \frac{q^3}{r^3} \qquad \text{dynamical equation,} \qquad (9.30)$$

where $\gamma, \mathcal{C} \in \mathbb{R}$ are constants computed from the observations, corresponding to γ_2, C_2 in the Gauss method, see (9.20) and to $1 - \Lambda_n, C$ in the Laplace method, see eqs. (9.26) and (9.7), reducing to $1, C$ in the geocentric approximation, see eq (9.8).

Equations (9.29) and (9.30) define surfaces of revolution around the axis $\hat{\mathbf{q}}$ passing through the center of the Sun and the observer. If the center of the Sun, the observer, and the observed body are not collinear at time $\bar{t}$, the observation line (also called the line of sight: a half-line from the observer position) and the axis $\hat{\mathbf{q}}$ define univocally a reference plane, which we shall use to study the intersection of these surfaces.

We introduce the **intersection problem**

$$\begin{cases} D(r, \rho) = (q\gamma - \mathcal{C}\rho)r^3 - q^4 = 0 \\ G(r, \rho) = r^2 - q^2 - \rho^2 - 2q\rho\cos\epsilon = 0 \\ r, \rho > 0, \end{cases} \qquad (9.31)$$

that is, given $(\gamma, \mathcal{C}, \epsilon) \in \mathbb{R}^2 \times [0, \pi]$ we search for pairs (r, ρ) of strictly positive real numbers, solutions of (9.30) and (9.29). For given values of $(\gamma, \mathcal{C}, \epsilon)$ the solutions of (9.31) correspond to the intersections of the observation line with the planar algebraic curve defined by (9.30) in the reference plane (see Figure 9.2).

We can perform elimination of the variable ρ by means of resultant theory (see Cox *et al.* 1996), thus from (9.31) we obtain the reduced problem

$$\begin{cases} P(r) = \mathbf{res}\,(D, G, \rho) = 0 \\ r > 0 \end{cases} \qquad (9.32)$$

where $\mathbf{res}(D, G, \rho)$ stands for the resultant of the polynomials $D(r, \rho)$ and $G(r, \rho)$ with respect to the variable ρ. The resultant computation gives

$$P(r) = \mathcal{C}^2 r^8 - q^2(\mathcal{C}^2 + 2\mathcal{C}\gamma\cos\epsilon + \gamma^2)r^6 + 2q^5(\mathcal{C}\cos\epsilon + \gamma)r^3 - q^8. \quad (9.33)$$

The reduced formulation (9.32) is suitable to obtain an upper bound for the maximum number of solutions, in fact $P(r)$ has only four monomials, thus by Descartes' sign rule there are at most three positive roots of $P(r)$, counted with multiplicity. Note that, if $r = \bar{r}$ is a component of a solution of (9.31), from (9.30) we obtain a unique value $\bar{\rho}$ for the other component and, conversely, from a value $\bar{\rho}$ of ρ we obtain a unique $\bar{r}$; there are no more than three values of ρ that are components of the solutions of (9.31).

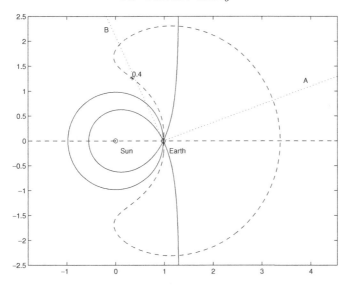

Fig. 9.2. Geometric view of the intersection problem: given a direction of observation, the solutions are the intersections with level curves of $\mathcal{C}$.

We define as a **spurious solution** of (9.32) a positive root $\bar{r}$ of $P(r)$ that is not a component of a solution $(\bar{r}, \bar{\rho})$ of (9.31) for any $\bar{\rho} > 0$, that is it gives a non-positive ρ through the dynamical equation (9.30).

The question is how many solutions are possible for the intersection problem, thus also for the preliminary orbit problem; from each solution of (9.31) a full set of orbital elements can be determined, in fact knowledge of the topocentric distance ρ allows us to compute the corresponding value of $\dot{\rho}$. In case of alternative solutions all of them should be used as a first guess for the differential corrections, not to miss the right solution.

9.6 Charlier theory

The **Charlier theory** describes the occurrence of alternative solutions in the problem defined by eqs. (9.8) and (9.4), with geocentric observations. Nevertheless, if we interpret ρ and q as the geocentric distance of the observed body and the heliocentric distance of the center of the Earth, then eq. (9.30) with $\gamma = 1$ corresponds to (9.8) and eq. (9.29) corresponds to (9.4). Therefore we shall discuss Charlier theory by studying the alternative solutions of (9.31) with $\gamma = 1$, and we shall see that in this case the solutions of (9.31) can be at most two. The discussion presented in this section is based on (Plummer 1918).

Charlier was the first to realize that *the condition for the appearance of another solution simply depends on the position of the observed body.* We stress that it assumes that the two-body model for the orbit of the observed body is exact and neglects the observation and interpolation errors in the parameters $\mathcal{C}, \epsilon$. The previous hypotheses imply the following assumption:

> the parameters $\mathcal{C}, \epsilon$ are such that the corresponding intersection problem with $\gamma = 1$ admits at least one solution. (9.34)

In real astronomical applications (9.34) may not be fulfilled and the intersection problem may have no solution. A reason for that is the presence of errors in the observations; these affect mostly the computation of $\mathcal{C}$. However we observe that condition (9.34) may hold also taking into account these errors, therefore it is more interesting for the applications.

For each choice of $\mathcal{C}, \epsilon$ the polynomial $P(r)$ in (9.32) has three changes of sign in the sequence of its coefficients; the coefficient of r^3 is positive because from (9.30) and (9.29) we have

$$\mathcal{C} \cos \epsilon + 1 = \frac{1}{2\rho^2 r^3} \left[(r^3 - q^3)(r^2 - q^2) + \rho^2(r^3 + q^3) \right] > 0 \, ,$$

thus the positive roots of $P(r)$ can be up to three. As $P(q) = 0$, there is always the physically meaningless solution corresponding to the center of the Earth, in fact, from the dynamical equation, $r = q$ corresponds to $\rho = 0$. Using (9.34), Descartes' sign rule and the relations

$$P(0) = -q^8 < 0 \, ; \qquad \lim_{r \to +\infty} P(r) = +\infty,$$

we conclude that there are always three positive roots of $P(r)$, counted with multiplicity. By (9.34) at least one of the other two positive roots r_1, r_2 is not spurious; if either r_1 or r_2 is spurious the solution of (9.31) is unique, otherwise we have two non-spurious solutions.

To detect the cases with two solutions we write $P(r) = (r - q)P_1(r)$, with

$$P_1(r) = \mathcal{C}^2 r^6 (r + q) + (r^2 + qr + q^2) \left[q^5 - (2\mathcal{C} \cos \epsilon + 1)q^2 r^3 \right]$$
$$P_1(q) = 2q^7 \mathcal{C} (\mathcal{C} - 3 \cos \epsilon).$$

From $P_1(0) = q^7 > 0$ and $\lim_{r \to +\infty} P_1(r) = +\infty$, it follows that if $P_1(q) < 0$ then $r_1 < q < r_2$, thus one root of $P_1(r)$ is spurious. Else, if $P_1(q) > 0$ either $r_1, r_2 < q$ or $r_1, r_2 > q$, and because of (9.34) both roots give meaningful solutions of (9.31). If $P_1(q) = 0$ there is only one non-spurious root of $P(r)$.

The dynamical equation gives us an expression of $\mathcal{C}$ as an algebraic function either in bipolar coordinates $\mathcal{C}(r, \rho)$ or in geocentric polar coordinates $\mathcal{C}(\rho, \epsilon)$; we shall plot the figures in a full plane, with $-\pi < \epsilon \leq \pi$, but the situation is symmetric with respect to the $\hat{\mathbf{q}}$ axis. Thus we can define two

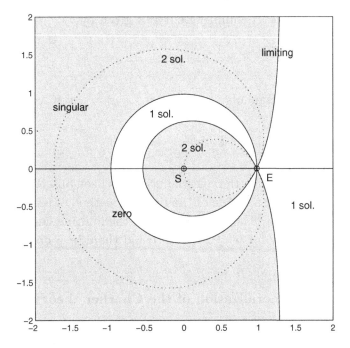

Fig. 9.3. The limiting curve and the zero circle divide the reference plane into four connected regions, two with a unique solution of (9.31) and two with two solutions (shaded in this figure). The singular curve (dotted) divides the regions with two solutions into two parts, with one solution each. The Sun and the Earth are labeled with S and E, respectively. We use heliocentric rectangular coordinates, and astronomical units (AU) for both axes. Reprinted from Gronchi (2009) with permission from Springer.

curves in this plane: the **zero circle** $\mathcal{C}(\rho, \epsilon) = 0$ and the **limiting curve** $\mathcal{C}(\rho, \epsilon) - 3\cos\epsilon = 0$, where

$$\mathcal{C}(\rho, \psi) = \frac{q}{\rho}\left[1 - \frac{q^3}{r^3}\right], \qquad r = \sqrt{\rho^2 + q^2 + 2q\rho\cos\psi}.$$

The limiting curve has a loop inside the zero circle and two unlimited branches with $r > q$. By the previous discussion the limiting curve and the zero circle divide the reference plane, containing the center of the Sun, the observer, and the observed body at time $\bar{t}$, into four connected components (see Figure 9.3), separating regions with a different number of solutions of the orbit determination problem. More precisely, given the position (ρ, ϵ) of a celestial body in the reference plane at time $\bar{t}$, eq. (9.30) with $\gamma = 1$ defines a value $\mathcal{C}$ such that the intersection problem defined by $\mathcal{C}$, ϵ and $\gamma = 1$ has the solution $(r, \rho) = (\sqrt{\rho^2 + q^2 + 2q\rho\cos\epsilon}, \rho)$ and, if the body is situated in a region with two solutions, we can find the second solution in the same region as the first. Using heliocentric polar coordinates (r, ϕ), with

$\rho^2 = r^2 + q^2 - 2qr \cos \epsilon$, the limiting curve is given by

$$4 - 3\frac{r}{q} \cos \epsilon = \frac{q^3}{r^3} \tag{9.35}$$

and, in heliocentric rectangular coordinates $(x, y) = (r \cos \phi, r \sin \phi)$, by

$$4 - 3\frac{x}{q} = \frac{q^3}{(x^2 + y^2)^{3/2}}.$$

Figure 9.3 shows that, when the celestial body has been observed close to the opposition direction, the solution of the Laplace method of preliminary orbit determination is unique. The two tangents to the limiting curve correspond to $\tan \epsilon_0 = 2$, thus only for $|\epsilon| \geq \simeq 63.43°$ could there be a double solution.

9.7 Generalization of the Charlier theory

In this section we consider the intersection problem (9.31) for a generic $\gamma \in \mathbb{R}$. Given a value of γ and the position (ρ, ϵ) of the observed body in the reference plane, in topocentric polar coordinates, eq. (9.30) defines a value of $\mathcal{C}$ such that the intersection problem defined by $(\gamma, \mathcal{C}, \epsilon)$ has the solution $(r, \rho) = (\sqrt{\rho^2 + q^2 + 2q\rho \cos \epsilon}, \rho)$. Therefore in the following we shall speak about the *intersection problem corresponding to, or related to, a fixed $\gamma \in \mathbb{R}$ and to a point of the reference plane.* We introduce the following assumption, that generalizes (9.34):

> the parameters $\gamma, \mathcal{C}, \epsilon$ are such that the corresponding intersection problem admits at least one solution. $\tag{9.36}$

Generically $r = q$ is not a root of $P(r)$, in fact

$$P(q) = q^8 (1 - \gamma) (2\mathcal{C} \cos \epsilon - (1 - \gamma)),$$

thus we cannot follow the steps of Section 9.6 to define the limiting curve. From the dynamical equation we define the function

$$C^{(\gamma)}(x, y) = \frac{q}{\rho} \left[\gamma - \frac{q^3}{r^3} \right], \tag{9.37}$$

where $\rho = \sqrt{(q - x)^2 + y^2}$ and $r = \sqrt{x^2 + y^2}$.

If $\gamma > 0$ we can also define the zero circle as the set of points such that $C^{(\gamma)}(x, y)$, that is $r = r_0 = q/\sqrt[3]{\gamma}$.

The topology of the level curves

For each $\gamma \in \mathbb{R}$

$$\lim_{\|(x,y)\|\to+\infty} C^{(\gamma)}(x,y) = 0, \qquad \lim_{(x,y)\to(0,0)} C^{(\gamma)}(x,y) = -\infty,$$

$$\lim_{(x,y)\to(q,0)} C^{(\gamma)}(x,y) \begin{cases} = -\infty & \text{for } \gamma < 1 \\ \text{does not exist} & \text{for } \gamma = 1 \\ = +\infty & \text{for } \gamma > 1. \end{cases}$$

The stationary points of $C^{(\gamma)}(x,y)$ have $y = 0$ and depend on γ as follows:

- for $\gamma \le 0$ there is only one saddle point, with $x \in (0, \frac{3}{4}q]$;
- for $0 < \gamma < 1$ there are three points: one saddle point inside the zero circle, one saddle and one maximum point outside.
- for $\gamma \ge 1$ there is a unique saddle point with $x < -r_0 = -q/\sqrt[3]{\gamma}$.

This result is useful to understand the topological changes in the level curves of $C^{(\gamma)}(x,y)$, see Figure 9.4 for all the significantly different cases, i.e. $\gamma \le 0$, $0 < \gamma < 1$, $\gamma = 1$, and $\gamma > 1$.

Table 9.1. Number of solutions at opposition.

# solutions	0	1	2	3
$\gamma \le 0$	$\mathcal{C} \ge 0$	$\mathcal{C} < 0$	/	/
$0 < \gamma < 1$	$\mathcal{C} > \mathcal{C}_{max}$	$\mathcal{C} \le 0$	$0 < \mathcal{C} \le \mathcal{C}_{max}$	/
$\gamma = 1$	$\mathcal{C} \le 0$ or $\mathcal{C} \ge 3$	$0 < \mathcal{C} < 3$	/	/
$\gamma > 1$	$\mathcal{C} \le 0$	$\mathcal{C} > 0$	/	/

In Table 9.1, for each value of γ, we describe the change with $\mathcal{C}$ in the number of solutions, when we observe in the opposition direction. $\mathcal{C}_{max}$ is the maximum value of $C^{(\gamma)}(x,y)$.

The singular curve

The function $C^{(\gamma)}(x,y)$ in topocentric polar coordinates (ρ, ϵ) is given by

$$C^{(\gamma)}(\rho, \epsilon) = \frac{q}{\rho}\left[\gamma - \frac{q^3}{r^3}\right], \qquad \text{where} \quad r = r(\rho, \epsilon) = \sqrt{\rho^2 + q^2 + 2q\rho\cos\epsilon}.$$

As the Jacobian of the transformation $(\rho, \epsilon) \mapsto (x,y) = (q + \rho\cos\epsilon, \rho\sin\epsilon)$ has determinant equal to ρ, the stationary points of $C^{(\gamma)}(\rho, \epsilon)$ just correspond

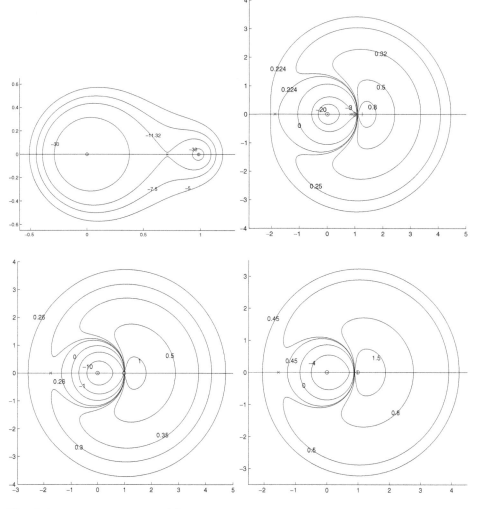

Fig. 9.4. The level curves of $C^{(\gamma)}(x, y)$. The Sun and the Earth are denoted by $\odot$ and $\oplus$, respectively. The saddle points are marked with $\times$ and the maximum point (present only for $0 < \gamma < 1$) with $+$. Top left: $\gamma = -0.5$. Top right: $\gamma = 0.8$. Bottom left: $\gamma = 1$. Bottom right: $\gamma = 1.5$. Reprinted from Gronchi (2009) with permission from Springer.

to those of $C^{(\gamma)}(x, y)$. For a given $\gamma \in \mathbb{R}$ we define

$$F(\mathcal{C}, \rho, \epsilon) = \mathcal{C}\frac{\rho}{q} - \gamma + \frac{q^3}{r^3(\rho, \epsilon)}.$$

The tangency points between the level lines of $C^{(\gamma)}(\rho, \epsilon)$ and the observation lines fulfill the equations

$$F(\mathcal{C}, \rho, \epsilon) = F_\rho(\mathcal{C}, \rho, \epsilon) = 0 \tag{9.38}$$

for each non-stationary value $\mathcal{C}$ of $C^{(\gamma)}(\rho, \epsilon)$, where

$$F_\rho(\mathcal{C}, \rho, \epsilon) = \frac{\mathcal{C}}{q} - 3\frac{q^3}{r^5}(\rho + q\cos\epsilon)$$

is the derivative of $F(\mathcal{C}, \rho, \epsilon)$ with respect to ρ. We can eliminate the dependence on $\mathcal{C}$ in (9.38) by considering the difference

$$F(\mathcal{C}, \rho, \epsilon) - \rho F_\rho(\mathcal{C}, \rho, \epsilon) = -\gamma + \frac{q^3}{r^3} + 3q^3\frac{\rho}{r^5}(\rho + q\cos\epsilon) \, ,$$

with $r = \sqrt{\rho^2 + q^2 + 2\rho q\cos\epsilon}$. The function $r^5(F - \rho F_\rho)$ in heliocentric rectangular coordinates becomes

$$\mathcal{G}(x, y) = -\gamma r^5 + q^3(4r^2 - 3qx), \qquad r = \sqrt{x^2 + y^2}.$$

We define the **singular curve** as the set

$$\mathcal{S} = \{(x, y) : \mathcal{G}(x, y) = 0\}.$$

Note that $\mathcal{S}$ contains all the points whose polar coordinates fulfill (9.38) plus $(x, y) = (0, 0)$. The shape of the singular curve for different values of γ, see Figure 9.5, is as follows. If $\gamma \neq 1$, the singular curve $\mathcal{S}$ contains a number of components, each a regular and simply closed curve:

- if $\gamma \leq 0$ it has a unique component, which is convex;
- if $0 < \gamma < 1$ it has two components inside and outside the zero circle;
- if $\gamma > 1$ it has a unique non-convex component intersecting the zero circle;
- if $\gamma = 1$ it has a self-intersection point at the observer position $(q, 0)$.

An even or an odd number of solutions

For $\gamma \leq 0$, eq. (9.30) has no real solution if $\mathcal{C} \geq 0$. Let us consider the polynomial $P(r)$ defined in (9.33). If $\mathcal{C} < 0$, from $P(0) < 0$ and $\lim_{r \to +\infty} P(r) = +\infty$ the number of roots of $P(r)$ in the interval $(0, +\infty)$, counted with their multiplicity, is odd and none of these roots is spurious.

Assuming $\gamma > 0$, let $r_0 = q/\sqrt[3]{\gamma}$ be the radius of the zero circle; we have

$$P(r_0) = \frac{\mathcal{C}^2 q^8}{\gamma^{8/3}}(1 - \gamma^{2/3}). \tag{9.39}$$

If $0 < \gamma < 1$ and $\mathcal{C} \neq 0$, from $P(0) < 0 < P(r_0)$ and $\lim_{r \to +\infty} P(r) = +\infty$ we obtain that in the interval $(0, r_0)$ the number of roots of $P(r)$ is odd, while in $(r_0, +\infty)$ it is even. By relation (9.30) the roots of $P(r)$ in $(0, r_0)$ are spurious iff $\mathcal{C} > 0$, those in $(r_0, +\infty)$ are spurious iff $\mathcal{C} < 0$. For $\mathcal{C} = 0$

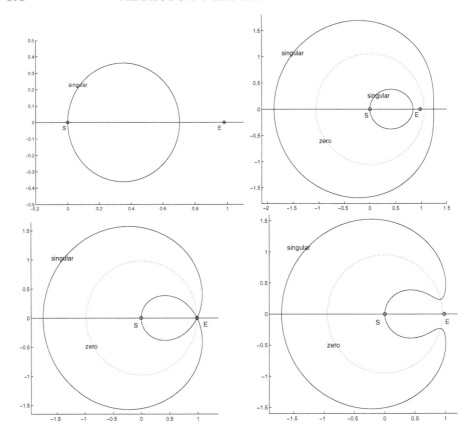

Fig. 9.5. Singular curve (continuous line) and zero circle (dotted). Top left: $\gamma = -0.5$. Top right: $\gamma = 0.8$. Bottom left: $\gamma = 1$. Bottom right: $\gamma = 1.1$. Note that the zero circle does not exist for $\gamma \leq 0$. Reprinted from Gronchi (2009) with permission from Springer.

the intersection problem (9.31) reduces to $r = r_0$, with ρ deduced from the geometric equation, and this solution is not spurious.

If $\gamma > 1$ and $\mathcal{C} \neq 0$, from $P(0), P(r_0) < 0$ and $\lim_{r \to +\infty} P(r) = +\infty$ we obtain that in the interval $(0, r_0)$ the number of roots of $P(r)$ is even, while in $(r_0, +\infty)$ it is odd. As in the previous case, the roots of $P(r)$ in $(0, r_0)$ are spurious iff $\mathcal{C} > 0$, those in $(r_0, +\infty)$ iff $\mathcal{C} < 0$. Since for $\gamma > 1$ we have $r_0 < q$, there is no solution for $\mathcal{C} = 0$ if $\cos \epsilon < \sqrt{q^2 - r_0^2}/q$, while if $\cos \epsilon \geq \sqrt{q^2 - r_0^2}/q$ the solutions are $(r, \rho) = \left(r_0, -q \cos \epsilon \pm \sqrt{q^2 \cos^2 \epsilon - (q^2 - r_0^2)} \right)$. We give a summary of this discussion in Table 9.2.

The bound on the solutions of the reduced problem implies that the solutions of the intersection problem cannot be more than three. In particular, for $(\gamma, \mathcal{C}, \epsilon)$ fulfilling (9.36) with $\gamma \neq 1$, when the number of solutions of (9.31) is even, they are two, when it is odd they are either one or three.

Table 9.2. *The table shows, for each value of $\gamma \neq 1$, the values of $\mathcal{C}$ allowing an even or an odd number of solutions of (9.31).*

	even	odd
$\gamma \leq 0$	$\mathcal{C} \geq 0$	$\mathcal{C} < 0$
$0 < \gamma < 1$	$\mathcal{C} > 0$	$\mathcal{C} \leq 0$
$\gamma > 1$	$\mathcal{C} \leq 0$	$\mathcal{C} > 0$

The limiting curve

Indeed the assertion by Charlier that the occurrence of alternative solutions depends only on the position of the observed body cannot be generalized to the Gauss method of preliminary orbit determination or to the modified Laplace method, with topocentric observations; in fact the position of the body defines different intersection problems for different $\gamma \in \mathbb{R}$. Actually for each fixed value of $\gamma \in \mathbb{R}$ we shall divide the reference plane into connected components such that, if a solution of an intersection problem lies in one of these components, then we know how many solutions occur in that problem, and all of them lie in the same component.

In the applications of this theory the parameters $\gamma, \mathcal{C}, \epsilon$ are computed from the three observations, thus there is no guarantee that assumption (9.36) holds. The failure of this assumption can occur for different reasons: mostly due to the unavoidable errors in the observations, and to the error made in considering three observations of different objects as belonging to the same.

For $\gamma \neq 1$ we define, with $r = \sqrt{x^2 + y^2}$, the sets

$$\mathcal{D}_2(\gamma) = \begin{cases} \emptyset & \text{if } \gamma \leq 0 \\ \{(x,y) : r > r_0\} & \text{if } 0 < \gamma < 1 \\ \{(x,y) : r \leq r_0\} & \text{if } \gamma > 1 \end{cases}$$

and $\mathcal{D}(\gamma) = \mathbb{R}^2 \setminus (\mathcal{D}_2(\gamma) \cup \{(q,0)\})$. To use a simpler notation, we shall suppress the dependence on γ in $\mathcal{D}(\gamma), \mathcal{D}_2(\gamma)$. For a fixed $\gamma \neq 1$, if we consider a point in $\mathcal{D}_2$ and if (9.36) holds for the parameters $(\gamma, \mathcal{C}, \epsilon)$ of the corresponding intersection problem, then there are two solutions of (9.31), both contained in $\mathcal{D}_2$. We shall also say that $\mathcal{D}_2$ is a region with two solutions of (9.31). Our aim is to divide the complementary set $\mathcal{D}$ into two connected regions, each with the same number of solutions of (9.31). Let $\mathfrak{S} = \mathcal{S} \cap \mathcal{D}$ be the portion of the singular curve $\mathcal{S}$ contained in $\mathcal{D}$. Note that $\mathfrak{S}$ is connected. In $\mathcal{D}$ the solutions of (9.31) are one or three, and the solutions lying on the singular curve have intersection multiplicity ≥ 2, therefore for each point $P \in \mathfrak{S}$ the related intersection problem must have three solutions.

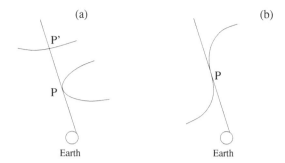

Fig. 9.6. We sketch the tangent intersection between an observation line and a level curve of $C^{(\gamma)}(x, y)$, giving rise to a residual point in the region $\mathcal{D}$ with an odd number of solutions. Case (a) is a generic situation, with P corresponding to a solution with multiplicity two and P′ corresponding to the third solution (the residual point). Case (b) is non-generic: P is a self-residual point, with intersection multiplicity equal to three. Reprinted from Gronchi (2009) with permission from Springer.

There are two cases, sketched in Figure 9.6 with labels (a), (b). Case (a) is the generic situation: we have $F_{\rho\rho}(\mathcal{C}, \bar{\rho}, \bar{\psi}) \neq 0$ for $(\bar{\rho}, \bar{\psi})$ corresponding to P and $\mathcal{C}$ such that $F(\mathcal{C}, \bar{\rho}, \bar{\psi}) = 0$, thus P corresponds to a solution of (9.31) with multiplicity two and there is another point P′ $\neq$ P corresponding to the third solution of (9.31). In case (b) we have $F_{\rho\rho}(\mathcal{C}, \bar{\rho}, \bar{\psi}) = 0$, so that in P the observation line is tangent to both the singular curve and to the level curve $C^{(\gamma)}(x, y) = \mathcal{C}$, and it corresponds to a solution with multiplicity three of the related intersection problem. For $\gamma \neq 1$ there are only two points of the reference plane, outside the x-axis, corresponding to solutions with multiplicity three (Gronchi 2009).

Let us fix $\gamma \neq 1$ and let $(\bar{\rho}, \bar{\psi})$ correspond to a point P $\in \mathfrak{S}$. If $F_{\rho\rho}(\mathcal{C}, \bar{\rho}, \bar{\psi}) \neq 0$, we call the **residual point** related to P the point P′ $\neq$ P lying on the same observation line and the same level curve of $C^{(\gamma)}(x, y)$ (see Figure 9.6 (a)). If $F_{\rho\rho}(\mathcal{C}, \bar{\rho}, \bar{\psi}) = 0$ we call P a **self-residual point**, i.e., we consider P as a residual point related to itself. We agree that the point $(x, y) = (q, 0)$, corresponding to the observer position, is the residual point related to $(x, y) = (0, 0)$, when the latter belongs to $\mathfrak{S}$. For $\gamma = 1$ each point of the singular curve has the observer position as residual point.

Let $\gamma \neq 1$. The limiting curve $\mathcal{L}$ is the set of all the residual points related to the points in $\mathfrak{S}$. By the symmetry of $\mathfrak{S}$ and of the level curves of $C^{(\gamma)}(x, y)$, the limiting curve is also symmetric with respect to the x axis. If the point $(q, 0)$ is in $\mathcal{L}$, it is not isolated. It has the following properties:

• (**separating property**): for $\gamma \neq 1$ the limiting curve $\mathcal{L}$ is a connected simple continuous curve, separating $\mathcal{D}$ into two connected regions $\mathcal{D}_1, \mathcal{D}_3$;

$\mathcal{D}_3$ contains the whole portion $\mathfrak{S}$ of the singular curve. If $\gamma < 1$ then $\mathcal{L}$ is a closed curve, if $\gamma > 1$ it is unbounded;

- **(transversality)**: the level curves of $C^{(\gamma)}(x, y)$ cross $\mathcal{L}$ transversely, except for the two self-residual points and where $\mathcal{L}$ meets the x axis;

- **(limiting property)**: for $\gamma \neq 1$ the limiting curve $\mathcal{L}$ divides the set $\mathcal{D}$ into two connected regions $\mathcal{D}_1, \mathcal{D}_3$: the points of $\mathcal{D}_1$ are the unique solutions of the corresponding intersection problem; the points of $\mathcal{D}_3$ are solutions of an intersection problem with three solutions, all lying in $\mathcal{D}_3$.

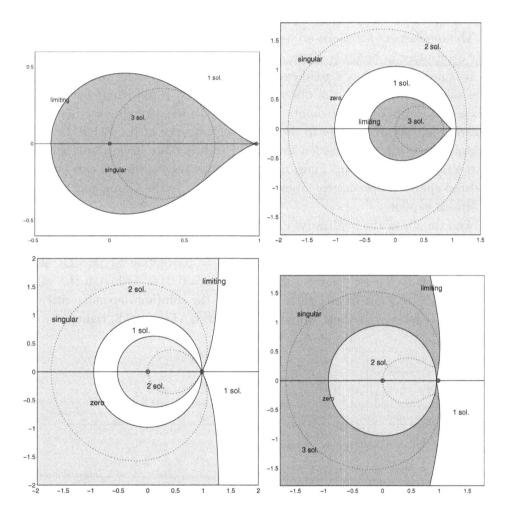

Fig. 9.7. Summary of the results on alternative solutions for all the qualitatively different cases. The regions with a different number of solutions are shaded: we use *light gray* for two solutions, *dark gray* for three solutions. Top left: $\gamma = -0.5$. Top right: $\gamma = 0.8$. Bottom left: $\gamma = 1$ (Charlier case). Bottom right: $\gamma = 1.1$. Reprinted from Gronchi (2009) with permission from Springer.

Figure 9.7 summarizes the results for all the significantly different cases: we distinguish among regions with a unique solution of (9.31) (white), with two solutions (light gray), and with three solutions (dark gray). For $\gamma = -0.5$ (top left) there are only two regions, with either one or three solutions. For $\gamma = 0.8$ (top right) in the region outside the zero circle there are two solutions; the region inside is divided by the limiting curve into two parts, with either one or three solutions. On the bottom left, we have the Charlier case ($\gamma = 1$), discussed in Section 9.6. For $\gamma = 1.1$ (bottom right) inside the zero circle there are two solutions; the region outside can contain either one or three solutions. In each case the singular curve separates the regions with alternative solutions into parts with only one solution each.

The numbers of solutions are generically different from those of Charlier: the solutions can be up to three, and up to two close to the opposition.

It is not easy to find a good example with three solutions; in many cases the solution nearest to the observer has distance ρ too small for the heliocentric two-body approximation to be applicable. A value $\rho = 0.01$ AU corresponds approximately to the sphere of influence of the Earth, i.e., the region where the perturbation from the Earth is more important than the attraction from the Sun. Thus, a solution with such a small ρ can be considered spurious because the approximation used in the Gauss and Laplace methods is poor.

An example of the qualitative changes when $\gamma_2 \neq 1$, based upon real data, is given by the first three nights of observation of asteroid 2002 AA$_{29}$: with $\epsilon \sim 79°$, there is only one solution with $\rho = 0.045$ (see Figure 9.8, left) which leads to a least squares solution with $\rho = 0.044$. Although the value of γ_2 is not very far from 1 the existence of the solution depends critically on $\gamma_2 \neq 1$: for $\gamma_2 = 1$ there would be no solution (Figure 9.8, right).

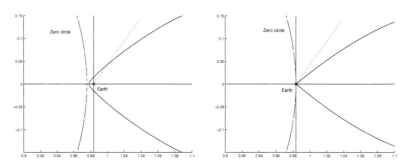

Fig. 9.8. For the preliminary orbit of 2002 AA$_{29}$ we show the relevant level curve ($\mathcal{C}_2 = 1.653$) and the zero circle; the observation direction is dotted. Left: using the actual value $\gamma_2 = 1.025$. Right: using a value of $\gamma_2 = 1$ not accounting for topocentric observations. Reprinted from Milani *et al.* (2008) with permission from Elsevier.

10

WEAKLY DETERMINED ORBITS

In most cases of population orbit determination the fit parameters $\mathbf{x}$ are just the initial conditions for an orbit, thus $\mathbf{x} \in \mathbb{R}^6$. We assume that there are at least enough observations to compute an attributable; if the arc is short, an approximate rank deficiency can occur, with order 1 or at most 2. In this chapter we discuss the special techniques which can be used to handle this kind of weak orbit determination, how to sample large confidence regions, the origin of such weakness, typically in a too short observations time span, and the impact on the quality of the orbit solution. This chapter is based on our papers (Milani *et al.* 2005c, Milani *et al.* 2007, Milani *et al.* 2008).

10.1 The line of variations

Given any point $\mathbf{x}$ in the space of initial conditions, we can compute the 6×6 normal matrix $C(\mathbf{x})$. Even when the inverse $\Gamma(\mathbf{x})$ cannot be computed, or is numerically unstable because of a large conditioning number, we can define

$$Z_L(\sigma) = \{\mathbf{y} | (\mathbf{y} - \mathbf{x}) \cdot C(\mathbf{x}) \, (\mathbf{y} - \mathbf{x}) \leq \sigma^2\}$$

which is an ellipsoid if $C(\mathbf{x})$ is positive definite. The main assumption in this chapter is that the observation information is above the minimum required to compute an attributable,[1] to the point that the matrix $C(\mathbf{x})$ has rank > 4. Still it could have rank 5, with a zero eigenvalue, or rank 6, but with a very small eigenvalue.

Ellipsoid long axis and weak direction

Let $\mathbf{v}_1(\mathbf{x})$ be an eigenvector of $C(\mathbf{x})$ with the smallest eigenvalue $\lambda_1(\mathbf{x}) \geq 0$:

$$C(\mathbf{x}) \, \mathbf{v}_1(\mathbf{x}) = \lambda_1(\mathbf{x})\mathbf{v}_1(\mathbf{x}).$$

---

[1] This implies that the number of scalar observations is > 4.

The other eigenvalues $\lambda_j(\mathbf{x}), j = 2, 6$, are assumed strictly $> \lambda_1(\mathbf{x})$. In the extreme case $\lambda_1(\mathbf{x}) = 0$, $Z_L(\sigma)$ is a cylinder with axis parallel to $\mathbf{v}_1(\mathbf{x})$. If $\lambda_1(\mathbf{x}) > 0$, the longest semiaxis of the confidence ellipsoid is in the direction $\mathbf{v}_1(\mathbf{x})$ and has length $k_1(\mathbf{x}) = 1/\sqrt{\lambda_1(\mathbf{x})}$ for $\sigma = 1$. $\mathbf{v}_1(\mathbf{x})$ is also an eigenvector of $\Gamma(\mathbf{x}) = C^{-1}(\mathbf{x})$ with the largest eigenvalue $1/\lambda_1(\mathbf{x}) = k_1^2(\mathbf{x})$, thus it defines the **weak direction** of the least squares fit.

If the fit were linear, the nominal solution $\mathbf{x}^*$ could be found from $C\,\mathbf{x}^* = D$ (see Section 5.2), without iterations, and the target function would be just

$$m\,Q(\mathbf{y}) = (\mathbf{y} - \mathbf{x}^*) \cdot C\,(\mathbf{y} - \mathbf{x}^*) + m\,Q^*.$$

Let H be the hyperplane spanned by the other eigenvectors $\mathbf{v}_j(\mathbf{x}), j = 2, \ldots, 6$. The tip $\mathbf{x}_1 = \mathbf{x}^* + k_1\,\mathbf{v}_1$ of the longest axis of the confidence ellipsoid is the minimum point of the target function restricted to the affine hyperplane $\mathbf{x}_1 + H$ and is also a local minimum point of the target function restricted to the sphere $|\mathbf{y} - \mathbf{x}^*| = k_1|\mathbf{v}_1|$. These properties, equivalent in the linear regime, are not equivalent in general.

The weak direction vector field

For each $\mathbf{x}$, let us select the eigenvector $\mathbf{v}_1(\mathbf{x})$ to be a unit vector. Then

$$\mathbf{F}(\mathbf{x}) = k_1(\mathbf{x})\,\mathbf{v}_1(\mathbf{x}), \tag{10.1}$$

with $k_1(\mathbf{x}) = 1/\sqrt{\lambda_1(\mathbf{x})}$, is a vector field. The unit eigenvector $\mathbf{v}_1$ is not uniquely defined, $-\mathbf{v}_1$ is also a unit eigenvector. Thus $k_1(\mathbf{x})\,\mathbf{v}_1(\mathbf{x})$ is what is called an **axial vector**, with well-defined length and direction but an arbitrary sign. However, given an axial vector field defined over a simply connected set, there is always a way to define a true vector field $\mathbf{F}(\mathbf{x})$ such that the function $\mathbf{x} \mapsto \mathbf{F}(\mathbf{x})$ is continuous. At an initial point we can select the sign according to some rule, e.g., such that the directional derivative of the semimajor axis a is positive in the direction $\mathbf{v}_1(\mathbf{x})$. Then the orientation is maintained by imposing that $\mathbf{v}_1(\mathbf{x})$ is continuous.

Problems could arise, for some value of $\mathbf{x}$, if either $\lambda_1(\mathbf{x}) = 0$ or the normal matrix $C(\mathbf{x})$ had its smallest eigenvalue of multiplicity 2. If the normal matrix is degenerate, see the discussion in Section 6.1. The exact equality of two eigenvalues does not occur generically, and even an approximate equality is rare in applications, as we have found from a large set of examples. Anyway, whenever the two smallest eigenvalues are of the same order of magnitude this method has serious limitations, as discussed in Section 10.3.

Given the vector field $\mathbf{F}(\mathbf{x})$ defined above, the differential equation

$$\frac{d\mathbf{x}}{d\sigma} = \mathbf{F}(\mathbf{x}) \qquad (10.2)$$

has a unique solution for each initial condition, because the vector field is smooth. If a nominal solution $\mathbf{x}^*$ has been found, let us select the initial condition $\mathbf{x}(0) = \mathbf{x}^*$, that is $\sigma = 0$ corresponds to the nominal solution, and let us denote with $\mathbf{x}(\sigma)$ the unique solution with such initial value. In the linear approximation, the solution $\mathbf{x}(\sigma)$ is one tip of the major axis of the confidence ellipsoid $Z_L(\sigma)$. Without approximations $\mathbf{x}(\sigma)$ is indeed curved and can be computed by numerical integration of the differential equation.

This approach could be used to define a special curve in the initial conditions space. However, such a definition may not be a constructive one, because of two problems. First, the definition cannot be used unless the nominal solution $\mathbf{x}^*$ is known. Second, there is a numerical instability in the algorithm to compute it. As an intuitive analogy, for weakly determined orbits the graph of the target function is like a very steep valley with an almost flat river bed at the bottom. The river valley is steeper than any canyon you can find on Earth; so steep that a very small deviation from the stream line sends you up the valley slopes by a great deal. This problem cannot be efficiently solved by brute force, that is by increasing the order or decreasing the step-size in the numerical integration of the differential equation. The only way is to slide down the steepest slopes until the river bed is reached again, which is the intuitive analog of the definition below.

Constrained differential corrections

Where the vector field $\mathbf{v}_1(\mathbf{x})$ is defined, the orthogonal hyperplane $H(\mathbf{x})$ is

$$H(\mathbf{x}) = \{\mathbf{y} \mid (\mathbf{y} - \mathbf{x}) \cdot \mathbf{v}_1(\mathbf{x}) = 0\}.$$

Given an initial guess $\mathbf{x}$, it is possible to compute one step of the differential corrections constrained to $H(\mathbf{x})$ by defining the $5 \times m$ matrix $B_{\mathbf{h}}(\mathbf{x})$ with the partial derivatives of the residuals with respect to the coordinates of the vector $\mathbf{h}$ of $H(\mathbf{x})$. Then the constrained normal equation is defined by the constrained normal matrix $C_{\mathbf{h}}$, which gives the restriction of the linear map associated to C to the hyperplane $H(\mathbf{x})$, and by the right-hand side $D_{\mathbf{h}}$, which is the projection of the vector D along the hyperplane:

$$C_{\mathbf{h}} = B_{\mathbf{h}}^T B_{\mathbf{h}}, \qquad D_{\mathbf{h}} = -B_{\mathbf{h}}^T \xi, \qquad C_{\mathbf{h}} \Delta \mathbf{h} = D_{\mathbf{h}}$$

with solution

$$\Delta\mathbf{h} = \Gamma_\mathbf{h}\, D_\mathbf{h}, \qquad \Gamma_\mathbf{h} = C_\mathbf{h}^{-1}$$

where the constrained covariance matrix $\Gamma_\mathbf{h}$ is not the restriction of the covariance matrix Γ to the hyperplane (see Section 5.4). The computation of $C_\mathbf{h}$, $D_\mathbf{h}$ can be performed by means of a rotation to a basis with $\mathbf{v}_1(\mathbf{x})$ as the first vector; then $C_\mathbf{h}$ is obtained by removing the first row and the first column of C, $D_\mathbf{h}$ by removing the first coordinate from D.

The constrained differential corrections process gives the corrected $\mathbf{x}' = \mathbf{x} + \Delta\mathbf{x}$ where $\Delta\mathbf{x}$ coincides with ΔH along $H(\mathbf{x})$ and has zero component along $\mathbf{v}_1(\mathbf{x})$. Then the weak direction $\mathbf{v}_1(\mathbf{x}')$ and the hyperplane $H(\mathbf{x}')$ are recomputed, and the next correction is constrained to $H(\mathbf{x}')$. This procedure is iterated until convergence.[2] At the convergence value $\overline{\mathbf{x}}$, $D_\mathbf{h}(\overline{\mathbf{x}}) = \mathbf{0}$, that is the right-hand side $D(\overline{\mathbf{x}})$ of the unconstrained normal equation is parallel to the weak direction $\mathbf{v}_1(\overline{\mathbf{x}})$. This equation is equivalent to the following property: the restriction of the target function to the hyperplane $H(\overline{\mathbf{x}})$ has a stationary point in $\overline{\mathbf{x}}$. The constrained corrections correspond to the intuitive idea of "falling down to the river".

Thus we can define the **line of variations** (LOV) as the set

$$\{\mathbf{x} \mid D(\mathbf{x}) = s\,\mathbf{v}_1(\mathbf{x}) \text{ for some } s \in \mathbb{R}\}, \qquad (10.3)$$

where the gradient of the target function is in the weak direction; if there is a nominal solution $\mathbf{x}^*$ with $D(\mathbf{x}^*) = 0$, it also belongs to the LOV. However, the LOV is defined independently of the existence of a local minimum of the target function. The definition by eq. (10.3) does not give the same curve as that resulting from the solutions of eq. (10.2), unless the problem is linear. For a discussion of different possible definitions of LOV, see (Milani *et al.* 2005c, Appendix A).

Parameterizing and sampling the LOV

The equation $D(\mathbf{x}) = s\,\mathbf{v}_1(\mathbf{x})$ corresponds to five scalar equations in six unknowns, thus it has generically a smooth one-parameter set of solutions, i.e., a differentiable curve. However, we have an analytical expression neither for the points of this curve nor for its parameterization (e.g., by the arc length).

An algorithm to compute the LOV by continuation from one of its points $\mathbf{x}$ is the following. The vector field $\mathbf{F}(\mathbf{x})$, deduced from the weak direction vector field $\mathbf{v}_1(\mathbf{x})$, is orthogonal to $H(\mathbf{x})$. A step in the direction of $\mathbf{F}(\mathbf{x})$, such

---

[2] In a numerical procedure, convergence is defined as having the last iteration with a small enough correction; in this context, the following properties are satisfied only approximately.

as an Euler step of the solution of the differential equation $dx/d\sigma = \mathbf{F}(\mathbf{x})$, that is $\mathbf{x}' = \mathbf{x} + \delta\sigma\,\mathbf{F}(\mathbf{x})$, does not provide another point on the LOV, unless the LOV itself is a straight line; this does not depend on the method employed to find a numerical solution of the differential equations (we normally use a second-order implicit Runge–Kutta–Gauss). However, $\mathbf{x}'$ will be close to another point $\mathbf{x}''$ on the LOV, which can be obtained by applying the constrained differential corrections algorithm, starting from $\mathbf{x}'$ and iterating until convergence, as shown in Figure 10.1.

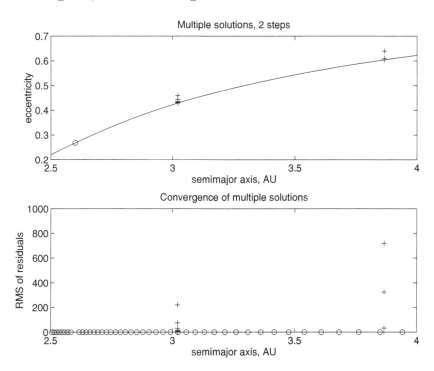

Fig. 10.1. The procedure to obtain multiple solutions; only two steps are shown projected onto the (a, e) plane. Top: starting from $\mathbf{x}^*$ (circle), the LOV solutions are obtained by propagation of eq. (10.2) followed by constrained differential corrections (each iteration is a cross); they converge to the "river" (continuous line), whose points have been computed by the same procedure with a much smaller step. Bottom: the RMS of the residuals is large at the starting point of each constrained differential corrections procedure, and rapidly converges towards the much smaller values obtained along the "river" line (circles). Reprinted from (Milani 1999) with permission from Elsevier.

If the LOV parameter of the starting point $\mathbf{x}$ is σ_0, we can set $\mathbf{x}'' = \mathbf{x}(\sigma_0 + \delta\sigma)$, which is an approximation (the value $\sigma_0 + \delta\sigma$ actually pertains to $\mathbf{x}'$). As an alternative, if we already know the nominal solution $\mathbf{x}^*$ and the local minimum value of the target function $Q(\mathbf{x}^*)$, we can compute the χ parameter as a function of the value of the target function at $\mathbf{x}''$ by $\chi = \sqrt{m \cdot [Q(\mathbf{x}'') - Q(\mathbf{x}^*)]}$. In the linear regime, the two definitions are related by $\sigma = \pm\chi$, but this is by no means the case in strongly nonlinear conditions. Thus we can adopt the definition $\sigma_Q = \pm\chi$, where the sign is taken to be the same as that of σ, for an alternative parameterization of the LOV. If we assume that the probability density at the initial conditions $\mathbf{x}$ is an exponentially decreasing function of χ, as in the Gaussian distributions of Chapter 3, then it is logical to terminate the sampling of the LOV at some value of χ, that is, to use the intersection of the LOV with the nonlinear confidence region $Z(b)$, where b is the maximum χ value.

This algorithm to compute the LOV can be used when a nominal solution $\mathbf{x}^*$ is known and when it is unknown, even non-existent. If $\mathbf{x}^*$ is known, then we can set $\mathbf{x}^* = \mathbf{x}(0)$ as the origin of the parameterization and proceed by using either σ or σ_Q as parameters for the other points computed with the alternating sequence of numerical integration steps and constrained differential corrections. Else, when a nominal solution is not available, we must first reach some point on the LOV by constrained differential corrections starting from some initial condition (a preliminary orbit). Once on the LOV, we can begin moving along it as was done from the nominal solution. In such cases, we set the LOV origin $\mathbf{x}(0)$ to whatever point $\overline{\mathbf{x}}$ of the LOV we have found first. Then we compute the other points as above and use the parameterization σ with arbitrary origin. The parameterization σ_Q cannot be computed point by point: it can be derived a posteriori.

10.2 Applications of the constrained solutions

There are two classes of applications. First, a single LOV solution can be used as intermediary for further orbit determination (or for identification). This may stabilize the procedure of orbit determination and/or identification, allowing us to increase its efficiency. Second, **multiple solutions** sampling the LOV can be computed in the attempt of representing all the possible orbits within some confidence region. For example, $2p+1$ LOV solutions $\mathbf{x}_k$ can be computed, with $-p \le k \le p$, with $\mathbf{x}_0$ the nominal solution if available, and with a fixed step $\Delta\sigma$ in the σ LOV parameter between two consecutive ones. Possibly the most important application of confidence region sampling along the LOV is *impact monitoring*, discussed in Chapter 12.

Orbit determination

A procedure for computing an orbit from the astrometric data, by the methods of this and the previous chapter, consists of the following steps, depending upon the quality control parameter Σ to be selected:

(i) if no orbit is already available, compute a preliminary orbit $\mathbf{x}$ with either the Gauss method or with the methods of Chapter 8;

(ii) compute constrained differential corrections by using $\mathbf{x}$ as first guess;

(iii) if constrained differential corrections converge to a LOV solution $\bar{\mathbf{x}}$ (with RMS of the residuals $\leq w \, \Sigma$, for some margin $w > 1$), then attempt differential corrections by using $\bar{\mathbf{x}}$ as first guess;

(iv) if the full differential corrections converge to a nominal solution $\mathbf{x}^*$ (with RMS of the residuals $\leq \Sigma$) then adopt this as orbit, with uncertainty described by its covariance;

(v) if the full differential corrections fail, then adopt $\bar{\mathbf{x}}$ as orbit if it has RMS $\leq \Sigma$;[3]

(vi) if the constrained differential corrections fail to converge, attempt full differential corrections with the preliminary orbit $\mathbf{x}$ as first guess.

After obtaining a least squares orbit, be it a nominal or just a LOV solution, we can apply the continuation algorithm of the previous section for multiple LOV solutions. By this procedure, it is possible to obtain a significantly larger number of orbits, which can be the starting point for the applications which follow; this increase results from

- LOV solutions in case nominal solutions are not available;
- nominal solutions computed starting from LOV orbits, when the iterations starting from the preliminary orbits diverge;
- multiple LOV solutions computed from the nominal ones;
- multiple LOV solutions computed from one of them, without a nominal.

Multiple ephemerides and recovery

Sampling along the LOV is a very useful tool when the predictions have a large uncertainty and are extremely nonlinear. This happens when the confidence region is very large, either at the initial epoch (because of very limited observational data) or at some subsequent time, after the propagation has stretched the confidence region preferentially in the along-track direction, as it is already clear from the model problem of Section 5.6.

A typical use of multiple solutions is to compute observation predictions, that is **ephemerides**. For each observation epoch t, we can compute the $2p+1$ points $\mathbf{y}_k = F(\mathbf{x}_k(t))$ on the celestial sphere, and plot the line joining these points, as in Figure 10.2. This method (Milani 1999) is comparable to

---

[3] In this case, the covariance provides uncertainty information but it does not define a confidence region, because the minimum value of the target function is not known (may not exist). There is also a problem in the use of the same quality control parameter Σ for orbits with a different number of free parameters, see Section 11.5.

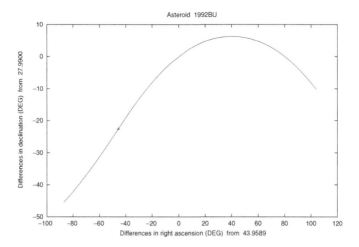

Fig. 10.2. Multiple solutions for the asteroid 1992 BU give multiple ephemerides 31 years before, when 4161 PLS was discovered. The actual observation of 4161 PLS is marked with a cross. Reprinted from Milani (1999) with permission from Elsevier.

that of the semilinear confidence boundary when the orbit determination at an epoch near the observations is quasi-linear, but can work even for orbits which are very poorly determined at all times. This method has been found useful to search for precoveries in plate archives (Boattini *et al.* 2001).

Multiple orbit identification

Identifications could be achieved by comparing multiple solutions for two asteroids, observed during short and widely separated arcs. As an example, in Figure 10.3 we show the already mentioned case of 4161 PLS = 1992 BU studied in this way. The two curves, plotted in the (a, e) plane (the inclination and node are typically better determined), are the multiple solutions computed for both single opposition orbits (the gaps correspond to the nominal). The two lines cross in only one point; we select, among the multiple solutions computed, the two which are closest to this intersection point. From them by the linear identification formula (7.1) we compute the first guess for the least squares fit to all the observations of both arcs; the differential corrections converge to an orbit, shown with a cross.

A better method computes orbit identification penalties from (7.2) for each couple of multiple solutions, finds the minimum of the $(2p+1)^2$ penalties, for a given couple of objects, and proposes the couple as identification if this minimum is below some control value. An effective method requires us to reduce the computational complexity for N objects below $\mathcal{O}(N^2(2p+1)^2)$.

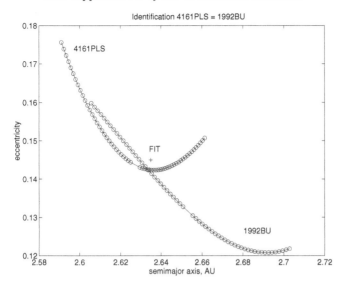

Fig. 10.3. Multiple solutions (up to $\sigma = 3$) for the asteroids 4161 PLS and 1992 BU: the projections of the LOV on the (a, e) plane have a single intersection point which is close to the least squares fit (with the observations of both arcs). These two asteroids could have been identified by this method, although they were actually identified with a different procedure (Sansaturio *et al.* 1996). Reprinted from Milani (1999) with permission from Elsevier.

Milani *et al.* (2005c, Section 5) discuss the systematic application of this class of methods to a large data set of asteroid orbits, with considerable success ($\simeq 1500$ confirmed identifications found in a single run).

Recursive attribution

In the linkage problem, after computing the preliminary orbits, e.g., with the methods of Section 8.5 and/or those of Section 8.6, the next step is to compute, starting from the preliminary orbits, least squares solutions. However, in most cases the observational data available are very limited even after the identification, e.g., just enough to compute two attributables. Thus constrained differential corrections are necessary as the first step, and in most cases the LOV solutions are the only ones achievable.

Figure 10.4 summarizes a hypothetical procedure of linkage for the Centaur (31824) Elatus: it was discovered in October 1999, given the designation 1999 UG$_5$, and then followed up until a good orbit could be computed, allowing us to attribute to it prediscovery observations from October 1998. The test consists in linking the data from the discovery night with the precovery data one year earlier, with the methods of Section 8.5. The figure shows the admissible region computed from the October 1999 attributable and the

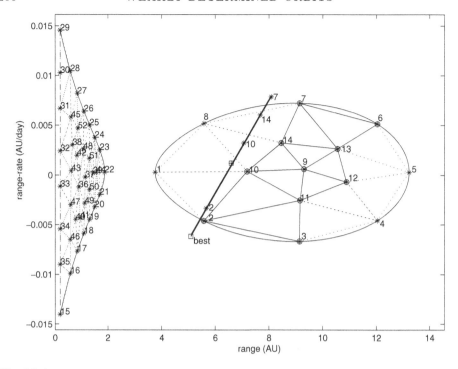

Fig. 10.4. (31824) For 1999 UG$_5$, linkage of the four discovery observations with one night of data one year earlier. The continuous lines join the nodes with identification penalty $K^i < 5^2$: none of them belongs to the first connected component of the admissible region, the one closer to the Earth. The nominal least squares solution, marked "best", is a hyperbolic orbit: it is not close to the true solution, which is marked with a crossed square near the node number 10. The LOV has been approximated by a straight line best fitting all the LOV points obtained in this way. Reprinted from Milani *et al.* (2005a) with permission from Elsevier.

Delaunay triangulation to create a swarm of virtual asteroids, with the same conventions as Figure 8.5. The nodes for which the attribution penalty is below some control have been used to compute preliminary orbits, in turn used as a first guess of constrained differential corrections: the points in the $(\rho, \dot{\rho})$ plane corresponding to the LOV solutions fit well to a straight line. The LOV also contains a nominal solution, which is hyperbolic: it has been computed by using Cartesian coordinates; it would not exist in Keplerian/equinoctial elements. Then the LOV solutions have been used to attribute another one-night arc from September 14, 1999, and this provides a three-night full least squares solution, marked with a crossed square in the figure, which is very close to the line fitting the LOV, not close to the two-night nominal solution.

The above example is extreme, with a very long time span between the first two attributables. Thus it is a strong confirmation of the feasibility of **recursive attribution**, by which the data from single nights, providing attributables, are added one by one. The procedure starts from linkage, which in fact, with the virtual asteroids method, is also an attribution.

Qualitative analysis

The sampling along the LOV is also useful to understand the situation whenever the orbit determination is extremely nonlinear. The problem of nonlinearity in orbit determination is too complex to be discussed in full generality here. We shall show the use of the LOV sampling as a tool to understand the geometry of the nonlinear confidence region in a single case study, in which there are multiple local minima of the target function.

The asteroid 1998 XB was discovered in December 1998, at an elongation $\simeq 93°$ from the Sun. The first orbit published by the Minor Planet Center, with observation time span $\Delta t \simeq 10$ days, had $a = 1.021$ AU. In the following days the orbit was repeatedly revised by the MPC, with semimajor axis gradually decreasing to 0.989 AU for $\Delta t \simeq 13$ days. Then, with observations extending Δt to 16 days, the semimajor axis jumped to 0.906 AU.

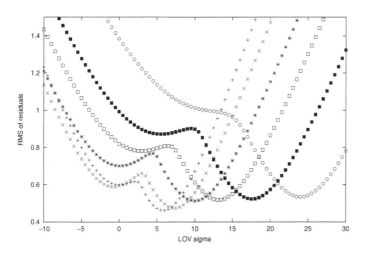

Fig. 10.5. The RMS of the residuals (in arcsec), for different observation data sets, as a function of the LOV parameter σ; $\sigma = 0$ has been fixed at one of the nominal solutions found for nine days. The lines are marked with plus signs (arc time span of nine days), crosses (10 days), stars (11 days), boxes (13 days), full boxes (14 days) and circles (16 days). Reprinted from Milani *et al.* (2005c) with permission from *Astronomy and Astrophysics*.

To understand this behavior we compute the LOV for different data sets, corresponding to $\Delta t = 9, 10, 11, 13, 14, 16$ days. Figure 10.5 shows a double minimum of the residuals RMS on the LOV; the secondary minimum moves in the direction of lower a as the data increase, but not as far as the location of the other minimum. The secondary minimum disappears with 16 days of data, then the differential corrections lead to the other solution.

As discussed in Chapter 9, the classical preliminary orbit methods can have two distinct solutions, especially when the elongation is below 116.5°. When applied with three observations selected in the shorter time spans, it can provide preliminary orbits close to both the primary and the secondary minimum. As an example, with data over 10 days we can compute with the Gauss method a preliminary orbit with $a = 0.900$, and from this a full least squares solution, which is a minimum of the target function, with $a = 0.901$ and RMS = 0.47 arcsec. There is also an alternative preliminary solution with $a = 1.046$, and from this we can compute another nominal solution, which is a local minimum, with $a = 1.032$ and RMS = 0.58 arcsec.

Indeed, if there were only three observations, both preliminary orbit solutions would correspond to very low RMS of residuals (not zero, because of planetary perturbations); the two local minima of the target function would be roughly at the same level. As the amount of data increases, the RMS increases for both the nominal solutions, but for one more than for the other. The existence of two local minima implies that there needs to be a **saddle point**, where the Hessian matrix of second derivatives has some negative eigenvalue, somewhere in the $\mathbf{x}$ space of orbital elements.[4] However, the LOVs for which the RMS is plotted in Figure 10.5 join two local minima without necessarily passing from the saddle, and even if they did, the LOV computation algorithm does not provide a way to find this out. Indeed, the normal matrices $C(\mathbf{x})$ used in the computation of the multiple solutions have eigenvalues ≥ 0; only the normal matrix of Newton's method, C_{new} of eq. (5.3), can have negative eigenvalues. Saddle points actually exist in asteroid orbit determination, but to find them it is necessary to use at least some approximation for the second partial derivatives of the residuals, and a more sophisticated optimization method (Sansaturio *et al.* 1996).

10.3 Selection of a metric

The eigenvalues λ_j of the normal matrix C are not invariant under a coordinate change. Thus a different weak direction and a different LOV would be obtained by using some other coordinates $\mathbf{y} = \mathbf{y}(\mathbf{x})$. This is true even when the coordinate change is linear $\mathbf{y} = S\,\mathbf{x}$: the normal matrix is transformed as $C_{\mathbf{y}} = \left[S^{-1}\right]^{T} C_{\mathbf{x}}\, S^{-1}$ and the eigenvalues are the same if $S^{-1} = S^{T}$, that is if the change of coordinates is isometric. Otherwise, the eigenvalues in

---

[4] Actually, the existence of saddle points may depend upon the coordinates used: as in Figure 10.4 the minimum is outside the $e = 1$ boundary, the same could occur for a saddle point.

the **y** space are not the same, and the eigenvectors are not the image by S of the eigenvectors in the **x** space. Thus the weak direction and the LOV in the **y** space do not correspond by S^{-1} to the weak direction and the LOV in the **x** space. A special case is scaling, a transformation changing the units along each axis, represented by a diagonal matrix S (see Section 6.4).

Coordinates to express initial conditions

A non-exhaustive list of coordinates used in orbit determination is:

- Cartesian heliocentric coordinates (position, velocity);
- cometary elements $(q, e, I, \Omega, \omega, t_p,$ with t_p the time of perihelion passage);
- Keplerian elements $(a, e, I, \Omega, \omega, \ell,$ with ℓ the mean anomaly);
- equinoctial elements $(a, h = e \sin(\varpi),$ $k = e \cos(\varpi),$ $p = \tan(I/2) \sin(\Omega),$ $q = \tan(I/2) \cos(\Omega),$ $\lambda = \ell + \varpi,$ with $\varpi = \Omega + \omega);$
- attributable elements $(\alpha, \delta, \dot{\alpha}, \dot{\delta}, \rho, \dot{\rho}).$

If the coordinate change is nonlinear, as it is for transformations between any two of the five types of orbital elements listed above, the covariance is transformed by the standard formula (5.5) with the Jacobian matrix

$$\mathbf{y} = \Phi(\mathbf{x}), \quad S(\mathbf{x}) = \frac{\partial \Phi}{\partial \mathbf{x}}(\mathbf{x}), \quad \Gamma_{\mathbf{y}} = S(\mathbf{x}) \, \Gamma_{\mathbf{x}} \, S(\mathbf{x})^T$$

and the constrained differential correction $\Delta \mathbf{y}$ can be computed accordingly.

If the computations are actually performed in the **x** coordinates, once the constrained differential correction $\Delta \mathbf{y}$ has been computed, we need to pull it back to the coordinates **x**. If $\Delta \mathbf{y}$ is small, as is typically the case when taking modest steps along the LOV, this can be done linearly:

$$\mathbf{x}' = \mathbf{x} + \left[\frac{\partial \Phi}{\partial \mathbf{x}}(\mathbf{x}) \right]^{-1} \Delta \mathbf{y}.$$

When the constrained differential corrections are large, as is likely to be when the initial point is not near the LOV, the corrections $\Delta \mathbf{y}$ must be pulled back to **x** nonlinearly, that is $\mathbf{x}' = \Phi^{-1}(\mathbf{y} + \Delta \mathbf{y}).$

Table 10.1 shows possible scalings for the five coordinate systems above, as proposed by Milani *et al.* (2005c). Cartesian position coordinates are measured in astronomical units (AU), but they are scaled as relative changes. Angle variables are measured in radians, but they are scaled in revolutions. Velocities in Cartesian coordinates are expressed in AU/day and are scaled as relative changes; angular velocities are scaled by the Earth mean motion $n_\oplus$. The range rate is scaled by $n_\oplus$ to get a parameter with the dimension of a length, thus commensurable to the range.

Table 10.1. Units and LOV scalings for different elements. r and v are the heliocentric distance and velocity. $n_\oplus \simeq k$ is the Earth mean motion in rad/day; $Z = 2\pi q^{3/2} n_\oplus^{-1} (1-e)^{-1/2}$ is a characteristic time for an orbit with large e.

Cartesian	x	y	z	v_x	v_y	v_z
Units	AU	AU	AU	AU/d	AU/d	AU/d
Scaling	r	r	r	v	v	v
Cometary	q	e	I	Ω	ω	t_p
Units	AU	–	rad	rad	rad	d
Scaling	q	1	π	2π	2π	Z
Keplerian	a	e	I	Ω	ω	ℓ
Units	AU	–	rad	rad	rad	rad
Scaling	a	1	π	2π	2π	2π
Equinoctial	a	h	k	p	q	λ
Units	AU	–	–	–	–	rad
Scaling	a	1	1	1	1	2π
Attributable	α	δ	$\dot{\alpha}$	$\dot{\delta}$	ρ	$\dot{\rho}$
Units	rad	rad	rad/d	rad/d	AU	AU/d
Scaling	2π	π	$n_\oplus$	$n_\oplus$	1	$n_\oplus$

Comparison of different LOVs

With the list of coordinates in Table 10.1, each with and without scaling, we can select 10 different LOVs. The question is to select the most effective for a specific usage. This question is complex, but we can state two rules of thumb. If the arc drawn on the celestial sphere by the apparent asteroid position is small, e.g., $\leq 1°$, the orbit determination is less nonlinear in the coordinate systems representing instantaneous initial conditions, such as the Cartesian and the attributable elements. In the latter the angular variables $\alpha, \delta, \dot{\alpha}, \dot{\delta}$ are well determined while $\rho, \dot{\rho}$ are very poorly determined.

On the contrary, orbital elements solving exactly the two-body problem perform better in orbit determination whenever the observed arc is comparatively wide, e.g., tens of degrees. The cometary elements avoid the discontinuity at the $e = 1$ boundary, the equinoctial ones avoid the coordinate singularity for $e = 0$ and for $I = 0$. The Keplerian elements are strongly nonlinear, because of nearby coordinate singularities for $e \simeq 0$, $e \simeq 1$ and $I \simeq 0$, thus they are not always suitable for orbit determination. Equinoctial elements modified by replacing a with n are especially suitable for orbit identification (see Section 7.4).

Figure 10.6 shows a comparison of the LOVs computed with different

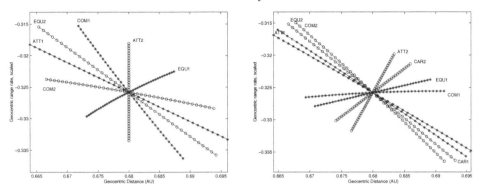

Fig. 10.6. For the asteroid 2004 FU$_4$ the computation of the LOV, by using only the first 17 observations, in different coordinates, without scaling on the left, with scaling on the right; the label denotes the coordinate system used and whether the line represents either the ordinary LOV or the second LOV. The Cartesian and attributable LOVs are indistinguishable on this plot and so only the attributable LOV is depicted. Reprinted from Milani *et al.* (2005c) with permission from *Astronomy and Astrophysics*.

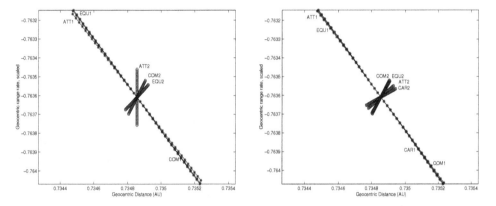

Fig. 10.7. For the asteroid 2002 NT$_7$ the computation of the LOV, by using only the first 113 observations, in different coordinates, without scaling on the left, with scaling on the right. The Cartesian and attributable LOVs are indistinguishable. Reprinted from Milani *et al.* (2005c) with permission from *Astronomy and Astrophysics*.

coordinate systems, without and with the scaling defined in Table 10.1, in the case of asteroid 2004 FU$_4$, observed only over a time span of $\simeq 3$ days, with an arc of only $\simeq 1°$. The data are projected on the $(\rho, \dot{\rho})$ plane, with $\dot{\rho}$ scaled by $n_\oplus$. For each coordinate system we show both the LOV, sampled with 41 VAs in the interval $-1 \leq \sigma_Q \leq 1$, and the *second LOV*, defined as the LOV but with the *second largest* eigenvalue λ_2 of the normal matrix and the corresponding eigenvector $\mathbf{v}_2$. The dependence of the LOV on the coordinates is very strong in this case. Note that the LOV of the Cartesian and attributable coordinates is closer to the second LOV, rather than to the first LOV, of the equinoctial coordinates. In such cases, with a very short observed arc, the confidence region has a two-dimensional spine, and the selection of a LOV in the corresponding plane is quite arbitrary. For example, in scaled Cartesian coordinates, the ratio of the two largest semiaxes of the confidence ellipsoid is $\simeq 2.4$. Then the best strategy to sample the confidence region would be either to use a number of LOVs, like in the figures, or to use a fully two-dimensional sampling, as in Section 8.2.

Figure 10.7 shows the comparison of the LOVs in the case of asteroid 2002 NT$_7$ when the available observations spanned 15 days, forming an arc almost $9°$ wide. In this case the ratio of the two largest semiaxes (in scaled Cartesian) is $\simeq 7.3$ and the LOVs computed with different coordinates are very close. As the confidence region becomes smaller, but also narrower, the long axis becomes less dependent upon the metric. Note that the attributable and the Cartesian coordinates have LOVs quite close in all cases. This can be understood knowing that the $(\rho, \dot{\rho})$ plane of the attributable coordinates corresponds exactly to a plane in Cartesian coordinates.

Uncertainty of curvature

Given the explicit formulae (9.2) and (9.3) we can compute the covariance matrix of the quantities $(\kappa, \dot{\eta})$ by propagation of the covariance matrix of the angles and derivatives with the matrix of partial derivatives for κ, $\dot{\eta}$ with respect to the 6-vector $V = (\alpha, \delta, \dot{\alpha}, \dot{\delta}, \ddot{\alpha}, \ddot{\delta})$

$$\Gamma_{(\kappa, \dot{\eta})} = \frac{\partial(\kappa, \dot{\eta})}{\partial V} \Gamma_V \left[\frac{\partial(\kappa, \dot{\eta})}{\partial V} \right]^T. \tag{10.4}$$

The covariance matrix Γ_V for the angles and their first and second derivatives is obtained by the procedure of least squares in a fit to the individual observations as a quadratic function of time. The partials of κ and $\dot{\eta}$ are given in (Milani *et al.* 2008, Section 5.1): we list just the last four of these partials, the 2×2 matrix $\partial(\kappa, \dot{\eta})/\partial(\ddot{\alpha}, \ddot{\delta})$, because they contribute to the principal part of the covariance of $(\kappa, \dot{\eta})$ for short arcs, as shown below.

$$\frac{\partial \kappa}{\partial \ddot{\alpha}} = -\frac{\dot{\delta} \cos \delta}{\eta^3}; \quad \frac{\partial \kappa}{\partial \ddot{\delta}} = \frac{\dot{\alpha} \cos \delta}{\eta^3}; \quad \frac{\partial \dot{\eta}}{\partial \ddot{\alpha}} = \frac{\dot{\alpha} \cos^2 \delta}{\eta}; \quad \frac{\partial \dot{\eta}}{\partial \ddot{\delta}} = \frac{\dot{\delta}}{\eta}. \tag{10.5}$$

We use a full computation of the covariance matrix to assess the significance of curvature with the formula

$$\chi^2 = \begin{bmatrix} \kappa \\ \dot{\eta} \end{bmatrix}^T \Gamma_{(\kappa, \dot{\eta})}^{-1} \begin{bmatrix} \kappa \\ \dot{\eta} \end{bmatrix} \tag{10.6}$$

and we assume that the curvature is *significant* if $\chi^2 > \chi^2_{min}$.

The infinite distance limit

Low values of $\mathcal{C}$, see eq. (9.30), can occur in two ways: near the zero circle and for large values of both ρ and r. The size of the deviations from a great circle will depend upon the length of the observed arc (both in time Δt and in arc length $\sim \eta \Delta t$). Thus for short observed arcs it may be the case that the curvature is not significant, and the preliminary orbit algorithms will yield orbits which may fail as starting guesses for differential corrections.

We will now focus on the case of distant objects. We would like to estimate the magnitude of the uncertainty in the computed orbit with respect to the small parameters ν, τ, b where ν is the astrometric accuracy of the individual observations (in radians) and $\tau = n_\oplus \Delta t$, $b = q_\oplus/\rho$ are small for short observed arcs and for distant objects, respectively. Note that the proper motion η for $b \to 0$ has principal part $n_\oplus b$ – the effect of the motion of the Earth. The uncertainty in the angles (α, δ) and their derivatives can be estimated as follows (Crawford *et al.* 1930, page 68):

$$
\begin{aligned}
\text{RMS}(\alpha) \simeq \text{RMS}(\delta) &= \mathcal{O}(\nu), \\
\text{RMS}(\dot\alpha) \simeq \text{RMS}(\dot\delta) &= \mathcal{O}(\nu\tau^{-1}), \\
\text{RMS}(\ddot\alpha) \simeq \text{RMS}(\ddot\delta) &= \mathcal{O}(\nu\tau^{-2}).
\end{aligned}
$$

The uncertainty of the curvature components $(\kappa, \dot\eta)$ should be estimated by the propagation formula (10.4) but it can be shown that the uncertainty of $(\alpha, \delta, \dot\alpha, \dot\delta)$ contributes with higher order terms. Thus we use the estimates based upon the partials in (10.5)

$$
\frac{\partial(\kappa, \dot\eta)}{\partial(\ddot\alpha, \ddot\delta)} = \left[\begin{array}{cc} \mathcal{O}(b^{-2})\, n_\oplus^{-2} & \mathcal{O}(b^{-2})\, n_\oplus^{-2} \\ \mathcal{O}(1) & \mathcal{O}(1) \end{array} \right]
$$

and obtain

$$
\Gamma_{(\kappa, \dot\eta)} = \nu \left[\begin{array}{cc} \mathcal{O}(b^{-4}\tau^{-2}) & \mathcal{O}(b^{-2}\tau^{-2})n_\oplus^2 \\ \mathcal{O}(b^{-2}\tau^{-2})n_\oplus^2 & \mathcal{O}(\tau^{-2})n_\oplus^4 \end{array} \right].
$$

To propagate the covariance to the variables $(\rho, \dot\rho)$ we use the implicit equation connecting $\mathcal{C}$ and ρ obtained by eliminating r from (9.4) and (9.30):

$$
F(\mathcal{C}, \rho) = \mathcal{C}\,\frac{\rho}{q} + \frac{q^3}{(q^2 + \rho^2 + 2q\rho\cos\varepsilon)^{3/2}} - 1 + \Lambda_n = 0. \tag{10.7}
$$

For $b \to 0$ we have $\mathcal{C}\, b^{-1} \to 1 - \Lambda_n$; thus $\mathcal{C} \to 0$ and is of the same order as b. Although $\mathcal{C}$ depends upon all the variables $(\alpha, \delta, \dot\alpha, \dot\delta, \ddot\alpha, \ddot\delta)$, by the approximation with the Laplace method (see Section 9.4) $\mathcal{C} \simeq C$. C is given by eq. (9.27); it contains $\kappa\eta^2$ and has an uncertainty mostly depending upon the uncertainty of κ and, ultimately, upon the difficulty in estimating the second derivatives of the angles. Thus we can use the derivatives of the implicit function $\rho(\kappa)$; assuming $\cos \varepsilon, \eta, \hat{\mathbf{n}}$ to be constant and keeping only the term of lowest order in q/ρ, we find

$$\frac{\partial \rho}{\partial \kappa} = -\frac{\eta^2 \, q^4}{\mu \, \hat{\mathbf{q}}_\oplus \cdot \hat{\mathbf{n}}} \frac{\rho}{q_\oplus \, C} + \mathcal{O}\left(\frac{q^3}{\rho^3}\right) = q_\oplus \, \mathcal{O}(1).$$

In the same way from (9.28) we deduce $\dot{\eta} = n_\oplus^2 \, \mathcal{O}(b)$ and obtain the estimates

$$\frac{\partial \dot{\rho}}{\partial \kappa} = n_\oplus \, q_\oplus \, \mathcal{O}(1), \qquad \frac{\partial \dot{\rho}}{\partial \dot{\eta}} = \frac{q_\oplus}{n_\oplus} \, \mathcal{O}(b^{-2}).$$

For the covariance matrix,

$$\Gamma_{(\rho,\dot{\rho})} = \frac{\partial(\rho,\dot{\rho})}{\partial(\kappa,\dot{\eta})} \, \Gamma_{(\kappa,\dot{\eta})} \, \left[\frac{\partial(\rho,\dot{\rho})}{\partial(\kappa,\dot{\eta})}\right]^T,$$

we compute the main terms of highest order in b^{-1}, τ^{-1} as

$$\Gamma_{(\rho,\dot{\rho})} = \nu \, b^{-3} \, \tau^{-2} \begin{bmatrix} q_\oplus^2 \, \mathcal{O}(1) & q_\oplus^2 \, n_\oplus \, \mathcal{O}(1) \\ q_\oplus^2 \, n_\oplus \, \mathcal{O}(1) & q_\oplus^2 \, n_\oplus^2 \, \mathcal{O}(1) \end{bmatrix}. \qquad (10.8)$$

In conclusion, if $(\rho, \dot{\rho})$ are measured in the appropriate units (AU for ρ and $n_\oplus$ AU for $\dot{\rho}$) their uncertainties are of the same order: this confirms the scaling of Table 10.1. In the scaled $(\rho, \dot{\rho})$ plane the weak direction, thus the LOV, can be in any direction, as can be seen from Figures 8.5 and 10.4.

10.4 Surface of variations

When the confidence region is not elongated in one direction much more than in the others, as in the example of Figure 10.6, whatever LOV we select may not be representative of the entire confidence region. If we use for a short arc of observations the attributable elements $(A, \rho, \dot{\rho})$, where A is the attributable, the confidence region is a "thin" shell surrounding a subset of the admissible region (see Section 8.4). We define as **surface of variations** the set S of the points where the target function has a local minimum with respect to changes of A, for each fixed $(\rho, \dot{\rho})$, with minimum RMS of the residuals below some control Σ. S is, generically, a two-dimensional manifold. Under the conditions assumed in this chapter, that is when there is little information beyond A, S is parameterized by $(A(\rho, \dot{\rho}), \rho, \dot{\rho})$, defined on a subset B of the $(\rho, \dot{\rho})$ plane; B is an open set, not necessarily connected. Then the surface S can be computed point by point: for each $(\rho_0, \dot{\rho}_0)$ we can correct only A, i.e., perform "doubly constrained" differential corrections, with normal equation

$$C_A \, \Delta A = D_A, \quad C_A = B_A^T \, B_A, \quad D_A = -B_A^T \, \xi, \quad B_A = \partial\xi/\partial A. \qquad (10.9)$$

If these corrections converge to a point of minimum $A(\rho_0, \dot\rho_0)$, and if the RMS of the residuals at this minimum is $< \Sigma$, the point $(A(\rho_0, \dot\rho_0), \rho_0, \dot\rho_0)$ belongs to the surface of variations, and $(\rho_0, \dot\rho_0)$ belongs to B.

Since the attributable elements are well defined also for hyperbolic orbits, B does not need to be a subset of the admissible region $\mathcal{D}(A)$. Indeed Figures 8.5 and 10.4 clearly show that B pokes out of the admissible region; in the latter case it is clear that the LOV would extend much further, if it was continued from the LOV points found by recursive attribution. Thus the choice depends upon the purpose: if we are aiming at sampling as much as possible the confidence region, even the hyperbolic orbits, a sampling of B is more useful than a sampling of $\mathcal{D}(A)$. On the contrary, if the goal is to discover the largest number of real objects, we can take into account that the number density of objects being discovered with $e \gg 1$ is very small (possibly zero, since none is known) and use a sampling of $\mathcal{D}(A) \cap B$, which can be obtained by discarding the nodes of a triangulation of $\mathcal{D}(A)$ for which the doubly constrained differential corrections give a residual RMS too high.

Thus to compute the surface of variations it is not required to compute the admissible region: we can start from a set of points sampling the $(\rho, \dot\rho)$ plane in any convenient way, e.g., a rectangular grid.[5] This and closely related methods are widely in use, e.g., (Chesley 2005, Tommei 2005). Another class of methods selects sample points in a two-dimensional space at random: this could be done in the space $(\rho, \dot\rho)$, but also in the space (ρ_1, ρ_2) of the distances at two epochs t_1, t_2; then the sample orbits could be selected according to some criteria, like discarding the hyperbolic ones but also preferentially exploring some portion of the phase space (Virtanen *et al.* 2001).

10.5 The definition of discovery

The quality of the least squares orbits improves as new observations, extending the observed arc time span, are added. The problem is how to find an algorithm to classify the observed arcs in quality classes with some predictive value on the information content of the orbit. This is connected with the definition of discovery: how many observations are enough to consider

---

[5] This method is described by Tholen and Whiteley in an unpublished paper. It is implemented in the free software KNOBS, available from D. Tholen, IfA, University of Hawaii, with the capability of sampling $\mathcal{D}(A) \cap B$ by discarding grid points corresponding to $e > 1$.

that a new Solar System object has been discovered? An obvious requirement would be to have enough information to be able to decide the nature of the object, e.g., to discriminate between a near-Earth asteroid (NEA), a main belt asteroid (MBA), a Jupiter trojan, a trans-neptunian object (TNO), a long periodic comet, and so on. In the definition of discovery there are a number of legal and science policy aspects, which are interesting, but not to be discussed here: we shall only give the mathematical background which should be used to build one such definition on rigorous grounds.

In modern surveys, the observations of Solar System objects are singled out among star images as moving objects: a number m of digital images of the same portion of the sky is taken within a short interval, on the same night,[6] and they are "digitally blinked", that is, computer programs remove the images stationary with respect to the stellar reference frame and assemble the transients into groups which could belong to a single moving object. One such group is called a **tracklet**, containing astrometric observations assembled without computing an orbit, by either a linear fit or a quadratic one with upper limits on the curvature (Kubica *et al.* 2007). Since $m \geq 2$, from a tracklet it is always possible to compute an attributable, but in most cases there is little, if any, curvature information.

We would like to capture in a definition an observed arc that does not allow us to compute a useful least squares orbit. Unfortunately such a definition would not be a clear-cut and operational one: the orbits computed from a given observational data set depend upon the algorithms used. For example, the complex procedure presented in Section 10.2 often succeeds in computing some orbit from a very limited data set, but then it is often a LOV orbit, or anyway with a very badly conditioned covariance matrix.

To provide a definition which is computable in all cases and is independent of the methods used in the orbit determination, we define as **too short arc** (TSA) an observed arc with no significant curvature, measured by eq. (10.6). The main problems with such a definition are two: the choice of the control value χ^2_{min}, and the fact that some observed arcs may contain enough information for a significant third derivative of the angles with respect to time. If the latter is the case, the residuals of a quadratic fit have a characteristic Z shape (Milani *et al.* 2007, Figure 1), which is significant if the standard deviation of the (normalized) residuals is larger than some RMS$_{min}$. Such definitions depend upon the error model used, see Section 5.8; the covariance matrix used in (10.6) contains the weights, and the RMS of the residuals after a quadratic fit needs to be the normalized one.

---

[6] Typical time intervals are between 15 min and 2 hours.

Because tracklets are typically formed with observations within the same night, most of them are also TSA. In some cases an observed arc may already be the result of the identification of ≥ 2 tracklets, and still be a TSA; this is often the case for TNO.

We call an **arc of type N** an observed arc which can be split into exactly N disjoint TSA, in such a way that each couple of TSA consecutive in time, if joined, would show a significant curvature. To obtain in all cases a unique value for the arc type it is necessary to specify the method by which observed arcs with significant curvature are to be split. A *recursive procedure* to compute the arc type is as follows: if the arc has significant curvature and/or a significant Z shape, it is split into two arcs by selecting the largest time gap between two observations. If the two subarcs have no significant curvature and no significant Z shape, the arc type is 2. Otherwise, the same procedure is applied to the two subarcs, and the arc type is the sum of the arc types of the subarcs. The recursion terminates because the number of observations in the subarcs decreases, and a subarc with < 3 observations can have neither curvature nor Z shape.

This definition, with predictive value on the orbit quality for all orbital classes of objects, should replace the currently used definition of *M-nighter*, an arc with observations belonging to M distinct nights. For TNOs, two nights of observations in most cases form an arc of type 1. For NEAs discovered near the Earth a single night of observations often is an arc of type ≥ 2, then an orbit with moderate uncertainty can be computed for a one-nighter.

Test of possible definitions

Milani *et al.* (2007) discusses the outcome of large-scale tests of possible definitions of discovery based on the arc type. We have used all the public asteroid astrometry available in March 2006: 9.2 million observations, including 185 296 observed arcs with time span < 190 days (single opposition). The values $\chi^2_{min} = 9$ and $\mathrm{RMS}_{min} = 4$ were used.[7] The error model for weighting the astrometry was based on (Carpino *et al.* 2003). A summary of the results is presented graphically in Figure 10.8. Going from type 2 to type 3 there is a sharp improvement in the quality of the orbit, allowing us in almost all cases to discriminate between MBA and NEA. As shown by Milani *et al.* (2007, Table IV), the arc type 3 does not correspond to three-nighters, in particular for MBA three *non-consecutive* nights are required.

A formal decision on the definition of discovery is not just mathematics; however, the above results suggest the mathematical properties which should

---

[7] Weaker definitions with lower controls gave less satisfactory results.

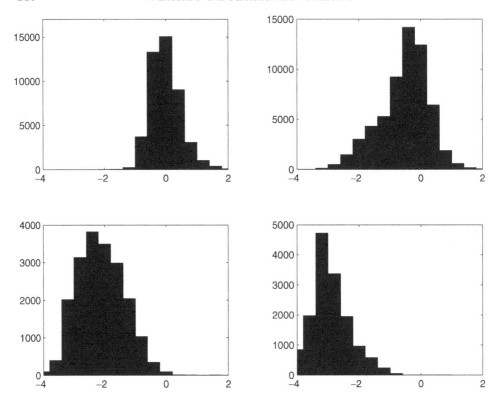

Fig. 10.8. Distribution of the RMS of the perihelion ($\log_{10}$ RMS(q) in AU) for observed arcs of different types. Top: type 1 (left), type 2 (right). Bottom: type 3 (left), type 4 (right). Reproduced with permission of Springer from Milani *et al.* (2007).

be required from a data set to form a discovery. A **discovery** of a Solar System object should include an arc of type N with $N \geq 3$. Moreover, the information supplied should contain a unique least squares orbit (full, with six free parameters), fitting the data with residuals compatible with the current error model. The data should also contain enough photometric information to fit an absolute magnitude, constraining the size of the object.

The object also needs to be new, not yet discovered according to the same definition, but this becomes complicated, both from the legal and the mathematical point of view, if the same discovery contains data from different observatories. Even the definition of arc type needs to be changed in case two tracklets, from different observatories, are overlapping in time, to take into account the topocentric corrections in a smooth way. One way is to split the arc by tracklets, not by observations. Anyway it is necessary to use information on the time when the data have been made public.

11

SURVEYS

This chapter is devoted to population orbit determination, that is not just computing the orbit for a single object, but compiling a catalog of orbits given a large number of observations. A survey is a project aiming at collecting observations of the largest and most representative sample of objects possible. We deal here only with the case in which the target population belongs to the Solar System; of course an astronomical survey may target simultaneously extrasolar populations. We deal with Earth satellites in Section 8.7. This chapter is based on our papers (Milani *et al.* 2005a, Milani *et al.* 2008, Milani *et al.* 2006) and ongoing research, in particular that in preparation for **Pan-STARRS**, a next-generation survey.

11.1 Operational constraints of Solar System surveys

The following three arguments should be taken into account in the definition of an identification/orbit determination procedure for a modern sky survey.

First, **Moore's law** tells us that the number of elements in an electronic chip grows exponentially with time; the doubling time has been around 18 months for more than 30 years. There is no indication that this trend might slow down; although in the last few years it has no longer been possible to increase the clock frequency, the increase in the complexity of the chips is now used to produce "multicore" CPUs. Assuming the multicores are used in an efficient parallelization procedure, the practical performance of computers continues to increase by a factor of 4 every three years. The CCD cameras used to produce astronomical images have also increased their number of pixels according to Moore's law,[1] thus the capability of surveys to produce

---

[1] The last few years have seen the assemblage of many individual CCD chips into very large arrays, with a reading time for the array comparable to that of the individual chips; this is another form of parallel data processing.

astrometric data is also growing exponentially with time. To exploit this technology is more cost effective than increasing the telescope size.

Second, the surveys have interest in decreasing the signal-to-noise (S/N) ratio at which the observations are considered, because state-of-the-art CCDs have less noise and because a gain in the acceptable S/N gives an increase in the limiting magnitude (at which objects can be considered detected) without an increase in the telescope diameter. Then some false observations are unavoidably accepted. The problem is that the false data may propagate through the orbit determination procedure and degrade the results.[2]

Third, the surveys have interest in scheduling their observations to get a minimum number of observations for each Solar System object to be discovered. This is due to the preference given to multidisciplinary surveys, trying to obtain results for very different astronomical, astrophysical, and cosmological investigations with the same telescope and even the same images. To obtain orbits for Solar System objects there is a requirement of multiple observations at constrained time intervals. This can generate a conflict of interest between the Solar System subtask and the requirement of other investigations to be conducted simultaneously. This implies the need for a more aggressive orbit determination, using less data for each object.

These three considerations lead to choices, in the design of a survey, which cannot be reversed just because we would like orbit determination to be easier. Thus we must not neglect the following implications.

First, it is essential to keep under control the **computational complexity** of the procedure used for identification and orbit determination; if the number of objects which could be discovered is N and the number of observations is M, the algorithms need to use a number of operations of order smaller than quadratic, e.g. $\mathcal{O}(N \log N)$, $\mathcal{O}(N \log M)$, or $\mathcal{O}(M \log M)$. Any algorithm with a quadratic complexity $\mathcal{O}(N^2)$, $\mathcal{O}(M^2)$, or $\mathcal{O}(N M)$ becomes impossible to use when $N \simeq 10^7$ and $M \simeq 10^9$ (including false), as it is planned for the surveys starting at the time this book is being written: 3 nanoseconds used in an $N \times M$ loop correspond to about a year of CPU time. It is important to be aware that we cannot brush away this constraint by waiting for more powerful computers of the next generation, because the next generation surveys will produce more data with more or less the same increase in performance: a quadratic algorithm would become more and more inadequate as time goes by. Note that traditional procedures, e.g., computing Gauss preliminary orbits with all triples of observations from

---

[2] *False facts are highly injurious to the progress of science, for they often endure long...*, C. Darwin, *The Origin of Man*, 1871.

different nights, an $\mathcal{O}(M^3)$ algorithm, was adequate when it was invented, but cannot be used for the current surveys.

Second, there is the problem of **accuracy**, that is the fraction of the proposed identifications and orbits which are true. To keep the fraction of false under control without decreasing the **efficiency**, that is the fraction of the objects actually appearing in the images which can be declared discovered, is the most difficult task of population orbit determination. Individual detections presumed to be moving objects could be false (noise, statistical flukes, variable stars, etc.), each tracklet formed with them could be false (containing false detections and/or assembling detections from two different objects), and the identifications proposed among the tracklets can also be false (containing false tracklets and/or assembling tracklets from different objects). The false detections increase sharply with lower S/N, the false tracklets and false identifications grow quadratically (either $\mathcal{O}(M^2)$ or $\mathcal{O}(N^2)$, depending upon the methods used). To avoid this cascade of false conclusions, it is essential to use a very tight quality control on the final orbit; typically we use a **statistical quality control** on the residuals of the least squares fit. However, if the number of false tracks produced by the first stages of the procedure is $\mathcal{O}(N^2)$, this would be the complexity of the overall procedure.

Third, the orbit determination methods have to attempt to provide an orbit with much less data than the traditional ones, e.g., observations in only two nights, observations in two short arcs separated by more than one orbital period, and some observations at a low S/N mixed with more reliable ones. Some of the algorithms proposed in the previous chapters can be used for this. Of course this increases the difficulty of the verification procedure and of the statistical quality control to avoid false conclusions.

11.2 Identification and orbit determination procedure

The processing begins with the astrometric data,[3] containing the detections presumed to belong to moving objects above a given S/N threshold s. They include false detections, but s has to be chosen such that the false are not too many, e.g., the true/false ratio could be of the order of 1.

From detections to tracklets

The first step is to assemble each group of detections which could belong to one and the same moving object into a tracklet. Without an orbit, this can

---

[3] The work of image processing and astrometric reduction is also a critical segment of the pipeline, but it is beyond the scope of this book.

be done just by using spatial proximity between detections at different times. For two detection tracklets the number of possible tracklets is $\mathcal{O}(M^2)$, the most likely belonging to moving objects need to be selected by a smaller complexity algorithm, e.g. $\mathcal{O}(M \log M)$, as in (Kubica *et al.* 2007). The control values used to select the tracklets are the proper motion η and, if there are three detections, the curvature components $\kappa \eta^2, \dot{\eta}$ introduced in Chapter 9. Interesting objects passing close, mostly near-Earth asteroids, can have rather long and curved tracklets, thus the controls must be loose. Hence it is unavoidable to have a significant fraction of false tracklets. As we will see in Section 11.4, this does not result in a cascade of false identifications.

From tracklets to tracks

The second step is to assemble tracklets into **tracks**, or **proposed identifications**. A track is a list of tracklets which could belong to one and the same Solar System object. There are at least two different methods to select the tracks. First, the recursive attribution methods, described in Chapter 8 and Section 10.2. Second, the binary tree methods, described by Kubica *et al.* (2007). A possible third method uses the integrals of the Kepler problem, see Section 8.6, but this has not been fully tested yet, while the two previous methods have been shown to be very efficient in simulation tests. There are also other methods such as (Granvik *et al.* 2005).

The main difference is that binary trees are sophisticated algorithms to control the computational complexity, which remains $\mathcal{O}(M \log M)$ even to align triples, quadruples, and so on, of tracklets, but the controls used to select the tracks are just polynomial inequalities. No orbit is computed, although the controls with quadratic polynomials can be done in such a way to exploit the relationship between the distance ρ and $\eta^2 \kappa$ to exclude only objects nearer than those at the tiny object boundary (see Section 8.1).

On the contrary, in the recursive attribution methods the smooth function with which the data are compared is always the path on the celestial sphere computed by means of an orbit, maybe just a hypothetical one.

In this book we have discussed in detail the recursive attribution methods; for the binary tree methods we suggest reading (Kubica *et al.* 2007). In practice, the two classes of methods can be effectively used together in a multistage procedure, see (Milani *et al.* 2008, Section 6).

From tracks to identifications

The process to confirm a track and generate from it a **confirmed identification** can be conceptually decomposed into four steps:

(i) Compute a preliminary orbit, by using some information from the observations of all the tracklets, e.g., either by selecting one observation in each tracklet, or by using the attributable of each tracklet.

(ii) Use the preliminary orbit as first guess for a differential corrections procedure, which might contain several steps as in Section 10.2.

(iii) If the differential corrections converge, apply a statistical quality control to the final residuals. If the values of the **quality control parameters** are satisfactory, the track is accepted as identification.

(iv) Apply identification management and tracklet management to the global set of identifications and tracklets obtained, to remove duplications, contradictions, and incomplete identifications; see Section 11.4.

Only at the end of such a process is it possible to claim to have identified the real Solar System objects which have been observed. Those with a well-determined orbit could then be tested for the requirements of a definition of discovery (Section 10.5). Even after this rigorous procedure, a small fraction of false discoveries may remain, and special care needs to be taken to remove them from the results, thanks to the successive accumulation of data.

11.3 Controlling the computational complexity

There are two problems to be solved to find an appropriate algorithm for identification and orbit determination from the data of a large survey. First, the tracks have to be found with a fast algorithm, e.g., with a computational complexity $\mathcal{O}(M \log M)$, and a very high efficiency, that is only a small fraction of objects actually observed can be lost (a typical goal could be an efficiency 0.95–0.99). Second, the accuracy of the tracks has to be controlled; if the true tracks were just a fraction $\mathcal{O}(1/N)$ of the proposed ones, the tracks to be tested would be $\mathcal{O}(N^2)$ and the computational cost would be dominated by the confirmation procedure, mostly step (ii).

At first sight, these two requirements appear contradictory. The solution is to devise a somewhat more complex procedure, in which both efficiency and accuracy are achieved by successive steps, in such a way that the composite computational complexity of the pipeline is controlled. We will present here two possible solutions, which should not be considered alternative, but rather to be used together for best performance.

Binary tree method

If the binary tree method is used as the first step, to generate a list of tracks, a high efficiency can be achieved in finding the true tracks corresponding

SURVEYS

to Solar System objects observed over at least three different nights. The main problem is that they are mixed with many false tracks. Kubica *et al.* (2007) and Milani *et al.* (2008) describe full-scale simulations of a large survey with accuracies below 0.01 for the proposed tracks: this could lead to a large computational overhead in the confirmation procedure.

A solution is to use step (i) of the confirmation procedure, the preliminary orbit computation, as a filter to reduce the number of tracks to be submitted to differential corrections (which are much more computationally intensive). Since the binary tree method works with tracklets from three or more nights, the preliminary orbit method should be Gauss. A track can be refused on the basis of the preliminary orbit computation in two cases. The first one occurs when the hypothesis (9.36) is contradicted: if the values of $\gamma = \gamma_2, \mathcal{C} = \mathcal{C}_2, \epsilon$ obtained from the three observations selected among the ≥ 3 tracklets are such that there is no solution of the intersection problem of Section 9.7,[4] then the track should be false.

The second case of track rejection is when some preliminary two-body orbit can be obtained, but when the residuals of the observations from all tracklets are computed, they have a RMS value too large. The control value for this kind of track rejection has to be selected with care. The preliminary orbit is a two-body orbit, and even that is obtained with some truncation to some order in Δt, thus the accuracy cannot be the same as requested for a least squares solution with a full N-body model; e.g., in (Milani *et al.* 2008) values around 10 arcsec have been shown to be a good choice.

By combining both tests, it is possible to reduce the number of proposed tracks by an order of magnitude, thus the computational overhead of differential corrections, needed as the following step, becomes acceptable. Moreover, the procedure to confirm tracks is parallelizable by tracks, that is the proposed tracks may be split among a number of processors/cores.[5]

Recursive attribution method

Another procedure to compose tracks is recursive attribution, that is a $(j+1)$-track is formed by attribution to an orbit computed for a j-track, see Section 10.2. For $j = 1$ the procedure starts from virtual asteroids obtained by sampling the admissible region, as discussed in Sections 8.2 and 8.3. The main problem is how to avoid performing complex computations inside an $\mathcal{O}(M N)$ loop (the number of tracklets is $\mathcal{O}(M)$ and the number of orbits is

---

[4] Apart from spurious solutions, with either $\rho_2 < 0$ or ρ_2 so small that the attraction from the Earth cannot be neglected.

[5] The procedure to propose tracks uses a global algorithm and is not easy to parallelize, but it corresponds to a small fraction of the computational load.

$\mathcal{O}(N)$). From each tracklet an attributable is computed: however, each attributable A_i has its own mean time t_i. If we were to compute the predicted attributable at times t_i from each orbit E_k, the double loop with respect to i, k would contain a computationally expensive orbit propagation.

This problem can be solved by selecting a fixed time step, e.g., one day, and a discrete set of times, e.g., local midnight near where the telescopes are located. Each orbit is propagated to the local midnight of the N_{nig} nights in which there are observations to be attributed. Thus the number of orbit propagations is still $\mathcal{O}(N)$. Each attributable is propagated from time t_i to the local midnight t_0 by some simple formula, such as a linear extrapolation:

$$\alpha(t_0) = \alpha(t_i) + \dot{\alpha}(t_i)\,(t_0 - t_i), \quad \delta(t_0) = \delta(t_i) + \dot{\delta}(t_i)\,(t_0 - t_i).$$

If this approximation is too poor, it is possible to propagate to the time of each individual digital image (in modern asteroid surveys, a large number of objects is detected in each frame).

The extrapolated values $(\alpha(t_0), \delta(t_0))$ can be compared with the predictions for the same epoch from the orbit. This comparison is in a two-dimensional space, and the selection of the candidates for attribution can be done by using some simple two-dimensional metric K_2, e.g., by using either the confidence ellipse or the two-dimensional identification penalty. The prediction needs to be in a linear regime to do this with simple computations. The test $K_2 < K_{2max}$ can be used as a first filtering stage, with a very limited number of operations to be repeated $\mathcal{O}(N\,M)$ times.

This loop can be replaced by a faster one with computational complexity $\mathcal{O}(N \log M) + \mathcal{O}(M \log M)$; we can sort the $(\alpha_i(t_0), \delta_i(t_0))$ coordinates for each night by the value of the right ascension. The sorting has complexity $\mathcal{O}(M \log M)$ and can be performed with classical algorithms (Knuth 1998). Then for each of the N orbits we search the sorted list to find the attributable with the α_i nearest to the prediction and compare only those with neighboring α. This can be done by the binary search method, with complexity $\mathcal{O}(N \log M)$. This algorithm is simpler and less efficient than the binary tree, and other multidimensional sorting methods such as (Granvik *et al.* 2005), but it is good enough to reduce the computational complexity. Moreover, the procedure for attribution is parallelizable by orbits, that is the orbits may be split among a number of processors/cores.

The orbit-attributable couples passing the first filter are submitted to a second one, containing an orbit propagation to the exact time t_i, a prediction of an attributable at time t_i (with covariance) resulting into a four-dimensional vector of residuals (O-C), and the computation of the penalty K_4, see Section 8.4. If $K_4 < K_{4max}$, then the couple is proposed as an attribution.

Confirmation of a proposed attribution, the third filtering stage, is done by selecting a preliminary orbit as in Section 8.5.[6] This is used as first guess for differential corrections. If convergence is achieved, and the quality control of the residuals is satisfactory, the attribution is accepted, which is not the same as saying that the identification is true, see the next section.

11.4 Identification management

The procedure of **identification management** has the purpose of compiling a catalog of identifications, each with its orbit(s) and auxiliary information (covariance, residuals, quality control metrics), removing all kinds of duplications and contradictions.

Duplications may arise because the same identification may be obtained through different sequences, e.g., $((A, B), C)$ and $((A, C), B)$, where A, B, C are tracklets and the symbol $(\cdot, \cdot)$ denotes an identification. There are different kinds of contradiction: the most severe is of the form $((A, B), C)$ and $((A, D), E)$, i.e., two **discordant identifications** with a tracklet in common.

Both duplications and contradictions can be removed with a procedure of **identification normalization**, with an arbitrary list of identifications as input and as output a normalized list with only **independent identifications**, that is each observation belongs to only one of them.

The key issue is that normalization is a global procedure, which needs to be applied to all the identifications available, or at least to all those formed with a set of observations which may refer to the same objects. For example, if the survey covers, in a given lunation, a region near the opposition and one near quadrature (the so-called *sweet spots*, specially suited for detection of near-Earth asteroids), then we can apply the normalization procedure to the identifications formed with opposition data and, separately, to those with quadrature data, because they can be assumed to be independent.

Normalization procedure

We define the following relationships between two identifications: let $List(id_1)$, $List(id_2)$ be the lists of tracklets belonging to the identifications id_1, id_2

- $included(id_1, id_2) \iff List(id_1) \subsetneqq List(id_2)$
- $contains(id_1, id_2) \iff List(id_2) \subsetneqq List(id_1)$
- $independent(id_1, id_2) \iff List(id_1) \cap List(id_2) = \emptyset$

---

[6] If there are enough data, a Gauss preliminary orbit and even the previous least squares orbit may be used.

- $same(id_1, id_2) \Longleftrightarrow List(id_1) = List(id_2)$
- $discordant(id_1, id_2) = List(id_1) \cap List(id_2) \neq \emptyset, List(id_1), List(id_2).$

These five properties are mutually exclusive and they cover all possible cases. The goal of normalization is to select a subset with only independent identifications. Among non-independent identifications, the procedure needs to select those with more information and more likely to be true.

Our normalization procedure is defined as follows. The input list of identifications is sorted according to an ordering relationship called *better*. The definition we currently use is based upon the number nt of tracklets in the identification, and upon the RMS σ of the least squares fit residuals (if the identification has alternative orbit solutions, the lowest RMS value has to be used). An identification is better if it includes more tracklets or, for the same number of tracklets, if the residuals have a smaller RMS:

$$better(id_1, id_2) = (nt_1 > nt_2) \text{ OR } (nt_1 = nt_2 \text{ AND } \sigma_1 < \sigma_2).$$

Then the sorted list is scanned from the top: the "best" identification is inserted in the normalized list. We proceed as follows: For the following identifications id_k in the input list:

- if for each id_j in the normalized list $independent(id_k, id_j)$, then id_k is inserted in the normalized list;
- if there is a normalized id_j such that $included(id_k, id_j)$, then id_k is dropped;
- if there is a normalized id_j such that $same(id_k, id_j)$, then the solutions of id_k are added to those of id_j, and duplicate solutions (consistent within the uncertainty given by the covariance matrices) are removed.

Note that $contains(id_k, id_j)$ cannot occur: it would imply $better(id_k, id_j)$, while id_j comes from higher up in the sorted list.

The steps defined above are enough to remove from the normalized list all duplications and cases of included identifications: e.g., if $((A, B), C)$ is in the input list, both (A, B) and $((A, C), B)$ are removed, without losing track of possible double solutions with the same set of observations.

Discordant identifications

The critical step is how to handle a couple of discordant identifications. There are three appropriate choices: to keep in the normalized list only one if it is *much better* than the other, to discard both, or to try to "merge" them into a single identification.

The choice of the *much better* ordering relationship is critical. It should indicate that an identification is significantly more likely to be true than the

other, as measured from the quality control metrics. To "merge" two identifications requires fitting an orbit to all the observations belonging to both,
then to apply quality control to the resulting residuals. If neither of these
two cases applies, the only way to complete the normalization, removing all
discordancies, is to discard both. In doing this, as the simulations discussed
below show, we often sacrifice one true identification to remove a false one,
that is we privilege accuracy with respect to efficiency.[7]

In the tests we have done so far, we use a definition of *much better* based
only on the two parameters nt and σ:

$$much\_better(id_1, id_2) \Longleftrightarrow (nt_1 > nt_2) \text{ OR } (nt_1 = nt_2 \text{ AND } \sigma_1 + \delta_\sigma < \sigma_2),$$

with a control $\delta_\sigma > 0$ (we have used $\delta_\sigma = 0.25$). It would be possible to use
some of the other quality control parameters described in Section 11.5.

An example

To explain better the logic of the normalization procedure, let us use a
simple example. Let us assume that A, B, C, D, E, F are tracklets, and that
the output of the identification procedure is

2 *ids*	3 *ids*	4 *ids*
(A, B),	$((A, B), C)$,	$(((A, B), C), D)$,
(F, C),	$((E, F), C)$,	
(E, F).		

Let us assume the identification list, sorted by *better*, is

(1) $(((A, B), C), D)$
(2) $((A, B), C)$ *included* in (1)
(3) $((E, F), C)$ *discordant* with (1) which is *much better*
(4) (A, B) *included* in (1)
(5) (F, C) *discordant* with (1) which is *much better*
(6) (E, F) *independent* from (1).

Then the *normalized list* is

$$(((A, B), C), D), \qquad (E, F).$$

This example can also be used to show that the normalization must be done
globally, on all the identifications in the same set of data, not by adding
sequentially the tracklets and the identifications when they are available. Let

[7] An alternative would be to adopt a probabilistic approach: two discordant identifications with
comparable quality control metrics could each be given an estimated probability $\simeq 0.5$ of being
true. They should be both kept in a *weakly normalized* list, without duplications but with
some discordancies. This approach has not been fully tested yet.

us suppose the tracklet D has not been observed yet, and the normalization is started using as list of identifications the one above without (1). Then (2) and (3) are discordant: if none of the two is *much better*, both of them have to be discarded; the same may occur with (5) and (6), and the normalized list may be just (A, B). If D is added later, maybe $((A, B), D)$ is found, but the *much better* identification (1) is not available and the object corresponding to (E, F) is lost.

Controlling the computational complexity

The normalization procedure as outlined above does work, and gives an important contribution to increase the accuracy as discussed in Section 11.5. However, if N is the number of objects being discovered, the computational complexity of the normalization procedure as described is $\mathcal{O}(N^2)$. Thus for N large enough, the computational load of the normalization could exceed the load for obtaining the identifications.

An algorithm for normalization of complexity $\mathcal{O}(N \log N)$ is as follows. Whenever an identification id_j is added to the normalized list, all the tracklets in $List(id_j)$ are endowed with a pointer to id_j. Then, when another identification id_k from the original list is analyzed, we can assemble all the pointers to normalized identifications of the tracklets in $List(id_k)$; they define the subset of the normalized identifications id_r for which $independent(id_k, id_r)$ is false. This list is used in the normalization procedure as outlined above. The procedure for assembling the pointers has complexity $\mathcal{O}(N \log N)$ if binary search of the tracklets is used to access the pointers.

With this method, the computational load of the identification management becomes negligible, with respect to that for finding and confirming the identifications. The only caveat is that for large N a large random access memory (RAM) is needed to run the procedure globally, because of the long input list of identifications (possibly with duplications) and of the large set of pointers. If this results in using virtual memory, i.e., much slower access to disk, the performance could be severely impaired.[8]

Merging discordant identifications

We need to consider the possibility, in the case $discordant(id_k, id_j)$, of **merging identifications**, that is to look for an orbit which can fit all the observations belonging to the tracklets of $List(id_j) \cup List(id_k)$, with residuals passing the quality control. The track, hence the list of observations, is given; however, in the list of observations it is necessary to remove the duplications and check for contradictions (see under tracklet management).

---

[8] However, the available RAM also grows according to Moore's law.

The first guess orbit for differential corrections could be selected among the already known ones, an orbit of id_j being preferred to that of id_k because $better(id_j, id_k)$. Then differential corrections need to be applied: if they converge and the residuals pass the quality control, then the new identification id_m with $List(id_m) = List(id_j) \cup List(id_k)$ replaces id_j in the normalized list and id_k is dropped.

This algorithm has been shown by many tests to be very effective in assembling much larger identifications (with more tracklets) from smaller ones, but it may introduce serious problems in the overall procedure. First, the average computational complexity of the merging algorithm is hard to compute, but it is possible to generate fictitious examples showing that the worst case complexity could be terrible.[9] Second, the insertion of id_m in place of id_j destroys the work already done in the normalization procedure, in that id_m may well be discordant, e.g., merging (A, B) with (A, C) may result in discordancy with (C, D), which might have been already inserted into the supposedly normalized list. A solution to both problems is to use recursive attribution for as many steps as required to get to the identifications with M tracklets, where M is such that objects observed with more than M tracklets are exceptional cases, occurring for a small fraction of the population. For example, if there is just one tracklet per night for each object in the great majority of the cases, M could be the number of observing nights. Then identification merging allows us to find the best orbit even for the few "overobserved" objects. To obtain the normalized list, it is enough to run the normalization procedure twice, the first with merging, the second without.

Orbit identification

Once the normalized list of identifications for a given time span (e.g., a lunation) has been formed, it should be compared with those built previously. Since each identification has some fitting orbit, this problem could be solved with the methods of orbit identification of Chapter 7.

The tracklets observed in one lunation can also be attributed to the orbits computed for another lunation (be it the previous or the next): in this approach, the combination between the two sets of results is just a continuation of the recursive attribution procedure of Section 11.3, to be followed by identification management (for all the identifications obtained with the data

---

[9] Assume there are M tracklets all belonging to the same object and take all the possible 2-identifications in the input list. Depending upon the ordering by *better*, that is upon the values of the RMS, in the output of identification management there could be either a single merged identification growing to M tracklets in $\log_2 M$ steps, or many merged identifications, obtained with very redundant computations and discordant among them.

of both lunations). When the gap in time between the two observation time spans, in which the same objects might appear, is short, the attribution is a more efficient method. If the gap in time is long, and the orbits obtained are well determined (e.g., arc type ≥ 3, see Section 10.5), the orbit identification algorithms are very efficient. There is one difficult case, when an object has not been observed enough for a well-determined orbit, e.g., just two nights, and then may have been re-observed after a long time (years). There is currently no general solution for this case, but several methods are being tested, see Sections 7.4 and 8.6, and also (Granvik and Muinonen 2008).

Tracklet management

The composition of tracklets can have two problems. First, there can be *incomplete* tracklets, not including all the detections of an object in the same night: if a, b, c, d are detections of the same object, there could be tracklets $((a, b), c)$ and $((b, c), d)$. Second, there can be *false* tracklets, including false detections and/or detections of different objects. Thus there can also be **discordant tracklets**, with some (but not all) detections in common.

The problems with incomplete tracklets may result from uneven performance of the **scheduler**, the method used to select the sequence of telescope pointings to collect observations. Optimal scheduling belongs to the class of discrete optimization problems, known to be of non-polynomial complexity; this in practice means the perfect scheduler cannot be available. Uneven spacing between observations of the same area may result in several tracklets of the same object in one night, increasing the complexity of the identification procedure without increasing, in most cases, the quality of the orbits. The orderings used in the identification management, *better* and *much better*, may need to be redefined to compensate for this. Too long spacing between detections may result in failure to assemble some of them into tracklets.

One advantage of the identification management procedure is that a significant fraction of the problems with the tracklets can be solved a posteriori, after finding the identification. If two tracklets belong to the same identification, it does not matter if there are common detections, provided the duplicates are removed from the list of observations of the identification.[10] If there are two discordant tracklets, of which one is identified and the other is not, we can assume that the latter is false and discard it; this procedure is called **tracklet management**.

An important output of the identification management is the list of

---

[10] In case a track contains two detections on the same frame at different positions, the track itself has to be discarded.

leftover tracklets. It is obtained by removing from the input list the tracklets belonging to confirmed identifications and those discordant with the identified ones. This allows us to build a list of unidentified tracklets which is shorter and also contains fewer false tracklets.

To control the computational complexity of the method that finds the tracklets discordant with the identified ones, a list of discordant tracklets needs to be prepared with an $\mathcal{O}(M \log M)$ algorithm (where M is the total number of detections); this is possible by sorting the detections. After identification management this list is scanned, searching for identified tracklets.

11.5 Tests for accuracy

The best way to measure the efficiency and accuracy of an orbit determination method applied to a large survey is to run a full-scale simulation. Then we have a **ground truth**, that is a list of synthetic objects used in data simulation and their assumed orbits, with which the output of the procedure can be directly compared. While processing real data we have no way of knowing how many other identifications may be possible, and false identifications can be long lasting. The performance of the algorithms is strongly dependent upon the **number density** of detections; the future surveys are designed to achieve a much larger number density of detections than those of the currently available data. The results of these simulations depend upon so many assumptions, many of them implicit, that it is very hard to predict the performance of a future survey. The point is to show that the main limitation to the performance of the next-generation surveys, as far as Solar System objects concerned, is not due to the orbit determination task.

The main purpose of large-scale orbit determination simulation is to measure the accuracy of the procedure. However, efficiency and accuracy are not independent. The quality control parameters may be selected to favor efficiency, sacrificing accuracy. The identification management methods may be effective in removing false identifications, because they are discordant with one another, but in this way a true identification is often sacrificed to remove a false one, decreasing efficiency. Simulations are needed to test different sequences of algorithms, options, and control values to achieve the best possible compromise between efficiency and accuracy.

Quality control metrics

Accuracy can be increased with a statistical quality control based upon more than just one parameter, trying to capture information not only on the noise component (measured by the RMS) but also on systematic signals left in the residuals. For the full-scale simulations of Milani *et al.* (2008) we have used the following 10 metrics (control values in square brackets):

- normalized RMS of astrometric residuals (the assumed RMS of the observation errors was 0.1 arcsec) [1.0];
- RMS of photometric residuals in magnitudes [0.5];
- bias of the residuals in RA and in DEC [1.5];
- first derivatives of the residuals in RA and DEC [1.5];
- second derivatives of the residuals in RA and DEC [1.5];
- third derivatives of the residuals in RA and DEC [1.5].

To compute the bias and derivatives of the residuals we fit them to a polynomial of degree 3 and divide the coefficients by their standard deviation as obtained from the covariance matrix of the fit.[11]

There is a problem in comparing values of the above parameters with fixed control values, because the statistically expected values depend upon the number m of (scalar) observations and the number n of fit parameters, which could be 6, 5, and even 4 (see Section 10.4). One standard way to take this into account is to normalize the control parameters dividing by the factor $\sqrt{m/(m-n)}$ before comparing to a fixed control, independent of m, n.

Simulation results

As an example we give the output of the simulation of Milani *et al.* (2008), a full density simulation of a next-generation survey over an entire lunation, with limiting magnitude 24 and a large Solar System model (with 11 million objects). This simulation did not include false detections.

In a first iteration, the binary tree algorithm was used to form tracks, which were submitted to differential corrections. Then the resulting list of identifications was normalized. Table 11.1 summarizes the accuracy results, showing that identification management is necessary to reduce the false identifications to the desired very low level, while before normalization there were far too many. It also shows that the false tracklets are very seldom included in identifications. The identification management was also effective in removing false tracklets, e.g., at opposition the leftover tracklets were reduced in number by a fraction 0.743, among them the false tracklets were reduced by a fraction 0.794.

---

[11] When these algorithms are used on real data additional metrics should assess the outcome of outlier removal, see Section 5.8. For simulations this may not apply, depending upon the error model used to add noise and false detections to the simulated astrometry.

Table 11.1. Accuracy results, before (columns 2–4) and after (columns 5–7) normalization. For each case, the number of false identifications passing quality control, fraction of false, and number of true identifications with false tracklets.

Region	All Identifications			Normalized		
	False	Fraction	F.Tr.	False	Fraction	F.Tr.
Opposition	7 093	0.043	4	80	0.0005	1
Sweet spots	1 869	0.013	10	29	0.0002	0

Table 11.2. Overall and NEA-only identifications. Column 2: number of objects observed on three nights. Column 3: efficiency before normalization. Column 4: efficiency after normalization. Column 5: fraction lost in the first iteration and recovered in a second iteration. Column 6: combined efficiency from both iterations. Column 7: fraction of false identifications after both iterations.

	Total	Eff.	Eff. No.	Recovered	Eff. tot.	False Id.
Opposition	161 146	0.973	0.959	0.754	0.990	0.0006
NEAs	353	0.904	0.904	0.853	0.971	
Sweet spots	144 903	0.980	0.974	0.750	0.994	0.0002
NEAs	271	0.801	0.801	0.852	0.971	

A second iteration based on the recursive attribution algorithm was applied to the list of leftover tracklets, followed by normalization. Table 11.2 shows that the overall efficiency was already high in the first iteration, although the efficiency for the near-Earth asteroids was less so, especially at the sweet spots. Most of the objects observed over three nights and lost by the first iteration were recovered by the second one, especially among NEAs. The recursive attribution method also provided orbits for a fraction > 0.8 of the objects observed over two nights; these lower quality orbits may be used to be identified with similar orbits from the previous/next lunation, and also for recovery of low confidence detections, see the next section.

In conclusion, the algorithms described in Chapters 8, 9 and 10 and assembled in an identification procedure are adequate for the next-generation sky surveys, even with $N \simeq 300\,000$ objects discovered in each lunation.[12]

---

[12] The large number of discovered objects in these simulations shows that effects sensitive to the number N of objects, like the occurrence of false identifications, can be under control. They are not meant to be an estimate of the discovery rate of any specific survey.

11.6 Recovery of low confidence detections

There is one especially tricky case of attribution, the one in which the data to be attributed are not yet organized into attributables, but are just individual observations. This may occur in very deep surveys, in which the number density of observations per unit area on the sky is extremely high, to the point that pairing them to form tracklets is not easy. If this procedure is pushed to low levels of signal-to-noise ratio, a large fraction of the supposed observations may in fact be spurious. The question is how far down we can push the minimum acceptable S/N to avoid a *false identification catastrophe*, that is a sharp drop of the accuracy beyond a critical number density.

As a basic form of the problem, let us suppose we have an orbit and two images, taken at two different times t_1 and t_2; on each image there are M supposed observations, including the spurious ones. By computing from the orbit two predictions (α_i, δ_i) for the times $t_i, i = 1, 2$, we can compare them with the (supposed) observations in each image, computing a simple two-dimensional metric K_2. This allows us to select by the values of K_2 a subset of $M_i \ll M$ observations from the frames at time $t_i, i = 1, 2$.

If the numbers M_i are small, we can test each of the $M_1 \times M_2$ pairs of observations by forming an attributable (with covariance) and comparing it, by the K_4 metric, with the predicted attributable for time $\bar{t} = (t_1 + t_2)/2$. This procedure may fail if $|t_2 - t_1|$ is too large, because the actual path of the object on the celestial sphere might have enough curvature, so that the tangent vector $(\dot{\alpha}(\bar{t}), \dot{\delta}(\bar{t}))$ is significantly different from the velocity of the straight line approximation.[13] If $M_1 \times M_2$ is too large, we may use the sorting method to find an algorithm of complexity $\mathcal{O}(M_1 \log M_2)$.

The couples of observations giving a satisfactory value of K_4 can be submitted to differential corrections, trying to fit the data used to compute the orbit and the two additional observations. The quality control metrics of the fit with the additional observations should not be significantly worse than those without them. When this is achieved, the two observations can be attributed even if there was no way to know a priori that they belonged to an object whatsoever (they could have been spurious).

The computational complexity of the procedure described above is difficult to control. The main reason is that the first filter, based upon the two-dimensional metric K_2, must be applied searching frame by frame. The second filter with the four-dimensional metric K_4 has to be applied searching by orbit, that is the detections selected by the first filter have to be sorted in

---

[13] A better result could be obtained by comparing $[\delta(t_2) - \delta(t_1)]/(t_2 - t_1)$ with the average of the predictions $[\dot{\delta}(t_1) + \dot{\delta}(t_2)]/2$ (similarly for α).

an order different from that in which they have been computed. If all the data passing the first filter, let Z be their number, can be kept in memory, then binary sorting with computational complexity $\mathcal{O}(Z \log Z)$ can be used to sort by orbit. This results in an algorithm with considerable complexity.

Recovery simulations

The performance of the above algorithms to recover low confidence detections and promote them to tracklets attributed to available orbits has not been fully tested yet; some simulations are available, but it is difficult to model realistically the occurrence of false detections, which do not really follow simple statistics, like Poisson. One such simulation, with uniform probability density, indicates that the accuracy problem is the main one.

Including false detections with overall number density $\mu = 10^4$ per square degree, $\simeq 100$ times the number density of the real detections, accuracy in the attribution of a promoted low confidence tracklet in an additional night turns out to be 0.99 when the orbit is based on ≥ 3 nights of observations and 0.96 for orbits with just two nights of data. At this level the results may be useful, especially for upgrading the weak two-night orbits to three-night orbits, generally well determined, thus possibly complying with some definition of discovery such as the one of Section 10.5.

If $\mu = 10^5$ per square degree, $\simeq 1000$ times the real detections, we obtain 0.90 for the accuracy, with ≥ 3 nights of observations and 0.68 with just two nights of data. Such low reliability identifications cannot be used to claim discovery, especially those which would be most useful, i.e., the two-nighters promoted to three-nighters. They could provide candidate, or probabilistic, discoveries to be confirmed by targeted follow up.

A limitation to the number density cannot be avoided; if μ is the number density and $\Gamma_{(\alpha,\delta)}$ is the covariance matrix of the predicted observations α, δ, then the expected number of detections in the confidence ellipse $Z_L(\sigma)$ is $F = \mu \, \sigma^2 \, \pi \sqrt{\det \Gamma_{(\alpha,\delta)}}$, the number of tracklets formed with detections selected in a couple of frames is of the order of F^2, and spurious tracklets with accuracy of the order of $1/F$ in $\dot{\alpha}$ and $\dot{\delta}$ must often occur. The simulations indicate that the *false identification catastrophe* does not need to occur in the next-generation surveys, provided the required S/N is such that the number of false detections is not much larger than that of true detections.

12

IMPACT MONITORING

When an asteroid or a comet has just been discovered, its orbit is weakly constrained by the available astrometric observations and it might be the case that an impact on the Earth in the near future (within the next 100 years) cannot be excluded. If additional observations are obtained, the uncertainty of the orbit decreases and the impact may become incompatible with the available information. Thus, if we are aware that an impact is possible, it is enough to spread this information to the astronomers to convince them to follow up the object. On the contrary if this piece of information is not available, or is made available when the asteroid has been lost, the impact risk will remain until the same asteroid is accidentally recovered. This might occur too late for any mitigation action.

This problem can be solved if all the asteroids/comets, immediately after being discovered and before they can be lost, are "scanned" for possible impacts in the near future. If impacts are possible, this information has to be broadcast to the astronomers. This is the goal of **impact monitoring**.

It is somewhat surprising that this was not really possible until late 1999, when the first impact monitoring system, the CLOMON software robot of the University of Pisa, became operational. For many years, even after the risk of impacts of asteroids and comets on our planet had been identified and its probability estimated, even while dedicated surveys were scanning the sky to discover as many near-Earth asteroids (NEA) and comets as possible, the algorithms to scan a given, known object for possible impacts were not effective enough. By using the linear theory of impact prediction (see Section 12.1) it was possible to identify impact possibilities with comparatively high probability, of the order of 10^{-3}–10^{-4}. However, if the possible event was the impact of an asteroid with diameter exceeding 1 km, which would result in an explosion with a yield of more than 20 000 megatons, even a probability of the order of 10^{-6}–10^{-7} cannot be considered negligible,

237

and to omit to follow up such a dangerous asteroid would be a serious mistake. On the contrary, unfounded announcements of possible impacts, such as for asteroid 1997 XF_{11} in March 1998, can undermine the credibility of the scientific community involved and thus make it more difficult to obtain the resources necessary in a serious case.

In 1999 we were able for the first time to issue a warning of a possible impact for asteroid 1999 AN_{10} (Milani *et al.* 1999). The impact probability for the year 2039 was $\simeq 10^{-9}$, so small that it would not need to be cause of concern for the public, but the mathematical problem had been solved.[1] The new methods introduced for the 1999 AN_{10} case led to the impact monitoring system CLOMON later in the same year.

In 2002 the impact monitoring system CLOMON was replaced by the second generation CLOMON2 in Pisa (duplicated at the University of Valladolid) and by SENTRY at the NASA Jet Propulsion Laboratory. These two independent systems, whose output is carefully compared, now guarantee that the potentially dangerous objects are identified very early (within hours from the dissemination of the astrometric data) and followed up until the observations succeed in contradicting the possibility of an impact. During the time span over which these observations are obtained, the announcement that some asteroid has the possibility of impacting must be in full view of the public, and in practice it is posted on the web.[2] This is essential to communicate the need of observations to the astronomers and also reassures the public that no information on impact risk is withheld.

In case the impact possibility remains for a long time, as it is currently the case for asteroids (99942) Apophis and (144898) 2004 VD_{17}, which have been on the *risk pages* of CLOMON2 and SENTRY for years, it is reasonable to begin planning for the mitigation actions which may become necessary if the later observations were to confirm, rather than contradict, the impact. Although the impact probability is small for these cases, we need to have a technologically feasible method to deflect such asteroids which can be used if necessary, see Section 14.6. Otherwise, the practical utility of the surveys and of the impact monitoring itself would be cast into doubt.

The purpose of this chapter is to outline the mathematical methods used in impact monitoring. It is based on (Milani and Valsecchi 1999, Milani *et al.* 1999, Milani *et al.* 2000b, Gronchi 2002, Gronchi 2005, Chesley *et al.*

---

[1] The impact probability was later found to be higher for impacts in 2044 and 2046; a few months later, the precovery of 1999 AN_{10} in plates taken in 1955 allowed us to contradict the possibility of an impact in the twenty-first century.

[2] `http://newton.dm.unipi.it/neodys` and `http://unicorn.eis.uva.es/neodys` for CLOMON2, `http://neo.jpl.nasa.gov` for SENTRY.

2002, Valsecchi *et al.* 2003, Milani *et al.* 2005b, Gronchi and Tommei 2006, Gronchi *et al.* 2007).

12.1 Target planes

The geometry of the encounters with a planet can be described in terms of a target plane, a plane in 3-D space through the center of the target planet, e.g., the Earth, orthogonal to the direction of the relative velocity of the approaching small body. In this context, an **impact** can be described as an orbit containing a target plane point inside the planet cross-section.

There are two ways to define such a target plane. The simplest is the **modified target plane** (MTP) (Milani and Valsecchi 1999): it is obtained by considering the time $\bar{t}$ at which the small body orbit has a relative minimum of the distance from the planet center of mass (CoM). Let $\mathbf{d}$ and $\mathbf{v}$ be the planetocentric position and velocity vectors of the asteroid at the time $\bar{t}$: the distance being minimum, $\mathbf{d} \cdot \mathbf{v} = 0$. The MTP is the plane containing $\mathbf{0}$ (the CoM) and normal to $\mathbf{v}$. On this plane the point $\mathbf{d}$ represents the **close approach trace** on the MTP. A complete description of the close approach is obtained by assigning two coordinates on the MTP, two angles defining the orientation of the MTP, the size of the velocity $v = |\mathbf{v}|$, and the time $\bar{t}$. The cross-section of the planet on the MTP is a disk centered at $\mathbf{0}$ and with the radius R of the planet; if the minimum distance $d = |\mathbf{d}|$ at time $\bar{t}$ is less than R there is an impact.[3]

The other definition, called in the literature either just the **target plane** (TP) or **b-plane**, uses the same vectors $\mathbf{d}$ and $\mathbf{v}$ describing the state at the closest approach time $\bar{t}$ to compute a planetocentric two-body approximation of the orbit. If, as it is generally the case, such a two-body orbit is hyperbolic, then the TP is the plane containing $\mathbf{0}$ and orthogonal to the incoming asymptote of the hyperbola, corresponding to the limit vector $\mathbf{u}$ for $t \to -\infty$ of the planetocentric velocity along the hyperbolic trajectory; the size $u = |\mathbf{u}|$ is the velocity "at infinity" as used in astrodynamics. The point $\mathbf{b}$, representing the **trace of the close approach**, is the intersection of the asymptote with the TP; its size $b = |\mathbf{b}|$ is the **impact parameter**, larger than the minimum distance d by a factor

$$\frac{b}{d} = \sqrt{\frac{v^2 \, d}{v^2 \, d - 2\,GM}}$$

where GM is the gravitational constant multiplied by the planet mass. A

---

[3] This assumes the planet surface is a sphere; the oblateness of the planet is generally irrelevant for the possibility of an impact, although it may matter when predicting the point of impact.

complete description of the close approaching orbit is obtained by assigning
two coordinates ξ, ζ on the TP, two angles θ, ϕ defining the orientation of the
TP, the size of the escape velocity $u = |\mathbf{u}|$, and the time $\bar{t}$ (Greenberg *et al.*
1988). On the TP the **impact cross-section** is a disk of radius

$$B = R \sqrt{1 + 2\,GM/R\,u^2} \qquad (12.1)$$

larger than the radius R by a factor accounting for gravitational focusing.

The two planes are different, because the velocity $\mathbf{v}$ at the close approach
is rotated by an angle $\gamma/2$ around the axis of the planetocentric angular
momentum. The angle γ measures the total deflection from the incoming
to the outgoing asymptote and can be computed by

$$\sin(\gamma/2) = \frac{G\,M}{v^2\,d - G\,M}.$$

The transformation of coordinates rotating and rescaling the MTP into the
TP is not canonical, thus it is impossible to use the Hamiltonian formalism
including coordinates on the TP (Tommei 2006a, Tommei 2006b). Moreover, the choice of the coordinates on the two planes can be done in different
ways, and this has also to be accounted for in the transformation.

From an abstract point of view, it does not matter how we select a representative vector for a given close approach, provided it is a smooth function
of the orbit initial conditions: thus a smooth coordinate transformation is
acceptable. However, some coordinate systems are more equal than others, because the propagation of the uncertainty is easy in a linear approximation, by using the differential of the transformations, and a coordinate
change with large higher order derivatives introduces strong limitations in
the applicability of the linearization. Since gravitational focusing introduces
a deformation more nonlinear where gravity is stronger, that is near collisions, there is advantage in using the TP with respect to the MTP.

Linear predictions on target planes

For a given asteroid, and a set of orbital elements $\mathbf{x} \in \mathbb{R}^6$ at epoch t_0
there is a unique orbit, which can be accurately propagated for some time
span.[4] For each close approach to the Earth, occurring within this time
span, there is at least one point $\mathbf{y} \in \mathbb{R}^2$ which is the trace of this orbit on
the target plane. To avoid useless geometric complications, we consider as

---

[4] In the current impact monitoring systems, the orbits are generally propagated for 80–100 years.
Only for some orbits, determined in an especially accurate way, is it meaningful to propagate
for longer time spans; then non-gravitational perturbations, especially the Yarkovsky effect
(Section 14.2), have to be taken into account.

close approach only an encounter with a distance from the planet CoM not exceeding some value d_{max}; practical values for d_{max} range between 0.05 and 0.2 astronomical units (AU), thus the target planes are replaced by disks with a finite radius.[5]

Let us suppose the orbit determination solves only for the initial conditions $\mathbf{x} \in \mathbb{R}^6$ of the asteroid at some epoch t_0, and the differential corrections converge to the nominal solution $\mathbf{x}^*$, with normal and covariance matrices C, Γ (see Chapter 5). As the nominal solution $\mathbf{x}^*$ is surrounded by a six-dimensional confidence region of acceptable solutions, the trace point $\mathbf{y}^* = \mathbf{g}(\mathbf{x}^*)$ determined by the propagated nominal orbit on the target plane of some encounter is surrounded by a two-dimensional confidence region.

To compute an approximation, we use the differential of the map $\mathbf{g}(\mathbf{x})$ providing the TP trace (Milani and Valsecchi 1999). The trace point is reached at the time $t_c(\mathbf{x})$ of the target plane crossing for each orbit with initial conditions $\mathbf{x}$ in a neighborhood of $\mathbf{x}^*$. By using Cartesian geocentric coordinates ξ, η, ζ such that $\eta = 0$ is the target plane, the equation $\eta(t, \mathbf{x}) = 0$ implicitly defines the crossing time $t_c(\mathbf{x})$ as a differentiable function, thus $\xi(t_c(\mathbf{x}), \mathbf{x})$ and $\zeta(t_c(\mathbf{x}), \mathbf{x})$ are differentiable too. Using the differential $D\mathbf{g}(\mathbf{x}^*) = \partial(\xi, \zeta)(\mathbf{x}^*)/\partial \mathbf{x}$ we can compute the covariance and normal matrix of the $\mathbf{y}$ prediction by the linear covariance propagation formula

$$\Gamma_{\mathbf{y}} = D\mathbf{g}\, \Gamma_{\mathbf{x}}\, (D\mathbf{g})^T, \qquad C_{\mathbf{y}} = \Gamma_{\mathbf{y}}^{-1}$$

defining the confidence ellipse on the target plane

$$(\mathbf{y} - \mathbf{y}^*)^T\, C_{\mathbf{y}}\, (\mathbf{y} - \mathbf{y}^*) \leq \sigma^2 \qquad (12.2)$$

with the same confidence parameter σ used for the confidence ellipsoids. This formalism is applicable because the trace function is differentiable, but this does not imply that the quadratic approximation (12.2) is an accurate description of the confidence region on the target plane. However, if it is adequate, the **possibility of an impact** can be studied by looking for intersections of the confidence ellipses with the impact cross-section. By using a Gaussian probabilistic formalism, from the normal probability density $N(\mathbf{x}^*, \Gamma)$ we can define a probability density on the target plane. In the linear approximation corresponding to the differential $D\mathbf{g}(\mathbf{x}^*)$, $\mathbf{y}$ is Gaussian with density $N(\mathbf{y}^*, \Gamma_{\mathbf{y}})$. Then it is possible to estimate the impact probability by computing a probability integral on the impact cross-section.

The formalism above is well known for applications to the navigation of interplanetary spacecraft, a case in which the assumptions of small

---

[5] It is possible for a close approach to have multiple local minima of the distance to the CoM, thus multiple target plane trace points. Reducing d_{max} can often eliminate such complications.

confidence regions and therefore the applicability of linearization are well founded, for the reasons given in Section 1.4. To estimate the probability of impact of asteroids is much more difficult, due to nonlinearity.

12.2 Minimum orbital intersection distance

A convenient reference system $O\xi\eta\zeta$ for the geocentric position on the TP is obtained by aligning the negative ζ axis with the projection of the heliocentric velocity $\mathbf{v}_\oplus$ of the Earth, the positive η axis with the geocentric asymptotic velocity $\mathbf{u}$ (i.e., normal to the TP), and the positive ξ axis in such a way that the reference system is positively oriented. With this frame of reference the TP coordinates (ξ, ζ) indicate the cross-track and along-track miss distances, respectively. In other words, ζ is the distance by which the asteroid is early or late for the minimum possible distance encounter. The associated "miss time" of the target plane crossing ($\eta = 0$) is $\Delta t = -\zeta/(v_\oplus \sin\theta)$, where θ is the angle between $\mathbf{u}$ and $\mathbf{v}_\oplus$ and $v_\oplus = |\mathbf{v}_\oplus|$; a positive ζ implies that the asteroid is "late" at the date with the Earth, $\zeta < 0$ means the asteroid is early.

On the b-plane the ξ coordinate is the minimum distance that can be obtained by varying the timing of the encounter. This distance is closely related to the **orbit distance**, known as the **minimum orbital intersection distance** (MOID) in the literature (Bowell and Muinonen 1994), that is the minimum separation between the two osculating Keplerian orbits of the asteroid and the Earth as curves in three-dimensional space, without regard to the phase on each of the two. Note that the approximation of the MOID with the ξ coordinate is valid only in the linear approximation and can break down for distant encounters (e.g., beyond several lunar distances).

Stationary points of the Keplerian distance function

Two confocal Keplerian orbits can get close at more than a pair of points, for example near both the mutual nodes, thus it is useful to compute all the local minima of the **Keplerian distance function** d, the distance between two points on the two orbits, not only the absolute minimum. We compute these values as the stationary points of the function d^2, squared to be smooth also in case of *orbit crossing*, when the distance can be zero.

There are several papers in the literature on the computation of the minimum points of d, e.g. (Sitarski 1968, Hoots 1994). Recently some algebraic methods to compute all the stationary points of d^2 have been introduced, using *Gröbner bases* (Kholshevnikov and Vassiliev 1999) and *resultant*

theory (Gronchi 2002, Gronchi 2005). They are both based on a polynomial formulation of the problem. The algebraic formulation of the problem
allows us to search for all the solutions using the efficient methods of modern
computational algebra and gives us a bound for the maximum number of
stationary points, as discussed below.

Mutual geometry of confocal Keplerian orbits

The stationary points of d^2 have been proven to be ≤ 16 for the case of two
ellipses and at most 12 if one orbit is circular, except for very particular cases
with infinitely many stationary points (Gronchi 2002). From a large number
of numerical experiments we have found cases with at most 12 stationary
points of d^2 and at most four **local MOID**, that is local minima of d^2.

The statistics of the stationary and minimum points of the squared distance function d^2 using the orbit of the Earth and those of the known NEAs
shows that most mutual orbit configurations give two local minima and one
maximum, among six stationary points: this is the most intuitive case, with
a simple geometry. There are also several cases with only one local minimum. No real asteroid has been found so far with four minimum points.

When there is a **crossing** between the orbits (MOID $= 0$) and the mutual
inclination is not zero the minimum point of d corresponds to a mutual node.
It is not always the case that at least one local minimum point of d is close
to a mutual node: there are examples of real NEAs with two minima, both
far enough from the mutual nodes. Such cases arise from orbits with low
mutual inclination.

Uncertainty of the MOID

The role of the MOID in impact monitoring is to select, among the large
number (thousands, even tens of thousands) of close approaches possible for
a given asteroid, the ones which could be very close. If the TP coordinates
have a small value of ξ and a large value of ζ, then the encounter has not
been close, but another orbit with different orbital phase might get in time
to the date with the Earth at the local MOID point. If the value of ζ has
a large enough uncertainty, such a phase change could be compatible with
the available observations. In a linear approximation, applicable to very
close encounters, the confidence ellipse has a major axis almost parallel to
the ζ axis and a minor axis almost parallel to the ξ axis, that is expressing
the uncertainty of the local MOID value.

Let $(\mathbf{e}, v)$ be a set of orbital elements such that $\mathbf{e} = (e_1, \ldots, e_5)$ describes
the geometric configuration of the orbit and v is a parameter along the

trajectory, e.g., the true anomaly. The least squares solution gives us a nominal orbit $(\mathbf{e}^*, v^*)$, together with its uncertainty represented by the 6×6 covariance matrix

$$\Gamma_{(\mathbf{e},v)} = \begin{pmatrix} \Gamma_{\mathbf{ee}} & \Gamma_{\mathbf{e}v} \\ \Gamma_{v\mathbf{e}} & \Gamma_{vv} \end{pmatrix},$$

which is the inverse of a normal matrix $C_{(\mathbf{e},v)}$. The 5×5 submatrix $\Gamma_{\mathbf{ee}}$ gives the marginal covariance of the five elements $\mathbf{e}$, independently from the value of the sixth one v, and $C^{\mathbf{ee}} = \Gamma_{\mathbf{ee}}^{-1}$ is the marginal normal matrix.

The minimal distance maps and their singularities

Let $(\mathbf{e}, v)$ be the orbital elements of the asteroid and $(\mathbf{e}_\oplus, v_\oplus)$ those of the Earth, supposedly known with negligible errors. For each configuration $\mathbf{e}$ we consider the minimum points $(v, v_\oplus) = \mathbf{v}_h(\mathbf{e})$ of the Keplerian distance function (assuming $\mathbf{e}_\oplus$ as fixed parameters) and we define the maps

$$d_h(\mathbf{e}) = d(\mathbf{e}, \mathbf{v}_h(\mathbf{e})) \qquad \text{local minimal distance,}$$

$$d_{min}(\mathbf{e}) = \min_h d_h(\mathbf{e}) \qquad \text{orbit distance (MOID).}$$

where h is an index with a finite number of values.

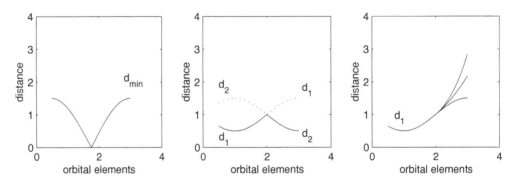

Fig. 12.1. The singularities of the maps d_h and d_{min} can occur in three forms, as described in the text. Reproduced from Gronchi and Tommei (2006).

In Figure 12.1 we show the singularities of d_h and of d_{min}, which belong to three types. Left: d_h and d_{min} are not differentiable where they vanish. Center: in a neighborhood of an orbit configuration $\mathbf{e}^*$, two local minima can exchange their role as absolute minimum; then d_{min} can lose its regularity even without vanishing. Right: when a bifurcation occurs the definition of the maps d_h may become ambiguous after the bifurcation point. Note that this ambiguity can occur only where the 2×2 Hessian matrix of $d^2(\mathbf{e}, \mathbf{v})$ is degenerate and does not occur for the d_{min} map.

Computation of the uncertainty of d_h and d_{min}

The errors in the orbit also affect the computation of the local minima of d and it is important to estimate the size of this effect. Let us consider the orbit distance map d_{min}; the same method can be applied to the minimal distance maps d_h. For a given $(\mathbf{e}^*, \mathbf{e}_\oplus)$, the nominal orbit configuration $\mathbf{e}^*$ being endowed with its covariance matrix $\Gamma_{\mathbf{ee}}$, we can compute the covariance of $d_{min}(\mathbf{e}^*)$ by a linear propagation of the matrix $\Gamma_{\mathbf{ee}}$:

$$\Gamma_{d_{min}}(\mathbf{e}^*) = \left[\frac{\partial d_{min}}{\partial \mathbf{e}}(\mathbf{e}^*)\right] \Gamma_{\mathbf{e}}(\mathbf{e}^*) \left[\frac{\partial d_{min}}{\partial \mathbf{e}}(\mathbf{e}^*)\right]^T. \qquad (12.3)$$

The possibility of crossings between the orbits produces a singularity in this computation because the partial derivatives $\partial d_{min}/\partial \mathbf{e}$ do not exist at $\mathbf{e}^*$ when $d_{min}(\mathbf{e}^*) = 0$, e.g., when the two orbits in the configuration $(\mathbf{e}^*, \mathbf{e}_\oplus)$ intersect each other. Moreover the uncertainty of a non-zero but small orbit distance may allow meaningless negative values of the distance. Note that we are interested in knowing the uncertainty just when the orbit distance can be small or vanishing, that is when a collision or a close approach is possible. Thus the classical covariance propagation formula to compute the uncertainty of the MOID is applicable only when it is not very useful.

Regularization of the minimal distance maps

We introduce a regularization of the maps d_h, d_{min}, generalizing the approach by Wetherill (1967) and Bonanno (2000). Let us take into account the map d_{min}, the same method can also be applied to d_h. It is possible to make d_{min} locally analytic even where its value is zero, simply by changing its sign according to some properties of the orbit configuration.

The idea of the regularization can be illustrated by a simple example. Let us consider the positive function, defined on the whole plane, $f(x, y) = \sqrt{x^2 + y^2}$ and the function $\tilde{f}$, defined on a smaller domain,

$$\tilde{f}(x, y) = \begin{cases} -f(x, y) & \text{for } x > 0 \\ f(x, y) & \text{for } x < 0. \end{cases}$$

The directional derivative of f in $(x, y) = (0, 0)$ does not exist for every choice of the direction. The regularized function $\tilde{f}$, extended by continuity to the origin $(0, 0)$, has all the directional derivatives in $(x, y) = (0, 0)$. How to extend such a method to the problem at hand is discussed below.

Geometric definition of the regularization

Let $\boldsymbol{\tau}_1$, $\boldsymbol{\tau}_2$ be the tangent vectors to the orbits at the minimum point and let $\boldsymbol{\Delta}_{min}$ be the vector joining the two tangency points ($|\boldsymbol{\Delta}_{min}| = d_{min}$). If $\boldsymbol{\tau}_1$ is not parallel to $\boldsymbol{\tau_2}$ we can define the non-zero vector $\boldsymbol{\tau_3} = \boldsymbol{\tau_1} \times \boldsymbol{\tau_2}$. Due to the stationary points properties, if $\boldsymbol{\Delta}_{min} \neq 0$, $\boldsymbol{\Delta}_{min}$ is parallel to $\boldsymbol{\tau_3}$. We define the regularized map $\tilde{d}_{min}$ by setting $|\tilde{d}_{min}| = d_{min}$ and choosing the sign $+$

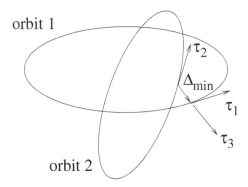

Fig. 12.2. Geometry of the regularization: the orientation of the two parallel vectors $\boldsymbol{\Delta}_{min}$, $\boldsymbol{\tau_3}$ is the key to defining a regular map $\tilde{d}_{min}$ by simply changing the sign of d_{min} on selected configurations $\mathbf{e}$. Reproduced from Gronchi and Tommei (2006).

for $\tilde{d}_{min}$ if $\boldsymbol{\Delta}_{min}$ and $\boldsymbol{\tau_3}$ have the same orientation, the sign $-$ otherwise. This sign is well defined, with the only exception of the cases in which $\boldsymbol{\tau_1}$ and $\boldsymbol{\tau_2}$ are parallel. Then we extend the definition domain to most crossing orbits setting $\tilde{d}_{min} = 0$ if $d_{min} = 0$. The orbit configurations with parallel tangent vectors to minimum points are also excluded from the definition domain even if they are not crossing points. The resulting map $\mathbf{e} \mapsto \tilde{d}_{min}(\mathbf{e})$ is locally analytic almost everywhere, without excluding a neighborhood of most orbit configurations $\mathbf{e}$ such that $d_{min}(\mathbf{e}) = 0$. In particular the partial derivatives can be computed as

$$\frac{\partial \tilde{d}_{min}}{\partial e_k}(\mathbf{e}^*) = \left\langle \frac{\boldsymbol{\tau_3}(\mathbf{e}^*)}{|\boldsymbol{\tau_3}(\mathbf{e}^*)|}, \frac{\partial \boldsymbol{\Delta}}{\partial E_k}(\mathbf{e}^*, \mathbf{v}_{min}(\mathbf{e}^*)) \right\rangle \qquad k = 1\ldots5 \qquad (12.4)$$

where $\mathbf{v}_{min}(\mathbf{e}^*)$ is the absolute minimum point and $\boldsymbol{\Delta}(\mathbf{e}, \mathbf{v})$ is the vector joining the points corresponding to v and $v_\oplus$ on the orbit of the asteroid and of the Earth respectively.

Thus it becomes possible to use for the smooth function $\tilde{d}_{min}(\mathbf{e})$ the standard covariance propagation formula, applicable only to differentiable functions, including the interesting low MOID cases. For each nominal orbit configuration $\mathbf{e}^*$, with covariance matrix $\Gamma_{\mathbf{ee}}$, we can compute the variance of $\tilde{d}_{min}(\mathbf{e}^*)$ as

$$\Gamma_{\tilde{d}_{min}}(\mathbf{e}^*) = \left[\frac{\partial \tilde{d}_{min}}{\partial \mathbf{e}}(\mathbf{e}^*) \right] \Gamma_{\mathbf{e}}(\mathbf{e}^*) \left[\frac{\partial \tilde{d}_{min}}{\partial \mathbf{e}}(\mathbf{e}^*) \right]^T \qquad (12.5)$$

by using the smooth partial derivatives of eq. (12.4).

The above statement needs to be taken with some caution. It is necessary to check that the singular case (τ_1 parallel to τ_2) does not occur at $\mathbf{e}^*$ and is not even within the confidence ellipsoid. We need to check the variance of the determinant of the Hessian matrix to look for possible bifurcations of the stationary points. Last but not least, the propagation of the covariance by the linear formula of eq. (12.5) may be mathematically consistent, but to assume $\tilde{d}_{min}(\mathbf{e})$ is a Gaussian random variable is a good approximation only provided the function $\tilde{d}_{min}$ is quasi-linear, which does occur when the uncertainty on $\mathbf{e}$ is small, see (Gronchi *et al.* 2007).

Potentially hazardous asteroids

Bowell and Muinonen (1994) define a **potentially hazardous asteroid** (PHA) as an asteroid having MOID ($= |\tilde{d}_{min}|$) ≤ 0.05 AU and absolute magnitude $H \leq 22$; these are the most relevant objects for impact monitoring. However, this definition refers only to the nominal orbits and does not take into account the uncertainty. To be complete, we should consider all the **virtual hazardous asteroids**, that is asteroids that have a significant probability of being a PHA, taking into account the joint probability density function of the variables $(\tilde{d}_{min}, H)$. The use of the regularized minimal distances $\tilde{d}_h$ is essential for this purpose; for small nominal values of d_h also negative values > -0.05 have to be considered.

The probability that $\tilde{d}_h$ belongs to $[-0.05\,\text{AU}, 0.05\,\text{AU}]$ is

$$\mathcal{P}\left(|\tilde{d}_h| \leq 0.05\,\text{AU}\right) = \frac{1}{\sqrt{2\pi}} \int_{z_1}^{z_2} \exp(-z^2/2)\,dz \qquad (12.6)$$

with

$$z_i = \frac{x_i - \tilde{d}_h(\mathbf{e}^*)}{\sigma_{\tilde{d}_h}(\mathbf{e}^*)} \qquad (i = 1, 2)$$

where $x_1 = -0.05$, $x_2 = +0.05$ and $\sigma_{\tilde{d}_h}(\mathbf{e}^*)$ is the standard deviation of $\tilde{d}_h$, defined by

$$\sigma_{\tilde{d}_h}(\mathbf{e}^*) = \sqrt{\Gamma_{\tilde{d}_h(\mathbf{e}^*)}}.$$

The variance Γ_{HH} of the absolute magnitude depends both on the photometry and astrometry, because it is computed from the apparent magnitudes by a formula involving the topocentric distance of the object. Given the variance V_{phot} of the photometry, assuming it is independent of the astrometry, we can decide if a celestial body is a virtual hazardous asteroid by looking at the 2×2 covariance matrix

$$\Gamma_{(\tilde{d}_{min}, H)}(\mathbf{e}^*) = \frac{\partial(\tilde{d}_{min}, H)(\mathbf{e}^*)}{\partial \mathbf{e}} \, \Gamma_{\mathbf{e}}(\mathbf{e}^*) \left[\frac{\partial(\tilde{d}_{min}, H)(\mathbf{e}^*)}{\partial \mathbf{e}} \right]^T + \begin{pmatrix} 0 & 0 \\ 0 & V_{phot} \end{pmatrix}$$

computed at the nominal orbit configuration $\mathbf{e}^*$.

12.3 Virtual asteroids

When an asteroid has been discovered only recently, or anyway has been observed only for a short time span, the orbit of the real object may belong to a large confidence region. Another way to describe our (lack of) knowledge is by using a swarm of virtual asteroids (VA), with slightly different orbits all compatible with the observations, that is belonging to the confidence region. The reality of the asteroid is shared among the virtual ones, in the sense that only one of them is real, but we do not know which one. Since the confidence region contains a continuum of orbits, each VA is in turn representative of a small region, i.e., its orbit is also uncertain, but to a much smaller degree. This smaller uncertainty enables us to use for each VA some local algorithms, which would not apply to the entire confidence region. Note that the nominal orbit is just one of the VAs, and is not extraordinary in this context.

The N-body problem not being integrable, there is no way to compute globally the totality of orbits of the confidence region; only a finite set of orbits can be numerically propagated. The reason for using a swarm of VAs is that they are a finite set of orbits, which can be propagated one by one, representing the totality of orbits compatible with the observations much better than the nominal solution alone. Moreover, by propagating together with the orbits the corresponding state transition matrices, we can use a linear approximation in a neighborhood of each VA: it is not easy to decide how many such points are needed to keep up with strong nonlinearities.

Thus the critical issue is how to sample the confidence region in an efficient way, that is with few orbits[6] but selected in such a way that they are as far as possible representative of the different possible orbits. There are two classes of sampling methods used in the selection of VA: the random, or **Monte Carlo** (MC) methods, and the **geometric sampling** methods, in which the sampling takes place on the intersection of a geometric object, a differentiable manifold, with the confidence region.

The MC methods directly use the probabilistic interpretation of the least

---

[6] In the impact monitoring practice with current computer hardware, this means between a few thousands and a few tens of thousands of orbits for each asteroid.

squares principle. Since the orbit determination process yields a probabilistic distribution in the space of orbital elements, the distribution can be randomly sampled to obtain a set of equally probable VAs. They will be more dense near the nominal solution, where the probability density is maximum, and progressively less dense as the RMS of the residuals increases (Chodas and Yeomans 1996). This can be implemented in different ways, by using a random number generator to sample an assumed probability density either in the space of the elements **x**, or in the space of all residuals ξ, or in some appropriate combination of the two as in *statistical ranging* (Virtanen *et al.* 2001).

When the computational resources are not an issue and the probabilistic error models are reliable, the MC methods are more rigorous and complete, thus they are often used for checking the results once a case of possible impact has been identified. If by impact monitoring we mean checking all newly discovered, or anyway re-observed, asteroids for the possibility of a future impact, then computational complexity is the main concern and the MC methods are too slow, thus the geometric sampling methods have to be used. In this chapter we will concentrate on the one-dimensional sampling methods, in which the geometric object is a smooth line sampled by a regular sequence of intervals. More complex sampling methods, such as two-dimensional ones using a surface of variations, have been proposed (Tommei 2005) but are not yet being used in operational impact monitoring.

The line of variations as geometric sampling

As discussed in Sections 5.6 and 7.3, some years after the epoch of initial conditions the confidence region becomes stretched in the along-track direction; for asteroids with low MOID, this effect is stronger because of the chaotic orbit, with a typical Lyapounov time of the order of the average time span between two close approaches to a major planet (Whipple 1995). Since the goal of impact monitoring is to find possible impacts with a long warning time (tens of years, longer than the Lyapounov time), the best way to sample the confidence region is by defining a curve which intuitively can be the "spine" of such an elongated confidence region.

The solution adopted by Milani *et al.* (1999) and used in the current impact monitoring systems is to use a sampling of the line of variations (LOV), see Section 10.1, as a set of virtual asteroids. The main advantage of this approach is that the set of VA has a geometric structure, that is they belong to a differentiable curve along which interpolation is possible. Thus the methods of impact monitoring are a version of **manifold dynamics**, in which a smoothly parameterized set of orbits is implicitly propagated.

The problem with the LOV is that it is not independent of the choice of coordinates in the space of initial conditions $\mathbf{x}$ (Section 10.3). Thus we have to choose the coordinate system which makes the LOV most representative of the set of orbits filling the confidence region, and this depends upon the purpose of the sampling. For impact monitoring we are interested in predictions for times much later than the initial conditions, then in most cases the important changes in the orbit elements are those in the semimajor axis, and the metric should be chosen accordingly.[7]

The LOV trace on the target planes

Once the LOV sampling has been computed, we have a set of VAs $\mathbf{x}_i$ for $1 \leq i \leq 2k+1$; let us assume the LOV points have been computed with a uniform spacing h in the LOV parameter σ, thus $\mathbf{x}_i$ corresponds to the value $\sigma_i = (i - k - 1) \cdot h$. By propagating each of the VA orbits for a given time span (80–100 years), we record for each VA all the close approaches to the Earth within the distance d_{max}. Each close approach is represented by at least one trace point $\mathbf{y} = (\xi, \zeta)$ on the TP. Up to this point the procedure is the same, whatever the sampling method.

Since the LOV sampling is not just a set of points, we can exploit the facts that they sample a smooth line and the trace of the LOV on the TP is also a smooth line. Let us suppose that two consecutive VAs, $\mathbf{x}_i$ and $\mathbf{x}_{i+1}$, have TP trace points $\mathbf{y}_i$ and $\mathbf{y}_{i+1}$ straddling the Earth impact cross-section, such that the trace point $\mathbf{y}_i$ is "early", that is $\zeta_i < 0$, while $\mathbf{y}_{i+1}$ is "late", $\zeta_{i+1} > 0$. Then there is one point $\mathbf{x}_{i+\delta}$ on the LOV (as a continuous curve) corresponding to the parameter $\sigma = (i - k - 1 + \delta) h$, with $0 < \delta < 1$, such that $\zeta_{i+\delta} = 0$; this must occur provided the trace of the LOV segment between $\mathbf{y}_i$ and $\mathbf{y}_{i+1}$ lies entirely within the distance d_{max} from the Earth CoM. This is the first instance of the principle of the simplest geometry we will further discuss in the next section: cases with strong nonlinearities, so that the function $\zeta(\sigma)$ is not defined in the interval $[\sigma_i, \sigma_{i+1}]$, are possible, but this is less frequent than the simple case in which the segment joining $\mathbf{y}_i$ to $\mathbf{y}_{i+1}$ behaves like a straight line.

The point $\mathbf{x}_{i+\delta}$ on the LOV, which was not among the original set of VAs, can be computed by using some iterative method such as regula falsi (see the

---

[7] With well determined NEA orbits, such as those with radar observations, impact monitoring can be extended to times > 100 years, and the most effective sampling is obtained by selecting a weak direction essentially along the gradient of the semimajor axis.

next section). If the TP trace $\mathbf{y}_{i+\delta}$ is inside the Earth impact cross-section, then around $\mathbf{x}_{i+\delta}$ there is a **virtual impactor** (VI), that is a connected set of initial conditions leading to an impact (at about the same date). If the point $\mathbf{y}_{i+\delta}$ is outside the impact cross-section, but the width w of the confidence ellipse computed by linearizing at $\mathbf{y}_{i+\delta}$ is large enough, there is an intersection between the confidence ellipse and the impact cross-section and there is anyway a VI, with initial conditions not belonging to the LOV.

By computing the probability density function with a suitable Gaussian approximation centered at $\mathbf{x}_{i+\delta}$ it is possible to estimate the probability integral on the impact cross-section, that is the **impact probability** (IP) associated with the given VI. These computations are approximate, but when the IP is low they are better than the estimates done with MC type sampling; the MC estimates based upon the number of impacting VA suffer from the uncertainties of small number statistic, e.g., a MC sampling is likely not to provide any impacting VA if the number of VAs is less than 1/IP. On the contrary, the geometric sampling methods described here can detect VIs with IP of the order of 10^{-7}–10^{-8} (and even less, as in the 1999 AN$_{10}$ case) starting from a few thousands of VAs on the LOV. The issue of completion in the searches for VIs is more complex and needs to be discussed in the context of the geometric theory of the next section.

12.4 Target plane trails

To understand the properties of the TP trace of the LOV we need to use the finite sample formed by the trace points $\mathbf{y}_i$ as markers of a geometric structure. To do this, after computing all the close approaches to the Earth for all the VAs $\mathbf{x}_i$, we sort them by the time of the closest approach. The recorded close approaches cluster around a discrete set of encounter times, associated to passages of the Earth through the point corresponding to the (local) MOID while the asteroid is neither very late nor very early at its (local) MOID point. Each of these clusters of close approaches forms a **shower**, and a shower is represented as a set of trace points on the TP.

In some cases, corresponding to comparatively slow encounters, the situation can be somewhat more complicated, see (Milani *et al.* 2005b, Figure 6), but let us assume this decomposition of the set of close approaches into showers has been obtained. Next we decompose each shower into contiguous LOV segments; this is easily obtained by sorting the shower according to the index i. A subset of a shower with contiguous indexes i is a **trail**. In some cases a trail is a **singleton**, formed by just one of the selected VAs.

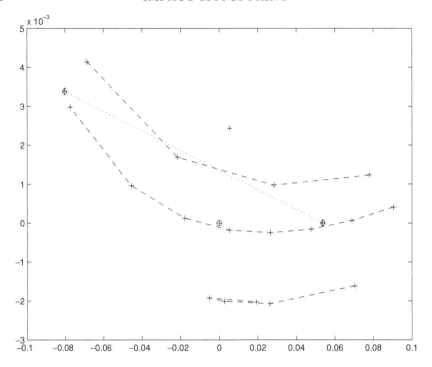

Fig. 12.3. A single shower with five trails for the asteroid 1998 OX$_4$ in January 2046. The Earth is depicted by the "⊕". The two diamonds that are connected by the dotted line actually correspond to a trail with collision, emphasizing that care must be taken when interpolating between solutions. Note the axes have different scales, thus the trails are very close to each other. These trails have been computed by using only the observations obtained in 1998; the asteroid was recovered in 2002 and the new observations have shown that there is no risk of impact. Reprinted from Milani *et al.* (2000b) with permission from Elsevier.

Figure 12.3 shows the trace on the TP for a shower containing five trails, including one singleton and a "pair" with just two TP points.

The principle of the simplest geometry

We can conjecture that a trail with $h \leq i \leq k$ corresponds to a continuous set, a segment of the LOV, with a corresponding curve segment of TP trace points joining $\mathbf{y}_h$ to $\mathbf{y}_k$. Because of the finite sampling, we cannot prove that this must be the case. This hypothesis could be verified by densifying the LOV sampling: if some of the new VAs miss the TP, that is do not have a close approach (within d_{max}) around the same date, we cannot exploit the differentiable structure of the LOV. However, if such a TP segment of a differentiable curve exists, we can draw very important conclusions.

Let us take as an example the "doubleton" of Figure 12.3. To interpolate linearly between the two extremes $\mathbf{y}_i$ and $\mathbf{y}_{i+1}$ (dotted line in the figure) is obviously a very poor approximation; the trails with more TP points do suggest a significant curvature of the TP trace curves. We can use additional information: the map from the LOV to the TP is differentiable and we can compute the tangent vector $\mathbf{s}(\sigma) = d\mathbf{y}(\sigma)/d\sigma$ to the TP trace curve at the point $\mathbf{y}_i = \mathbf{y}(\sigma_i)$; the length $s = |\mathbf{s}|$ is the **stretching**. Let $\mathbf{s}(\sigma_i) = \mathbf{s}_i$; if the angle between $\mathbf{s}_i$ and the vector to the origin $-\hat{\mathbf{y}}_i$ is $< \pi/2$, then the closest approach distance is decreasing for increasing values of the LOV parameter σ at the value σ_i corresponding to $\mathbf{x}_i$. The same computation can be done for $\mathbf{x}_{i+1}$, and if the angle between $\mathbf{s}_{i+1}$ and $-\mathbf{y}_{i+1}$ is $> \pi/2$ the closest approach distance increases with σ at the value σ_{i+1}. If the TP trace segment joining continuously $\mathbf{y}_i$ to $\mathbf{y}_{i+1}$ exists, then there is for some value of σ in the interval (σ_i, σ_{i+1}) a local minimum of the closest approach distance.

In conclusion, if we make the assumption that the behavior of the TP curve is simple, more exactly as simple as it is compatible with the existing decomposition of the shower into trails, we expect to have at least one local minimum of the closest approach distance for each trail. This is why we adopt the **principle of the simplest geometry**, by which the curve segment does not exit the TP disk of radius d_{max}; then there needs to be at least one minimum of the closest approach distance. We can define constructive algorithms for the determination of at least one minimum. Note that it is also an assumption that such a minimum is unique for each trail. In the case of the "doubleton" of Figure 12.3 the minimum is such that there is a VI, but this cannot be confirmed by using a linear approximation.

In fact, the shape of the TP trace curve between two given VAs can be much more complex than the simplest geometry, in which case the convergence of the minimum-finding algorithms and/or the uniqueness of the local minimum may fail. However, if the TP trace curve has extreme nonlinearities over a very short segment of the LOV, that implies the stretching is large and thus the IP is low. This is a robust qualitative argument, that is the VIs we can find by using this method have larger IP than those we can miss, but unfortunately we have not yet been able to transform it into a quantitative argument, that is into an estimate of the maximum IP of the missed VIs. Thus we do not yet have an analytical estimate of the maximum IP for which we can guarantee completion in the search for VIs.

Returns to close approach

Another application of the principle of the simplest geometry can be appreciated from the trail with 5 TP points near the bottom of Figure 12.3. Also in this case the first and last TP point of the trail, $\mathbf{y}_i$ and $\mathbf{y}_{i+4}$, correspond to decreasing and increasing closest approach distance, respectively. Thus, if the TP trace curve segment joins $\mathbf{y}_i$ and $\mathbf{y}_{i+4}$ without exceeding the distance d_{max}, then it must have at least one point with minimum closest approach distance. However, the behavior of the TP trace curve cannot be approximated by a straight line and the map between the confidence region for the initial conditions $\mathbf{x}$ and the TP points $\mathbf{y}$ is strongly nonlinear, because there is a **fold** line where the differential of the map $\mathbf{x} \mapsto \mathbf{y}$ is degenerate (Milani *et al.* 2005b). The closest approach distance decreases to a minimum, then increases again, because the TP trace "turns back".

A *return* is a trail with the additional condition that the VAs forming it have experienced another close approach between the times of the available observations and the times of the close approaches belonging to the trail. Among the returns there are those such that the intermediate close approach is to the same planet and has occurred near the same local MOID point of the close approaches belonging to the return (e.g., both near the ascending node for a high inclination orbit). That implies the Earth is near the same value of the mean anomaly in the intermediate and in the return encounter, i.e., the close approach occurs at about the same date in different years. Also the asteroid needs to be near the same anomaly to be close to the MOID point along its orbit. The time span Δt between the two encounters needs to be close to an integer multiple of the Earth orbital period and to an integer multiple of the asteroid period, thus the two orbits must be nearly resonant. This is the motivation for the name **resonant return** (Milani *et al.* 1999).

We have developed an analytical theory, based upon the Öpik formalism for planetary encounters, describing in a qualitative and approximately quantitative way the shape of the TP trace curves associated to returns, in particular for resonant returns (Valsecchi *et al.* 2003). This is possible because there is a comparatively simple analytical formula approximating the change in the asteroid semimajor axis resulting from the intermediate encounter, as a function of the coordinates (u, θ, ξ, ζ). Without going into long details, it is enough to point out that each intermediate encounter can generate as many as four "turning back points" in the TP trace curves of successive encounters. Non-resonant returns to the same planet (e.g., after a close approach near the other node) and even encounters with another planet can also generate reversals. Thus the behavior of the bottom trail of Figure 12.3 is by no means exceptional, rather generic, and the principle of the simplest geometry does not exclude it. We conclude that the algorithms to find local minima of the close approach distance for each trail must be able to cope with this case, thus they must not assume that the $\mathbf{x} \mapsto \mathbf{y}$ map is locally well approximated by a linear map.

Algorithms for the minimum close approach distance

By assuming the principle of the simplest geometry, each value of the LOV parameter σ belonging to a segment (σ_j, σ_k) corresponds to a point of the TP trace differentiable curve $\mathbf{y}(\sigma)$ joining the TP points $\mathbf{y}_j$ and $\mathbf{y}_k$. For each of these values of σ we can compute the squared distance to the center of the Earth on the TP, that is $b^2(\sigma) = \mathbf{y}(\sigma) \cdot \mathbf{y}(\sigma)$, and its derivative $f(\sigma) = db^2/d\sigma = 2\,\mathbf{s}(\sigma) \cdot \mathbf{y}(\sigma)$ with respect to the LOV parameter σ. Then we can search the set of TP points y_i, $j \leq i \leq k$, to find the couples of consecutive indexes $i, i+1$ where the signs are discordant, $f(\sigma_i) < 0$ and $f(\sigma_{i+1}) > 0$, then there is at least one value of $\sigma^* \in (\sigma_i, \sigma_{i+1})$ such that $f(\sigma^*) = 0$. The algorithm of **regula falsi** provides an approximate value of σ^*:

$$\sigma^* \simeq \sigma_i + \delta = \sigma_i - \frac{f(\sigma_{i+1}) - f(\sigma_i)}{\sigma_{i+1} - \sigma_i}.$$

A step of length δ is performed with the same formula $\mathbf{x}' = \mathbf{x}_i + \delta\,\sigma_1\mathbf{v}_1$ used for the original VA sampling, then constrained differential corrections are used to obtain a LOV point which is taken as $\mathbf{x}_{i+\delta}$. If the corresponding TP trace $\mathbf{y}_{i+\delta}$ is not the minimum distance point along the LOV trace, that is $f(\sigma_i + \delta)$ is significantly $\neq 0$, the procedure is iterated on the interval with extremes $\sigma_i + \delta$ and either σ_i or σ_{i+1}, in such a way that the signs of f at these extremes are discordant. At convergence we obtain a LOV point of (local) minimum of the close approach distance; if f is defined in the whole interval $[\sigma_i, \sigma_{i+1}]$, that is the TP trace is always within the disk of radius d_{max}, this regula falsi iteration has guaranteed convergence.

The difference with the procedure of the previous section is that no assumption needs to be done on the direction and curvature of the TP trace curve. Indeed, in resonant returns the TP trace curve may never cross the $\zeta = 0$ line, because it "turns back" before crossing it. Thus the minimum distance may be much larger than the local MOID, e.g., the two cases in Figure 12.3 of the "doubleton" and of the resonant return can be handled without problems. In other cases the curve may turn back *after* crossing the $\zeta = 0$ line, with a double minimum of the close approach distance. These different cases need to be handled with an adaptive algorithm, able to identify the simplest geometry of the TP trace curve compatible with the available sampling and to take the necessary action, that is selecting additional sampling points to be used as initial conditions for iterative procedures to reach multiple local minima. The case of a singleton TP trace $\mathbf{y}_j$ has to be handled with a different iterative procedure, which is a modification of Newton method on the variable σ with bounded increment (Milani *et al.* 2005b).

12.5 Reliability and completion of impact monitoring

Whenever one of the TP points $\mathbf{y}_0$ of local minimum (in the close approach distance along the LOV) is within the Earth impact cross-section, we have a **virtual impactor representative**, that is an explicitly computed set of initial conditions $\mathbf{x}_0$ compatible with the observations and leading to a collision at a given date. Whether the simpler algorithm, described in Section 12.3, or the more robust algorithm of Section 12.4 has been used does not matter: once found, the VI representative is a proof that the collision can occur, and the problem is how to associate an IP to the VI. Linearization at $\mathbf{y}_0$ of some Gaussian probability density is the only algorithm available and efficient enough to be used in operational impact monitoring, although targeted investigations with denser VA sampling and/or localized Monte Carlo are possible and are used in difficult cases, when there is significant nonlinearity in a neighborhood of the VI representative.

A delicate case is when $\mathbf{y}_0$ is outside the impact cross-section, but the TP confidence ellipse computed by linearizing at $\mathbf{x}_0$ does contain collisions. In this case an explicit representative of the VI is not available; the linearization can be of questionable accuracy, especially when the second largest eigenvalue of $\Gamma_{\mathbf{y}}$ is large, thus the **width** w of the TP confidence ellipse is large. Both including and excluding these cases from the list of VIs can be a mistake.

For the CLOMON2 impact monitoring system we have developed a method to confirm possible VIs by an iterative procedure which has shown the capability to converge to a VI representative, in most cases in which such a VI exists. It is based on a modified Newton method, first proposed by Milani *et al.* (2000c). If $\mathbf{y}_0$ is the point on the LOV TP trace with minimum distance from the Earth, corresponding to the initial condition $\mathbf{x}_0$, but $|\mathbf{y}_0| > B$, see equation (12.1), we select a point $\mathbf{y}'_1$ on the TP with $|\mathbf{y}'_1| = B$, e.g., by moving radially. Then we find the point $\mathbf{x}_1$ in the confidence region near $\mathbf{x}_0$ with the minimum penalty among those projecting into $\mathbf{y}'_1$ on the TP, using the differential of the $\mathbf{x} \mapsto \mathbf{y}$ map at $\mathbf{x}_0$; this is obtained by the same algorithm used in semilinear predictions (see Section 7.5). Then the TP trace $\mathbf{g}(\mathbf{x}_1) = \mathbf{y}_1$ is computed, and it is not $\mathbf{y}'_1$ because of the nonlinearity, but by iterating this procedure the convergence to a VI representative is possible. The difficult point is to define a criterion to terminate the above iterative procedure when convergence is not achieved. Such "divergence" should provide a good indication that the intersection of the linear confidence ellipse and the impact cross-section was a "spurious" VI, see (Milani *et al.* 2005b).

Generic completion

How complete is the search for VIs? This is a difficult question to answer rigorously, but it can be useful to understand how efficient a system could be under idealized circumstances. Specifically, we can ask what is the highest impact probability of a VI which could possibly escape detection if the associated trail on the TP is fully linear; we also assume that the width is small with respect to the size of the Earth impact cross-section. This we call the **generic completion** level of the system.

To obtain this maximum IP, we assume the VI is a very narrow strip passing through the origin of the TP, with a length $2 B_\oplus > 2 R_\oplus$, depending upon the gravitational focusing. If a sufficient number of VAs intersects the target plane disk of radius d_{max}, then the methods described in the previous subsection will reveal the VI. Otherwise, the VI will be missed in the scan.

The generic completion level is given by

$$IP^* = \frac{\delta p \, 2 \, B_\oplus}{\Delta TP}$$

where δp is the probability integral between TP trace points and ΔTP is the maximal separation between TP trace points for which a VI detection is assured; $\Delta TP \simeq s \, \Delta\sigma$, where s is the stretching and $\Delta\sigma$ is the separation of the consecutive VAs in terms of the LOV parameter σ.

For CLOMON2, $d_{max} = 0.2$ AU, and, under the linearity hypothesis, only one point on the TP is required to detect a VI. Thus the spacing of consecutive VAs on the target plane cannot be more than 0.4 AU $\simeq 9400 R_\oplus$. By sampling 2400 VAs over the interval $|\sigma| \leq 3$ on the LOV, these points are separated by $\Delta\sigma = 0.0025$, and the maximum cumulative probability over one interval is $\delta p = \Delta\sigma/\sqrt{2\pi} \simeq 0.001$ (near the nominal, where the Gaussian probability density is $\simeq 1/\sqrt{2\pi}$). From this we can compute the CLOMON2 generic completion,

$$IP^* \simeq \frac{0.002 \, B_\oplus}{9400 \, R_\oplus} \simeq 2.1 \times 10^{-7} \frac{B_\oplus}{R_\oplus}.$$

Thus the generic completion level depends upon the amount of gravitational focusing, and can be higher for asteroids with a low velocity at infinity, u. A somewhat smaller value applies to SENTRY, currently not handling singletons but using more VAs. VIs with IP well below IP^* can be found, and are indeed found very often, but their detection is probabilistic, depending upon some VA crossing by chance the TP disk.

12.6 The current monitoring systems

The impact monitoring software robots CLOMON2 and SENTRY have been operational since early 2002. They handle on average about 100 cases per year: each case begins with the impact monitoring, either by CLOMON2 or by SENTRY, finding at least one VI for either a newly discovered or a re-observed asteroid. In most cases, these results are immediately posted on the web. In the most serious cases[8] the two impact monitoring systems cross-check their results before announcing the existence of a VI: this procedure takes typically just a few hours. Anyway, after the public announcement of a risk case, the astronomical community takes action by performing follow up observations until enough information is available to exclude the possibility of impacts in this century. In a fraction of cases, the asteroids have been lost while still having VIs; however, these are generally small asteroids ($<$ 100 m diameter), indicating that the telescopes available for this targeted follow up might not have been large enough. It is expected that this rate of VI cases will further increase when the next generation asteroid surveys will be operational and discover smaller asteroids.

One case has been a significant source of concern for the personnel of the monitoring systems: (99942) Apophis has been on the "risk pages" of CLOMON2 and SENTRY since December 2004, with a VI for 2029 with an estimated IP peaking at 1/37 and then declining as new and more accurate observations were received. Now the orbit is very well determined, also with radar observations: the 2029 approach will be very close (the asteroid will be visible to the human eye) but there is no possibility of impact for that year. The estimated IP for the VI resulting from the 2036 resonant return is low but it is difficult to contradict this impact possibility. This is because the current optical observations are not accurate enough to improve the orbit, and also because of the extreme sensitivity of the outcome in 2036 with respect to the exact conditions of the 2029 encounter, to the point that non-gravitational perturbations cannot be neglected. This case has for the first time raised the public issue of planning for mitigation action: knowing there is some risk does not solve the problem, if the risk is not contradicted by follow up observations. Thus a dedicated space mission, such as the one discussed in Section 14.6, might be needed.

---

[8] The priority is assigned to the most serious cases by using a numeric metric, the *Palermo Scale*, taking into account the impact probability, the impact energy, and the time to the impact date (Chesley *et al.* 2002).

Part IV

Collaborative Orbit Determination

13

THE GRAVITY OF A PLANET

The equation of motion for a satellite is dominated by the gravity field of the central body, be it the Earth or another major planet. This chapter discusses the mathematical properties of the gravity field of an extended body, how to parameterize it, and how to represent it as a function of both Cartesian coordinates and orbital elements.

13.1 The gravity field

The equation of motion for artificial satellites has the monopole attraction of the planet as the main term; the main perturbation is the effect of the planet shape and internal mass distribution. Thus we need to discuss the gravity field of an extended body and the way to parameterize it. Expressive parameters are the coefficients of the spherical harmonic expansion, which can be included in the fit parameters for **satellite gravimetry**.

Gravity of point masses

The gravity field of accelerations generated by a *point mass* with mass M and located in $\mathbf{p} \in \mathbb{R}^3$, computed at $\mathbf{x}$, is

$$\frac{d^2\mathbf{x}}{dt^2} = \mathbf{v}(\mathbf{x}) = \frac{GM}{|\mathbf{x} - \mathbf{p}|^3}(\mathbf{p} - \mathbf{x}) = \frac{GM}{r^3}(\mathbf{p} - \mathbf{x})$$

where $r = |\mathbf{p} - \mathbf{x}|$ is the distance and G is the **universal gravitational constant**, whose value depends only upon the system of units used; e.g., in the CGS system, $G = 6.6726 \times 10^{-8}\,\mathrm{cm}^3/\mathrm{s}^2\,\mathrm{g}$. Note that in the orbit determination problem the constant G always appears in the combination GM and cannot be solved independently. The field $\mathbf{v}(\mathbf{x})$ is defined and smooth at every point $\mathbf{x} \neq \mathbf{p}$; it is a conservative vector field, that is rot $\mathbf{v} =$

0 and it can be obtained from a **gravitational potential**

$$U(\mathbf{x}) = GM/r, \quad \mathbf{v}(\mathbf{x}) = \operatorname{grad} U(\mathbf{x})$$

determined up to a constant, conventionally selected by imposing $\lim_{r \to +\infty} U = 0$. The vector field $\mathbf{v}(\mathbf{x})$ is divergence free: $\operatorname{div} \mathbf{v} = 0$, thus by the **divergence formula**[1] for every volume $W \subset \mathbb{R}^3$, with as boundary the oriented surface S, the flow of $\mathbf{v}(\mathbf{x})$ across S (from the inside to the outside of W) can be computed as the volume integral of the divergence

$$\int_S \mathbf{v} \cdot \mathbf{n} \, dS = \int_W \operatorname{div} \mathbf{v}(\mathbf{x}) \, d\mathbf{x}$$

where $d\mathbf{x}$ is the volume element in $\mathbb{R}^3$, dS the element of surface on S, and $\mathbf{n}$ the unit vector normal to S pointing to the exterior of W. If the attracting point mass is not in W the volume integral and the flow across S are zero. If $\mathbf{p} \in W$ the divergence formula does not apply because of the singularity at $\mathbf{x} = \mathbf{p}$. The flow of $\mathbf{v}(\mathbf{x})$ does not change under a deformation of the surface S, if $\mathbf{p}$ remains inside; the flow can be computed by using the sphere $S(r)$ of radius r centered in P, with normal vector $\mathbf{n} = (\mathbf{x} - \mathbf{p})/r$:

$$\int_{S(r)} -\frac{GM}{r^3}(\mathbf{x} - \mathbf{p}) \cdot \mathbf{n} \, dS = \int_{S(r)} -\frac{GM}{r^2} \, dS = -4\pi \, r^2 \frac{GM}{r^2} = -4\pi \, GM.$$

A simple generalization is the N-body gravity field: given a finite number of point masses $M_1, M_2, \ldots, M_n$ located at $\mathbf{p}_1, \mathbf{p}_2, \ldots, \mathbf{p}_n$, by the **superposition principle** the gravity field and the potential are the sums

$$\mathbf{v}(\mathbf{x}) = \sum_{i=1}^n \frac{GM_i}{|\mathbf{x} - \mathbf{p}_i|^3}(\mathbf{p}_i - \mathbf{x}), \quad U(\mathbf{x}) = \sum_{i=1}^n \frac{GM_i}{|\mathbf{x} - \mathbf{p}_i|}.$$

The flow across a surface S is the sum of the flows, thus for every surface S corresponding to the (oriented) boundary of some volume $W \subset \mathbb{R}^3$

$$\int_S \mathbf{v} \cdot \mathbf{n} \, dS = -4\pi G \sum_k M_k$$

where the sum is extended to the point masses $\mathbf{p}_k \in W$, inside S. To represent a solid body – a planet – as a swarm of point masses is possible, but this is neither physically intuitive nor computationally efficient.

---

[1] Also known as the **Gauss divergence formula**, but in this book there are so many results due to Gauss.

Mass and gravity of an extended body

To describe the gravity field of an extended body it is better to abandon the unphysical mathematical model of point masses and to use a continuous mass distribution, defined by a **mass density function** $\rho(\mathbf{p}) \geq 0$ which is positive only on a limited subset $W \subset \mathbb{R}^3$, the support of the mass distribution. Then the **total mass** is given by the volume integral

$$M = \int_W \rho(\mathbf{p}) \, d\mathbf{p}, \tag{13.1}$$

well defined under some regularity conditions; e.g., the total mass is well defined if S is a smooth surface and ρ is continuous.[2] The gravity field generated by the mass density ρ is

$$\mathbf{v}(\mathbf{x}) = \int_W \frac{G\rho(\mathbf{p})}{|\mathbf{x} - \mathbf{p}|^3} (\mathbf{p} - \mathbf{x}) \, d\mathbf{p} \tag{13.2}$$

where the integral is over the points $\mathbf{p} \in W$, with $\mathbf{x}$ fixed.

The extended body may well move, thus the mass density may also be time dependent, satisfying a mass conservation equation. If we use the Newtonian approximation, by which the gravity field acts instantaneously (with infinite propagation speed), the gravity field can be computed at each instant of time for every point $\mathbf{x} \in \mathbb{R}^3$ with the same formula.

Also the gravitational potential of an extended body can be defined by a volume integral:

$$U(\mathbf{x}) = \int_W \frac{G\rho(\mathbf{p})}{|\mathbf{x} - \mathbf{p}|} \, d\mathbf{p}. \tag{13.3}$$

By exchanging the operations of integral over $\mathbf{p}$ and differentiation with respect to the components of $\mathbf{x}$, we can obtain eq. (13.2) by applying the gradient operator to the integral above.[3]

The conventional definition of the (Riemann) integral used in eq. (13.1) is obtained by partitioning the support W by parallelograms W_k defined by intervals for each of the coordinates of $\mathbf{p}$. The contribution for each k to the sums approximating the integral is just the volume of W_k times the value of ρ in some point $\mathbf{p} \in W_k$. This corresponds to the intuitive idea that the total mass of the extended body can be obtained by cutting it into small bricks and by estimating the density of each brick, e.g., by weighing it. However, this is certainly not an operational definition of the mass of a planet.

---

[2] Discontinuous jumps in the mass density are common in geophysical models, e.g., at the transition between core and mantle, but the volume integrals such as (13.1) are still well defined.
[3] This is simpler when $\mathbf{x}$ is outside W, thus the points $\mathbf{x}$ and $\mathbf{p}$ are never the same. Equations (13.2) and (13.3) hold also for $\mathbf{x} \in W$, but this requires some results from integration theory.

To obtain a more practical definition, we can use the divergence formula. By exchanging derivatives and integrals, also for an extended body div $\mathbf{v}(\mathbf{x}) = 0$ for all $\mathbf{x}$ outside W, where there is no mass. Inside the body div $\mathbf{v}(\mathbf{x}) = -4\pi G\rho(\mathbf{x})$; the proof requires some non-trivial steps.

Let S be an oriented surface, containing the planet, that is the set W where $\rho(\mathbf{p}) > 0$, inside. Then the mass of the planet can be measured by the flow of its gravity field across S

$$M = \int_W \rho(\mathbf{x})\, d\mathbf{x} = -\frac{1}{4\pi G} \int_W \operatorname{div} \mathbf{v}(\mathbf{x})\, d\mathbf{x} = -\frac{1}{4\pi G} \int_S \mathbf{v} \cdot \mathbf{n}\, dS.$$

This formula suggests that the mass of the planet can be determined by scanning some closed surface, either at the physical surface of the planet or above it, with a **gravimeter** measuring the gravity field vector.

As an historical example, let us approximate a planet like our Earth by a sphere of radius $R_\oplus$, with the gravity field at the surface everywhere normal to the sphere and of constant size $|\mathbf{v}| = g$; we are also neglecting the apparent forces due to the rotation of the planet. In this approximation the mass

$$M = -\frac{1}{4\pi G} \int \int_S -g\, dS = \frac{1}{4\pi G}\, g\, 4\pi\, R_\oplus^2 = \frac{g\, R_\oplus^2}{G}$$

can be estimated if g and G are known.[4]

Harmonic functions

By combining the equations $\mathbf{v}(\mathbf{x}) = \operatorname{grad} U(\mathbf{x})$ and div $\mathbf{v}(\mathbf{x}) = -4\pi G\rho(\mathbf{x})$ we obtain the **Poisson equation**

$$\operatorname{div}(\operatorname{grad} U(\mathbf{x})) = -4\pi G\rho(\mathbf{x})$$

for the points $\mathbf{x} \in W$ where there is source mass density, and the **Laplace equation**

$$\operatorname{div}(\operatorname{grad} U(\mathbf{x})) = 0$$

for $\mathbf{x}$ outside W, in empty space. The combined operator $\Delta U = \operatorname{div}(\operatorname{grad} U)$ can be expressed by means of the second partial derivatives of the potential

$$\Delta U = \operatorname{div}(\operatorname{grad} U) = \frac{\partial^2 U}{\partial x_1{}^2} + \frac{\partial^2 U}{\partial x_2{}^2} + \frac{\partial^2 U}{\partial x_3{}^2}.$$

A function $U(\mathbf{x})$ fulfilling the Laplace equation $\Delta U = 0$ is a **harmonic function**: the gravitational potential generated by an extended body with support W is harmonic in $\mathbb{R}^3 \setminus W$.

---

[4] This is why the first experiments to measure the gravitational constant G, performed by Newton and Cavendish, were described as measures of the mass of the Earth.

Harmonic functions have many important properties, including the following: they can have neither local maxima nor local minima (Evans 1998, Section 2.2). A function with continuous second derivatives and harmonic is smooth, that is it has continuous derivatives of whatever order. It is also analytic, that is with convergent Taylor series in a neighborhood of every point. This has an important implication in orbit determination, actually in all problems of celestial mechanics: the equation of motion and the corresponding general solution are always smooth, even analytic, for each body moving in empty space under the gravitational attraction of the other bodies. Only for non-gravitational perturbations might regularity problems occur (see Section 14.3).

Spherical symmetry

The simplest case in which the solutions of the Laplace equation can be explicitly computed is when the gravitational potential has spherical symmetry around the origin in the $\mathbf{x}$ space, that is $U(\mathbf{x}) = R(r)$ (where $r = |\mathbf{x}|$ and R is a smooth function). Then the Laplace operator Δ can be easily computed from the partials

$$\frac{\partial U}{\partial x_j} = \frac{dR}{dr}\frac{\partial r}{\partial x_j} = \frac{x_j}{r}\frac{dR}{dr},$$

$$\frac{\partial^2 U}{\partial x_j^2} = \frac{1}{r}\frac{dR}{dr} + \frac{x_j^2}{r}\left(\frac{1}{r}\frac{d^2 R}{dr^2} - \frac{1}{r^2}\frac{dR}{dr}\right)$$

and, by summing over $j = 1, 3$,

$$0 = \Delta U = 2\frac{1}{r}\frac{dR}{dr} + \frac{d^2 R}{dr^2} = \frac{1}{r^2}\frac{d}{dr}\left[r^2\frac{dR}{dr}\right];$$

we conclude that $r^2 dR/dr = -k$, with k an arbitrary constant. This gives all the possible spherically symmetric harmonic functions as $R(r) = \frac{k}{r} + const$. By selecting the additive constant to be 0, we find that the solution coincides with the gravitational potential of a point mass in $\mathbf{p} = \mathbf{0}$ with mass $M = k/G$.

This result has a very deep implication for all methods of gravimetry, including satellite geodesy. Let the mass density function be spherically symmetric, that is $\rho(\mathbf{p}) = \tilde{\rho}(|\mathbf{p}|)$ for some function $\tilde{\rho}$. The support W has necessarily spherical symmetry, and we say that the planet is spherically symmetric. Then the gravitational potential U is a spherically symmetric harmonic function in $\mathbb{R}^3 \setminus W$ and there is a positive constant M such that $U = GM/r$ outside W. By the divergence formula it can be shown that

M is indeed the mass as defined by eq. (13.1). Thus the gravity field of two spherically symmetric planets with equal mass are exactly the same, outside the support of both mass distributions; they are both the same as the potential of a point mass with the same M.

This implies that there is no way to determine the internal mass distribution of a planet, that is the mass density $\rho(\mathbf{p})$, by measuring the gravitational field either on or outside its surface. This limitation applies also to the methods of measurement using whatever consequence of the gravity field, including the orbits of satellites. Satellite gravimetry cannot solve for parameters describing the mass density function, unless such parameters are chosen in such a way that the "concentration" of the mass near the center depends upon the parameters to be measured by other means, e.g., by measuring the rotational properties of the planet (see Chapter 17).

13.2 Spherical harmonics

The example of spherically symmetric harmonic functions shows the advantage of using a system of coordinates adapted to the problem. To generalize this we use a coordinate system adapted to bodies with an approximate spherical shape, the **spherical polar coordinates** defined by

$$x_1 = r \cos\theta \cos\lambda, \qquad x_2 = r \cos\theta \sin\lambda, \qquad x_3 = r \sin\theta, \qquad (13.4)$$

that is, $r > 0$ is the distance from the selected center, θ the latitude $(-\pi/2 \leq \theta \leq \pi/2)$, and λ the longitude, an angle variable. On the Earth a reference system with the x_3 axis along the rotation axis and the (x_1, x_2) plane containing the equator is normally used.[5] Thus the gravitational potential can be expressed in polar coordinates: $U(x_1, x_2, x_3) = \Phi(r, \theta, \lambda)$.

To compute the Laplace operator in spherical polar coordinates we use the chain rule

$$\frac{\partial U}{\partial x_j} = \frac{\partial \Phi}{\partial r}\frac{\partial r}{\partial x_j} + \frac{\partial \Phi}{\partial \theta}\frac{\partial \theta}{\partial x_j} + \frac{\partial \Phi}{\partial \lambda}\frac{\partial \lambda}{\partial x_j},$$

and the derivatives of the inverse coordinate change (13.4)

$$\frac{\partial U}{\partial x_1} = \frac{\partial \Phi}{\partial r}\cos\theta\cos\lambda - \frac{\partial \Phi}{\partial \theta}\frac{1}{r}\sin\theta\cos\lambda - \frac{\partial \Phi}{\partial \lambda}\frac{\sin\lambda}{r\cos\theta}.$$

The second partial derivatives can be obtained by iterating the same procedure on the first derivatives and by summing we obtain an expression for

---

[5] The rotation axis of the Earth, as well as that of whatever planet, is not constant, thus the definition of the reference system requires some additional care.

ΔU containing only partial derivatives with respect to r, θ, λ:

$$r^2 \, \Delta U = \frac{\partial}{\partial r} \left(r^2 \frac{\partial \Phi}{\partial r} \right) + \Delta_S U, \tag{13.5}$$

where

$$\Delta_S U = \frac{1}{\cos \theta} \frac{\partial}{\partial \theta} \left(\cos \theta \frac{\partial \Phi}{\partial \theta} \right) + \frac{1}{\cos^2 \theta} \frac{\partial^2 \Phi}{\partial \lambda^2} \tag{13.6}$$

is called the **Laplace–Beltrami operator**; it is independent of r and can be applied to functions defined over a sphere $S(r)$ with fixed r, that is functions of (θ, λ) only.

Zonal spherical harmonics

We will first search for solutions of the Laplace equation $\Delta U = 0$ with **axial symmetry**, that is $U(x, y, z) = \Phi(r, \theta)$ (independent of λ). Then the Laplace operator in polar coordinates of (13.5) has the simpler expression

$$\Delta U = \frac{1}{r^2} \frac{\partial}{\partial r} \left[r^2 \frac{\partial \Phi}{\partial r} \right] + \frac{1}{r^2 \cos \theta} \frac{\partial}{\partial \theta} \left[\cos \theta \frac{\partial \Phi}{\partial \theta} \right] = 0.$$

We proceed to solve the partial differential equation above by **separation of variables**, that is by looking for special solutions as the product of a function of r and a function of θ: $U = \Phi(r, \theta) = R(r) \, F(\theta)$. Then

$$\Delta U = \frac{RF}{r^2} \left\{ \frac{1}{R} \frac{d}{dr} \left[r^2 \frac{dR}{dr} \right] + \frac{1}{F \cos \theta} \frac{d}{d\theta} \left[\cos \theta \frac{dF}{d\theta} \right] \right\} = 0.$$

The expression inside the bracket is the sum of a function of r and a function of θ. If the above equation is satisfied on an open set in the (r, θ) space, these functions are both constant. Let this constant be $\ell(\ell + 1)$; then

$$\frac{d}{dr} \left[r^2 \frac{dR}{dr} \right] = \ell(\ell + 1) \, R, \qquad \frac{d}{d\theta} \left[\cos \theta \frac{dF}{d\theta} \right] = -\ell(\ell + 1) \, F \cos \theta$$

are two ordinary differential equations to be satisfied by the functions $R(r)$ and $F(\theta)$. The first one has solutions of the form $R(r) = r^\gamma, \gamma \in \mathbb{R}$,

$$\frac{d}{dr} \left[r^2 \gamma r^{\gamma-1} \right] = \ell(\ell + 1) \, r^\gamma \iff \gamma(\gamma + 1) \, r^\gamma = \ell(\ell + 1) \, r^\gamma$$

with two possible solutions for γ: either $\gamma = \ell$ or $\gamma = -\ell-1$. By the standard existence and uniqueness theorem, the second-order ordinary differential equation for $R(r)$ has solutions depending upon two arbitrary constants A, B:

$$R(r) = A \, r^\ell + \frac{B}{r^{\ell+1}}.$$

To solve the equation for $F(\theta)$ we change variable: $\mu = \sin\theta$, $F(\theta) = f(\mu)$

$$\frac{1}{\cos\theta}\frac{d}{d\theta}\left[\cos\theta\,\frac{dF}{d\theta}\right] = \frac{d}{d\mu}\left[\cos^2\theta\,\frac{df}{d\mu}\right] = \frac{d}{d\mu}\left[(1-\mu^2)\,\frac{df}{d\mu}\right].$$

Thus the function $f(\mu)$ is a solution of a second-order linear equation, the **Legendre equation**:

$$(1-\mu^2)\frac{d^2 f}{d\mu^2} - 2\mu\frac{df}{d\mu} + \ell(\ell+1)\,f = 0. \tag{13.7}$$

The solutions of the Legendre equation are found by the method of **undetermined coefficients**, that is by representing $f(\mu)$ as a power series

$$f(\mu) = \sum_{k=0}^{+\infty} a_k\,\mu^k, \tag{13.8}$$

by substituting the series into (13.7) and by grouping terms according to their degree in μ:

$$0 = (1-\mu^2)\sum_{k=2}^{+\infty} a_k\,k(k-1)\,\mu^{k-2} - 2\mu\sum_{k=1}^{+\infty} a_k\,k\,\mu^{k-1} + \ell(\ell+1)\sum_{k=0}^{+\infty} a_k\,\mu^k$$

$$= \sum_{k=0}^{+\infty}\mu^k\left[a_{k+2}\,(k+2)(k+1) - a_k\,k(k-1) - 2\,k\,a_k + \ell(\ell+1)\,a_k\right].$$

The Legendre equation needs to be satisfied identically, for whatever θ in $-\pi/2 \le \theta \le \pi/2$, that is for $-1 \le \mu \le 1$, thus the coefficients of the power series above need to be all zero:

$$a_{k+2}\,(k+2)(k+1) - a_k\,[k(k-1) + 2k - \ell(\ell+1)] = 0$$

for every non-negative integer k. This is a second-order recursion formula, allowing us to solve for the unknown a_{k+2} provided a_k is known

$$a_{k+2} = \frac{k(k+1) - \ell(\ell+1)}{(k+2)(k+1)}\,a_k. \tag{13.9}$$

This recursion formula gives zero for $k = \ell$: this implies that we can find a solution with $f(\mu)$ polynomial in $\mu = \sin\theta$ provided ℓ is a non-negative integer.[6] If ℓ is even we set the initial conditions $a_0 \ne 0$ and $a_1 = 0$, and we obtain $f(\mu)$, a polynomial with only even degree monomials. If ℓ is odd we use $a_0 = 0$ and $a_1 \ne 0$, we obtain for $f(\mu)$ an odd polynomial. For example,

$$\ell = 0 \implies f(\mu) = a_0$$

---

[6] Polynomial solutions can also be obtained by selecting $-\ell = k+1$, but they are the same.

$$\ell = 1 \quad \Longrightarrow f(\mu) = a_1\,\mu = a_1\,\sin\theta$$

$$\ell = 2 \quad \Longrightarrow f(\mu) = -3\,a_0\,\mu^2 + a_0 = a_0\,(1 - 3\,\sin^2\theta).$$

By selecting for each integer ℓ a suitable constant factor we define a set of solutions of (13.7), the **Legendre polynomials** with argument $\sin\theta$:

$$P_\ell(\sin\theta) = \sum_{j=0}^{L} T_{\ell j}\,(\sin\theta)^{\ell - 2j}, \tag{13.10}$$

where L is the integer part of $\ell/2$ and $T_{\ell j}$ is a coefficient solution of an equation equivalent to (13.9) going backward, that is the value of the coefficient $T_{\ell 0}$ of the highest degree in $\sin\theta$ is assigned first (Kaula 1966, Chapter 1)

$$T_{\ell j} = -\frac{(\ell - 2j + 1)(\ell - 2j + 2)}{2j\,(2\ell - 2j + 1)}\,T_{\ell\,j-1}, \quad T_{\ell 0} = \frac{(2\ell)!}{(\ell!)^2\,2^\ell}. \tag{13.11}$$

The reason for this choice of $T_{\ell 0}$ will be explained later, by eq. (13.17). By combining the solution for $F(\theta)$ with the solutions for $R(r)$ we obtain two linearly independent solutions of the Laplace equation for each integer $\ell \geq 0$

$$P_\ell(\sin\theta)\,\frac{1}{r^{\ell+1}}, \quad P_\ell(\sin\theta)\,r^\ell,$$

where those with r^ℓ are smooth at the origin and unlimited for $r \to +\infty$, describing the gravity field *inside* a cavity surrounded by a mass distribution; they are **internal harmonics**. Those with $1/r^{\ell+1}$ are of interest for satellite orbits: they are singular for $r = 0$, and for $r \to +\infty$ they tend to 0, the **external harmonics**. In this book we shall consider only the external **zonal spherical harmonic** of degree ℓ. The Legendre polynomials have exactly ℓ real roots in the interval $-1 < \sin\theta < 1$ (Hobson 1931, pp. 18–19), that is the zonal spherical harmonics have as many zeros along each meridian as the degree ℓ. This can give an intuitive understanding of the shape corresponding to each harmonic; e.g., $\ell = 2$ corresponds to an oblate (or prolate) shape, with flattening along the x_3 axis, $\ell = 3$ to a pear shape giving different mass to the two hemispheres separated by the $x_3 = 0$ plane.

Tesseral spherical harmonics

To remove the assumption of axial symmetry, we look for solutions of the Laplace equation depending upon all the polar coordinates, by using again the separation of variables, that is assuming $U = \Phi(r,\theta,\lambda) =$

$R(r)F(\theta)\,G(\lambda)$. By eq. (13.5)

$$r^2\,\Delta\Phi = FG\,\frac{d}{dr}\left[r^2\,\frac{dR}{dr}\right] + R\,\Delta_S(FG) = 0$$

and dividing by $U = RFG$ we get an equation equivalent to Laplace:

$$\frac{r^2}{RFG}\,\Delta\Phi = \frac{1}{R}\frac{d}{dr}\left[r^2\frac{dR}{dr}\right] + \frac{\Delta_S(FG)}{FG} = 0.$$

The two terms in the equation being a function of r and a function of (θ,λ), they have to be constant, thus defining an ordinary differential equation for $R(r)$, the same as the zonal case, and the partial differential equation

$$\Delta_S(FG) = -\ell(\ell+1)\,FG,$$

that is, $F(\theta)G(\lambda)$ has to be an eigenfunction of the Laplace–Beltrami operator. The same argument can be applied to the two terms of

$$\frac{\cos^2\theta}{FG}\,\Delta_S(FG) = \frac{\cos\theta}{F}\frac{d}{d\theta}\left[\cos\theta\,\frac{dF}{d\theta}\right] + \frac{1}{G}\frac{d^2G}{d\lambda^2} = -\ell\,(\ell+1)\,\cos^2\theta.$$

By selecting a negative constant for the term containing λ

$$\frac{1}{G}\frac{d^2G}{d\lambda^2} = -m^2,$$

the equation for $G(\lambda)$ and its solutions are

$$\frac{d^2G}{d\lambda^2} + m^2\,G = 0 \iff G(\lambda) = C_{\ell m}\cos(m\lambda) + S_{\ell m}\sin(m\lambda),$$

pure trigonometric functions.[7] Thus the equation for $F(\theta)$ is

$$\frac{1}{F\cos\theta\,d\theta}\frac{d}{d\theta}\left[\cos\theta\,\frac{dF}{d\theta}\right] - \frac{m^2}{\cos^2\theta} = -\ell(\ell+1).$$

For $m = 0$ we obtain again eq. (13.7), that is the zonal harmonics. With the same arguments used in the previous section we can select for $R(r)$ only the solution $1/r^{\ell+1}$, to get the external harmonics. The equation for $F(\theta)$ is simplified by the change of variables $\mu = \sin\theta$: if we set

$$F(\theta) = (\cos\theta)^m\,f(\sin\theta),$$

the function $f(\mu)$ is a solution of the linear second-order differential equation

$$(1 - \mu^2)\frac{d^2 f}{d\mu^2} - 2(m+1)\,\mu\,\frac{df}{d\mu} + (\ell - m)(\ell + m + 1)\,f = 0\ . \qquad (13.12)$$

---

[7] If the constant for the λ term in the Laplace equation is positive, the solutions are a combination of exponentials; if m is not integer the trigonometric functions are not periodic of period 2π, in both cases not providing a smooth function of λ, an angle variable.

By using the series expansion (13.8) we obtain again a second-order recursion formula

$$a_{k+2} = \frac{k(k+2m+1) - (\ell-m)(\ell+m+1)}{(k+2)(k+1)} a_k. \qquad (13.13)$$

Again, as in the case of eq. (13.9), we can obtain solutions with a finite number of terms, a product of a polynomial in $\sin\theta$ of degree $k_{max} = \ell - m$ with a power of $\cos\theta$ can be obtained; the polynomial has only even powers if $\ell - m$ is even, odd powers if $\ell - m$ is odd. Thus a solution of the differential equation for $F(\theta)$ is the **Legendre associated function** of **harmonic degree** ℓ and **harmonic order** m

$$P_{\ell m}(\sin\theta) = (1 - \sin^2\theta)^{m/2} \sum_{j=0}^{L} T_{\ell m j}(\sin\theta)^{\ell-m-2j},$$

where L is the integer part of $(\ell-m)/2$, and the coefficients of the monomials in $\sin\theta$ are

$$T_{\ell m j} = -\frac{(\ell - m - 2j + 1)(\ell - m - 2j + 2)}{2j\,(2\ell - 2j + 1)} T_{\ell m\, j-1},$$

$$T_{\ell m 0} = (2\ell)!/\ell!\,(\ell - m)!\,2^\ell. \qquad (13.14)$$

By combining together the three functions $R(r)$, $F(\theta)$, and $G(\lambda)$, we find two **spherical harmonic** functions of degree ℓ and order m

$$\frac{P_{\ell m}(\sin\theta)}{r^{\ell+1}} \cos(m\lambda), \qquad \frac{P_{\ell m}(\sin\theta)}{r^{\ell+1}} \sin(m\lambda).$$

For $m = 0$ we obtain only one solution, which is a zonal spherical harmonic. For $m > 0$ we obtain **tesseral spherical harmonic** functions. Some qualitative properties of the spherical harmonics: the harmonic of degree ℓ and order m has $\ell - m$ zeros along each meridian and $2m$ zeros along each parallel (and zeros at the two poles for $m > 0$). For $\ell - m = 0$ we have **sectorial spherical harmonic** functions, independent of the latitude.

Expansion in spherical harmonics

The Laplace equation is linear, thus linear combinations of spherical harmonics are still solutions:

$$U = \frac{GM}{r} + \frac{GM}{r} \sum_{\ell=1}^{+\infty} P_\ell(\sin\theta) \frac{R_\oplus^\ell}{r^\ell} C_{\ell 0}$$

$$+ \frac{GM}{r} \sum_{\ell=1}^{+\infty} \sum_{m=1}^{\ell} P_{\ell m}(\sin\theta) \frac{R_\oplus^\ell}{r^\ell} [C_{\ell m} \cos(m\lambda) + S_{\ell m} \sin(m\lambda)]$$

where the length $R_\oplus$, to be interpreted as the equatorial radius of the Earth (or of the relevant planet), has been added to have adimensional coefficients $C_{\ell m}$, with $0 \le m \le \ell$, and $S_{\ell m}$ with $0 < m \le \ell$. M is the total mass of the planet as defined by eq. (13.1). By using the conventions $P_{\ell 0} = P_\ell$, $P_0 = 1$ we can use the more compact formula

$$U = \frac{GM}{r} \left\{ \sum_{\ell=0}^{+\infty} \sum_{m=0}^{\ell} P_{\ell m}(\sin\theta) \, \frac{R_\oplus^\ell}{r^\ell} \left[C_{\ell m} \cos(m\lambda) + S_{\ell m} \sin(m\lambda) \right] \right\}.$$

(13.15)

Another useful representation is by means of the set of harmonic functions on the sphere $r = R_\oplus$, which can be considered as functions of (θ, λ) only:

$$Y_{\ell m i} = P_{\ell m}(\sin\theta) \, \mathrm{trig}(m\lambda, i), \quad \mathrm{trig}(m\lambda, 1) = \cos(m\lambda), \quad \mathrm{trig}(m\lambda, 0) = \sin(m\lambda);$$

then the expansion (13.15) becomes

$$U = \sum_{\ell=0}^{+\infty} \frac{GM \, R_\oplus^\ell}{r^{\ell+1}} \sum_{m=0}^{\ell} \left[C_{\ell m} Y_{\ell m 1} + S_{\ell m} Y_{\ell m 0} \right].$$

(13.16)

What we need is a relationship between the expansions in series of harmonic functions, such as the one above, and the properties of the extended mass generating the gravity field, that is of the mass density function ρ. For this we need to restart from eq. (13.3) and to expand the kernel used to generate the potential U from the density ρ:

$$\frac{1}{|\mathbf{x} - \mathbf{p}|} = \frac{1}{|\mathbf{x}|} \left[1 - 2 \frac{|\mathbf{p}|}{|\mathbf{x}|} \cos\psi + \frac{|\mathbf{p}|^2}{|\mathbf{x}|^2} \right]^{-1/2}$$

(13.17)

where ψ is the angle between $\mathbf{x}$ and $\mathbf{p}$, that is $\cos\psi = \mathbf{x} \cdot \mathbf{p}/|\mathbf{x}| \, |\mathbf{p}|$. This inverse distance can be expanded as

$$\frac{1}{|\mathbf{x} - \mathbf{p}|} = \sum_{\ell=0}^{+\infty} \frac{|\mathbf{p}|^\ell}{|\mathbf{x}|^{\ell+1}} P_\ell(\cos\psi)$$

(13.18)

where P_ℓ are Legendre polynomials because $1/|\mathbf{x} - \mathbf{p}|$ is harmonic for $\mathbf{x} \ne \mathbf{p}$. However, the Legendre polynomials contain an arbitrary factor, thus we need to check that those appearing in eq. (13.18) have the coefficient we have selected in eq. (13.11). To confirm this we have to compute the coefficient of highest degree in $\cos\psi$ among the terms with factor $|\mathbf{p}|/|\mathbf{x}|$ in the expansion (13.17) and confirm it coincides with $T_{\ell 0}$:

$$\binom{-1/2}{\ell} (-2)^\ell = \frac{(-1)^\ell}{\ell!} \prod_{k=1}^{\ell} (2k - 1) = \frac{(2\ell)!}{(\ell!)^2 \, 2^\ell} = T_{\ell 0}.$$

By substituting the expansion in Legendre polynomials into eq. (13.3) we get

$$U(\mathbf{x}) = \int_W \frac{G\,\rho(\mathbf{p})}{|\mathbf{x}-\mathbf{p}|}\,d\mathbf{p} = \sum_{\ell=0}^{+\infty} \frac{G}{|\mathbf{x}|^{\ell+1}} \int_W \rho(\mathbf{p})\,|\mathbf{p}|^{\ell}\,P_{\ell}(\cos\psi)\,d\mathbf{p} \quad (13.19)$$

and comparing with (13.15), the part of U with factor $G/r^{\ell+1}$, where $r = |\mathbf{x}|$, is

$$\int_W \rho(\mathbf{p})\,|\mathbf{p}|^{\ell}\,P_{\ell}(\cos\psi)\,d\mathbf{p} = M\sum_{m=0}^{\ell} R_{\oplus}^{\ell}\,[C_{\ell m}\,Y_{\ell m 1} + S_{\ell m}\,Y_{\ell m 0}]. \quad (13.20)$$

Total mass and center of mass

As an example, let $\ell = 0$; $P_{\ell} = 1$ and the equation above reduces to

$$C_{00} = \frac{1}{M}\int_W \rho(\mathbf{p})\,d\mathbf{p} = 1.$$

Let $\ell = 1$; the spherical harmonics are

$$Y_{101} = P_{10} = x_3/r, \quad Y_{111} = P_{11}\cos\lambda = x_1/r, \quad Y_{110} = P_{11}\sin\lambda = x_2/r,$$

and eq. (13.20) multiplied by $r = |\mathbf{x}|$ becomes

$$\frac{1}{M}\,\mathbf{x}\cdot\int_W \rho(\mathbf{p})\,\mathbf{p}\,d\mathbf{p} = R_{\oplus}\,[C_{11}\,x_1 + S_{11}\,x_2 + C_{10}\,x_3],$$

the harmonic coefficients for $\ell = 1$ are related to the **center of mass**

$$\mathbf{c}^M = (1/M)\int_W \rho(\mathbf{p})\,\mathbf{p}\,d\mathbf{p};$$

in fact

$$C_{11} = c_1^M/R_{\oplus}, \quad S_{11} = c_2^M/R_{\oplus}, \quad C_{10} = c_3^M/R_{\oplus}.$$

As a consequence, if the origin of the coordinate system coincides with the planet center of mass, the coefficients of degree 1 are zero and the expansion of U after the point mass term begins with degree 2:

$$U = \frac{GM}{r}\left\{1 + \sum_{\ell=2}^{+\infty}\sum_{m=0}^{\ell} P_{\ell m}(\sin\theta)\left(\frac{R_{\oplus}}{r}\right)^{\ell}[C_{\ell m}\cos(m\lambda) + S_{\ell m}\sin(m\lambda)]\right\},$$

that is, the point mass approximation is accurate to $\mathcal{O}(R_{\oplus}^2/r^2)$.

Moments of inertia

Let $\ell = 2$; the Legendre functions are

$$P_{20} = \frac{3}{2} \sin^2 \theta - \frac{1}{2}, \quad P_{21} = 3 \sin \theta \cos \theta, \quad P_{22} = 3 \cos^2 \theta$$

and the spherical harmonics are

$$Y_{201} = P_{20} = (3 x_3^2 - r^2)/2 r^2, \qquad Y_{200} = 0,$$
$$Y_{211} = P_{21} \cos \lambda = 3 x_3 x_1/r^2, \qquad Y_{210} = P_{21} \sin \lambda = 3 x_3 x_2/r^2,$$
$$Y_{221} = P_{22} \cos(2\lambda) = 3 (x_1^2 - x_2^2)/r^2, \quad Y_{220} = P_{22} \sin(2\lambda) = 6 x_1 x_2/r^2,$$

and eq. (13.20) for $\ell = 2$ multiplied by $r^2/(M R_\oplus^2)$ becomes

$$\frac{r^2}{M R_\oplus^2} \int_W \rho(\mathbf{p}) |\mathbf{p}|^2 \frac{3 \cos^2 \psi - 1}{2} d\mathbf{p}$$
$$= \frac{C_{20}}{2} (2 x_3^2 - x_1^2 - x_2^2) + 3 C_{21} x_3 x_1 + 3 S_{21} x_3 x_2$$
$$+ 3 C_{22} (x_1^2 - x_2^2) + 6 S_{22} x_1 x_2.$$

The geophysical significance of the degree $\ell = 2$ coefficients can be understood by expressing them in terms of the integrals

$$A_{ij} = \frac{1}{M R_\oplus^2} \int_W \rho(\mathbf{p}) \, p_i \, p_j \, d\mathbf{p} \qquad i, j = 1, 2, 3.$$

Taking into account that

$$|\mathbf{x}|^2 |\mathbf{p}|^2 P_2(\cos \psi) = \frac{1}{2} \left[3 (\mathbf{x} \cdot \mathbf{p})^2 - |\mathbf{x}|^2 |\mathbf{p}|^2 \right]$$

and expanding as functions of the coordinates (x_1, x_2, x_3) and (p_1, p_2, p_3) we obtain the relationship between the integrals A_{ij} and the coefficients of harmonic degree $\ell = 2$

$$x_1^2 (2 A_{11} - A_{22} - A_{33}) + x_2^2 (2 A_{22} - A_{11} - A_{33}) + x_3^2 (2 A_{33} - A_{11} - A_{22})$$
$$+ 6 (x_1 x_2 A_{12} + x_2 x_3 A_{23} + x_1 x_3 A_{13})$$
$$= C_{20} (2 x_3^2 - x_1^2 - x_2^2) + 6 C_{21} x_3 x_1 + 6 S_{21} x_3 x_2 + 6 C_{22} (x_1^2 - x_2^2)$$
$$+ 12 S_{22} x_1 x_2.$$

The coefficients of polynomial degree 2 in the x_j variables provide a system of six equations, which is linear in the six integrals A_{ij} and in the five geopotential coefficients, and can be solved for the latter

$$C_{20} = A_{33} - \frac{A_{11} + A_{22}}{2}, \quad C_{22} = \frac{1}{4} (A_{11} - A_{22}),$$
$$C_{21} = A_{13}, \quad S_{21} = A_{23}, \quad S_{22} = \frac{1}{2} A_{12}.$$

If the origin of the coordinates, corresponding to $r = 0$, is the center of mass, then the integrals A_{ij} can be combined to provide the **inertia quadratic form** of the planet, represented by a symmetric positive definite 3×3 matrix with diagonal coefficients

$$I_{jj} = M\,R_\oplus^2\,(A_{ii} + A_{kk}) \quad \text{with } i \neq j \neq k \neq i \tag{13.21}$$

and off-diagonal

$$I_{ij} = -M\,R_\oplus^2\,A_{ij} \quad \text{with } i \neq j. \tag{13.22}$$

Then the degree 2 harmonic coefficients of the gravity field of the planet can be computed in terms of the inertia matrix

$$C_{20} = \frac{1}{M\,R_\oplus^2}\left[\frac{I_{11} + I_{22}}{2} - I_{33}\right], \quad C_{22} = \frac{1}{4\,M\,R_\oplus^2}\,(I_{22} - I_{11}),$$

$$C_{21} = \frac{-1}{M\,R_\oplus^2}\,I_{13}, \quad S_{21} = \frac{-1}{M\,R_\oplus^2}\,I_{23}, \quad S_{22} = \frac{-1}{2\,M\,R_\oplus^2}\,I_{12}. \tag{13.23}$$

It is possible to select a reference system diagonalizing the inertia quadratic form, that is such that $I_{ij} = 0$ for $i \neq j$; then C_{21}, S_{21}, S_{22} would be zero. The problem is that such a reference system is not known a priori: only information on the rotational state of the planet can be used to determine it. To solve for all the I_{jj} from the harmonic coefficients is not possible: some scale factor, e.g., the **concentration coefficient** $I_{max}/M\,R_\oplus^2$ where I_{max} is the largest eigenvalue of the inertia matrix, needs to be constrained by the rotation state information. This is another case of the following general property: the internal mass distribution cannot be determined by knowing only the gravity field outside of the body (see Section 13.1 and Chapter 17).

Recursion formulae

In the computation of the spherical harmonic functions at a given point with polar coordinates (r, θ, λ) it is not convenient to compute the coefficients $T_{\ell m j}$ from eq. (13.14), but rather the values of the Legendre polynomials and the associate functions could be computed by a **recursion formula**. There are many possible such formulae, we give here just one example: for the zonal harmonics (Hobson 1931, pp. 32–33)

$$\ell\,P_\ell(\mu) = (2\,\ell - 1)\,\mu\,P_{\ell-1}(\mu) - (\ell - 1)\,P_{\ell-2}(\mu) \tag{13.24}$$

and for the tesseral harmonics (Hobson 1931, pp. 107–108)

$$(\ell - m)\,P_{\ell m}(\mu) - (2\ell - 1)\,\mu\,P_{\ell-1m}(\mu) + (\ell + m - 1)\,P_{\ell-2m}(\mu) = 0. \tag{13.25}$$

Given the initial values

$$P_{00}(\mu) = 1, \qquad P_{10}(\mu) = \mu, \qquad P_{11}(\mu) = \sqrt{1 - \mu^2}$$

it is possible to compute all the Legendre polynomials and associate functions, up to a maximum degree ℓ_{max}, at some value of $\mu = \sin\theta$ by two-index recursion. The trigonometric functions $\sin(m\lambda), \cos(m\lambda)$ can also be computed recursively for a fixed λ by the trigonometric addition formulae. In this way it is possible to set up a very efficient algorithm to compute all the spherical harmonic functions up to $\ell = \ell_{max}$ at a given point.

To compute the partial derivatives of the potential, the gravity field components in the equation of motion and the second derivatives in the variational equation, it is efficient to use formulae expressing the derivatives of the spherical harmonics by means of combinations of Legendre polynomials, Legendre associated functions and trigonometric functions of λ. The derivatives with respect to r and λ are elementary; for those with respect to θ it is convenient to use the relationships between the Legendre polynomials and their derivatives (Hobson 1931, p. 32)

$$(\mu^2 - 1)\frac{dP_\ell(\mu)}{d\mu} = \ell\,[\mu\,P_\ell(\mu) - P_{\ell-1}(\mu)] \tag{13.26}$$

and for the associated Legendre functions (Wagner and Velez 1972, chapters 5–6)

$$\frac{dP_{\ell m}(\sin\theta)}{d\theta} = P_{\ell\,(m+1)}(\sin\theta) - m\,\tan\theta\,P_{\ell m}(\sin\theta) \tag{13.27}$$

where the first term $= 0$ for $m = \ell$ (we define $P_{\ell m} = 0$ for $m > \ell$).

13.3 The Hilbert space of the harmonic functions

What is the rigorous meaning of the infinite summation in eq. (13.15)? Are the **harmonic coefficients** $C_{\ell m}, S_{\ell m}$ uniquely defined for a given harmonic function $U(\mathbf{x})$? To solve these problems we need some additional properties of the spherical harmonics, and this requires some functional analysis.[8]

Orthogonality

The spherical harmonics on the sphere $Y_{\ell m i}$ have the property, which has been used in the procedure of separation of variables, of being eigenfunctions of the Laplace–Beltrami operator: $\Delta_S Y_{\ell m i} = -\ell(\ell+1)\,Y_{\ell m i}$. This implies

---

[8] Readers unfamiliar with functional analysis, such as Hilbert spaces, may skip this subsection and take for granted that eq.(13.15) is a uniquely defined, convergent series expansion.

that the functions $Y_{\ell m i}$ are orthogonal with respect to the scalar product defined by the surface integral over $S(1)$

$$\langle Y_{\ell m i}, Y_{\ell' m' i'} \rangle = \int_{S(1)} Y_{\ell m i}\, Y_{\ell' m' i'}\, d\,S = 0$$

unless $\ell = \ell', m = m', i = i'$. Indeed

$$-\ell(\ell+1)\,\langle Y_{\ell m i}, Y_{\ell' m' i'} \rangle = \langle \Delta_S\, Y_{\ell m i}, Y_{\ell' m' i'} \rangle = \int_{S(1)} \Delta_S\, Y_{\ell m i}, Y_{\ell' m' i'}\, d\,S$$

where the surface element in spherical polar coordinates is $d\,S = \cos\theta\, d\theta\, d\lambda$

$$\int_{S(1)} \Delta_S\, Y_{\ell m i}, Y_{\ell' m' i'}\, d\,S$$

$$= \int_0^{2\pi} d\lambda \int_{-\pi/2}^{\pi/2} [Y_{m'\ell' i'}\, \Delta_S\, Y_{\ell m i}]\, \cos\theta\, d\theta \qquad (13.28)$$

$$= I_\theta \int_0^{2\pi} tr(m'\lambda, i')\, tr(m\lambda, i)\, d\lambda = \pi\, \delta_{mm'}\, \delta_{ii'}\, I_\theta$$

where $\delta_{jk} = 0$ for $j \neq k$, $\delta_{jj} = 1$ and the integral I_θ over the variable θ can be computed by parts:

$$I_\theta = \int_{-\pi/2}^{\pi/2} \left[-\cos\theta\, \frac{\partial P_{\ell m}}{\partial\theta}\, \frac{\partial P_{\ell' m'}}{\partial\theta} - \frac{m^2}{\cos\theta}\, P_{\ell m}\, P_{\ell' m'} \right] d\theta.$$

The last step in (13.28) is due to the usual orthogonality of the sine and cosine functions over the interval $[0, 2\pi]$.

Given that $m = m'$, otherwise the other factor is zero, the integral I_θ is symmetric with respect to the exchange of (ℓ', m', i') with (ℓ, m, i), that is of the two spherical harmonics.[9] Then

$$-\ell(\ell+1)\,\langle Y_{\ell m i}, Y_{\ell' m' i'} \rangle$$
$$= \langle \Delta_S\, Y_{\ell m i}, Y_{\ell' m' i'} \rangle = \langle Y_{\ell m i}, \Delta_S\, Y_{\ell' m' i'} \rangle = -\ell'(\ell'+1)\,\langle Y_{\ell m i}, Y_{\ell' m' i'} \rangle$$

implying the scalar product is zero whenever $(\ell, m, i) \neq (\ell', m', i')$. Thus the spherical harmonic functions $\{Y_{\ell m i}\}$ are an orthogonal set and the coefficients $C_{\ell m}$ for $i = 1$, $S_{\ell m}$ for $i = 0$ define the corresponding components.

Normalization

The spherical harmonics $\{Y_{\ell m i}\}$ are not an orthonormal set, that is the squared L^2 norm on $S(1)$ is[10]

$$\langle Y_{\ell m i}, Y_{\ell m i} \rangle = \int_0^{2\pi} [trig(m\lambda, i)]^2\, d\lambda \int_{-\pi/2}^{\pi/2} \cos\theta\, [P_{\ell m}(\sin\theta)]^2\, d\theta$$

---

[9]   This symmetry applies to all functions on the sphere, that is the operator Δ_S is self-adjoint.
[10]   For this computation, see (Hobson 1931, p. 37) for $m = 0$, (Hobson 1931, p. 147) for $m > 0$.

$$\text{(for } m = 0, i = 1) \quad = \quad 2\pi \int_{-1}^{1} [P_\ell(\mu)]^2 \, d\mu = \frac{4\pi}{2\ell+1},$$

$$\text{(for } m > 0) \quad = \quad \pi \int_{-1}^{1} [P_{\ell m}(\mu)]^2 \, d\mu = \frac{2\pi}{2\ell+1} \frac{(\ell+m)!}{(\ell-m)!}.$$

Thus we can define the **normalized harmonics** and the normalized associate Legendre functions with unit quadratic mean on the sphere

$$\overline{Y}_{\ell m i} \quad = \quad \overline{P}_{\ell m}(\sin\theta) \, \text{trig}(m\lambda, i)$$

$$= \quad \sqrt{(2\ell+1)(2-\delta_{0m})\frac{(\ell-m)!}{(\ell+m)!}} \, Y_{\ell m i} = H_{\ell m} \, Y_{\ell m i}$$

($\delta_{0m} = 1$ for $m = 0$, $\delta_{0m} = 0$ otherwise). If we use $\{\overline{Y}_{\ell m i}\}$ as an orthonormal function set on $S(1)$ the expansion of the gravitational potential is

$$U = \frac{GM}{r} \left\{ \sum_{\ell=0}^{+\infty} \sum_{m=0}^{\ell} \overline{P}_{\ell m}(\sin\theta) \left(\frac{R_\oplus}{r}\right)^\ell \left[\overline{C}_{\ell m} \cos(m\lambda) + \overline{S}_{\ell m} \sin(m\lambda)\right] \right\}.$$

$$(13.29)$$

$\overline{C}_{\ell m}, \overline{S}_{\ell m}$ are *normalized harmonic coefficients* of degree ℓ and order m:

$$\overline{C}_{\ell m} = \frac{C_{\ell m}}{H_{\ell m}}, \quad \overline{S}_{\ell m} = \frac{S_{\ell m}}{H_{\ell m}}.$$

The degree $\ell = 1$ terms do not appear if the reference system has the center of mass as origin. For a given harmonic function $U = \Phi(r, \theta, \lambda)$, the normalized coefficients $\overline{C}_{\ell m}, \overline{S}_{\ell m}$ are uniquely defined by the harmonic function U, e.g., through the scalar products

$$\langle \overline{Y}_{\ell m 1}(\theta, \lambda), \Phi(R_\oplus, \theta, \lambda) \rangle \quad = \quad 4\pi \frac{GM}{R_\oplus} \overline{C}_{\ell m}$$

$$\langle \overline{Y}_{\ell m 0}(\theta, \lambda), \Phi(R_\oplus, \theta, \lambda) \rangle \quad = \quad 4\pi \frac{GM}{R_\oplus} \overline{S}_{\ell m}$$

$$\langle \overline{Y}_{\ell 0 1}(\theta), \Phi(R_\oplus, \theta, \lambda) \rangle \quad = \quad 4\pi \frac{GM}{R_\oplus} \overline{C}_{\ell 0}.$$

For the computation of the normalized spherical harmonics $\overline{Y}_{\ell m i}$ there are recursion formulae replacing those of the previous section. The reason for using them is that for ℓ large and m also large the L^2 norm of the unnormalized harmonic $Y_{\ell m i}$ grows to enormous values, producing computer overflows in the recursion formulae.[11] Balmino *et al.* (1990) give an algorithm in Cartesian coordinates (thus free from the singularities at the poles) and using normalized harmonic functions and coefficients.

---

[11] Long before overflow, the recursion is unstable and gives inaccurate results.

Convergence

Is the expansion (13.15) convergent? The series expansion (13.18) is uniformly convergent on each sphere $|\mathbf{x}| = r$ for every $r > |\mathbf{p}|$; this follows from the properties of the Taylor series of the function $(1 + z)^{-1/2}$ (holomorphic for $|z| < 1$). The formula (13.18) is a power series in $1/|\mathbf{x}|$, the integral is a continuous operator, thus it is convergent for $|\mathbf{x}| > R$ provided the support W of the mass density ρ is contained in the open ball $|\mathbf{x}| < R$.

A more subtle issue is what can be assumed about the convergence on the sphere $|\mathbf{x}| = R_\oplus$ if the support W touches it. If $\rho(\mathbf{x})$ is continuous, it is zero on the boundary of W, thus the potential is harmonic on $|\mathbf{x}| \geq R_\oplus$. However, a solid planet could be modeled with a mass density jumping discontinuously to zero at the surface. The empirical **Kaula rule** was introduced to somewhat model this behavior (Kaula 1966, Chap. 5):

$$\text{RMS}(\overline{C}_{\ell m}) = \text{RMS}(\overline{S}_{\ell m}) = K/\ell^2 \qquad (13.30)$$

where the coefficients of degree ℓ are taken to be random variables with standard deviation proportional to a constant $K \simeq 10^{-5}$ for the Earth, with other values for the other terrestrial planets (e.g., $K \simeq 10^{-4}$ for the Moon). In this way the series of spherical harmonics is convergent on the sphere $|\mathbf{x}| = R_\oplus$ in the L^2 norm, although only slowly, like the series $\sum 1/\ell^3$.

Completeness

The set of functions $\{\overline{Y}_{\ell m i}(\theta, \lambda)\}$ is a **complete orthonormal sequence** if it is a basis for the functional space of the harmonic functions on the sphere (Hobson 1931, p. 40–41). This requires that a function g orthogonal to every element of the sequence is identically zero on the sphere:

$$\langle g, \overline{Y}_{\ell m i} \rangle = \int_S g(\theta, \lambda) \, \overline{Y}_{\ell m i}(\theta, \lambda) \, dS = 0 \implies g(\theta, \lambda) = 0 \text{ for all } \theta, \lambda.$$

For a proof see (Hobson 1931, pp. 146–147) and (Albertella 1993, pp. 89–91).

The question arises of what is the place in this context of the solutions of the Legendre equation (13.7), and of the similar equation (13.12), obtained by allowing the power series expansion (13.8) to be infinite. It can be shown (Hobson 1931, p. 12 and Chapter V) that these infinite series of powers of $\mu = \sin \theta$ are not convergent for μ in the closed[12] interval $[-1, 1]$, thus their sums are not harmonic functions on any complete sphere $S(r)$.

A consequence of the convergence of the expansion and of the completeness of the basis $\{\overline{Y}_{\ell m i}\}$ is the solution of the **exterior Dirichlet problem**

---

[12] Although some are not uniformly convergent for μ in the open interval $\,]-1, 1[$.

with a spherical boundary: given assigned values on the sphere $r = R$, that is $\Phi(R, \theta, \lambda) = f(\theta, \lambda)$, where f is continuous on the sphere, the function Φ, harmonic outside of the sphere, exists and is uniquely determined. The expansion of Φ in normalized spherical harmonics is uniquely determined by the integrals on the sphere $\langle \overline{Y}_{\ell m k}(\theta, \lambda), f(\theta, \lambda) \rangle$, and then the sum of the series expansion such as eq. (13.29) is harmonic outside the sphere. It is unique because the difference of two such functions would have zero harmonic coefficients.

13.4 The gravity field along the orbit

We have expressed the gravity field as an expansion in spherical harmonics, functions of the satellite position $\mathbf{x}$. We would like to find an expression for the gravitational potential, and derived quantities such as the gravity field and the gravity gradient experienced by the satellite as a function of time, assuming the satellite follows an unperturbed two-body orbit. To this purpose we shall consider the gravitational potential U decomposed in harmonics

$$U = \frac{GM}{r} + \sum_{\ell=2}^{+\infty} \sum_{m=0}^{\ell} U_{\ell m},$$

where $U_{\ell m}$ is the component of degree ℓ and order m.

Equatorial orbit

We give an example to show that the potential can be expanded as a function of the orbital elements. Let us assume the satellite orbit is equatorial. The orbital elements are only (a, e, ϖ, l_0), with ϖ the angle between the inertial x_1 axis and the direction of the pericenter, l_0 the mean anomaly l at epoch t_0. Moreover, let us assume the planet is rotating around the axis x_3 with constant angular velocity $\Omega_\oplus$, with phase zero at $t = t_0$, thus the rotation phase is $\phi = \Omega_\oplus (t - t_0)$. Then

$$\begin{aligned} l &= n(t - t_0) + l_0, \qquad n = \sqrt{\frac{GM}{a^3}} \\ \lambda &= v(l) + \varpi - \phi, \qquad r = r(l) \end{aligned}$$

where the functions $v(l)$ (true anomaly) and $r(l)$ have to be computed by solving the Kepler equation. Since $I = 0$ the latitude $\theta = 0$ and the harmonic of the potential $U_{\ell m}$ of degree ℓ and order m along the satellite orbit is

$$U_{\ell m} = \frac{GM\,R_\oplus^\ell}{r^{\ell+1}} P_{\ell m}(0) \left[C_{\ell m} \cos(\psi_m) + S_{\ell m} \sin(\psi_m) \right] \tag{13.31}$$

where $\psi_m = m(v + \varpi - \phi)$. If the orbit is also circular, then $v = l$, $r = a$, $\psi_m = m[\varpi + (n - \Omega_\oplus)(t - t_0) + l_0]$, and the only frequency in the signal as a function of time is $m(n - \Omega_\oplus)$.

From this simple example we can already draw an interesting conclusion: for an equatorial circular orbit all the cosine harmonics with the same order m have the same dependence upon time, the same result holds for the sine terms. Moreover, the $C_{\ell m}$ and the $S_{\ell m}$ terms have the same spectrum of frequencies.

The observations do not directly measure the potential,[13] but many observables are obtained by either partial derivatives or time integrals from U, e.g., the gravity gradient can be measured by a gradiometer, and these observables can also be represented as a sum of spherical harmonics.

If the orbit were to remain circular and planar, the simple model above would imply that there is an exact rank deficiency such that only $2(\ell_{max} - 1)$ harmonic coefficients could be determined among those of degree $2 \leq \ell \leq \ell_{max}$. For example, the sectorial spherical harmonic coefficients $C_{\ell\ell}, S_{\ell\ell}$ could be solved, while all the others should be left among the consider parameters.

The orbit can remain neither circular nor planar because of the perturbations by the gravitational potential. However, the effects of the orbit perturbations on the perturbing potential are of second order in the small parameters $C_{\ell m}, S_{\ell m}$ and the above computations are correct to first order. Thus this model problem shows two important features of satellite geodesy: first, each spherical harmonic generates a signal in the observations with a characteristic frequency spectrum; second, there are spherical harmonics with different ℓ and m giving highly correlated signals, resulting in either exact or at least approximate rank deficiency.

Kaula expansion

For a general expansion of the gravitational potential as a Fourier series containing the orbital elements, we need to consider that in a reference system with the orbit plane as reference plane the expression of $U_{\ell m}$ is given by eq. (13.31). Thus we only need to perform a rotation of coordinates, from a reference system defined by the planet's equatorial plane and by some body-fixed direction in it to a reference system adapted to the Keplerian orbit (defined by the osculating orbital elements). The rotation from the (x, y, z) equatorial reference system to the (x', y', z') reference system with x' axis along the direction of the pericenter and z' axis along the angular

---

[13] Apart from the measurements of the ocean surface, a good approximation of an equipotential surface corresponding to the ocean mean potential, the **geoid**, by satellite altimetry.

momentum direction $\hat{\mathbf{c}}$ is almost the same as used in Section 6.5, with the
difference that the first rotation around the z axis is by an angle $\alpha = \Omega - \phi$

$$
\begin{bmatrix} x \\ y \\ z \end{bmatrix} = R_{\omega\hat{\mathbf{c}}}\, R_{I\hat{N}}\, R_{\alpha\hat{z}} \begin{bmatrix} x' \\ y' \\ z' \end{bmatrix} = R(\alpha, I, \omega) \begin{bmatrix} x' \\ y' \\ z' \end{bmatrix}
$$

where $\hat{N}$ is the current ascending node direction in the rotating equatorial plane; the angles (α, I, ω) play the role of Euler angles.[14] Thus
the composite rotation $R(\Omega, I, \omega)$, when applied to the scalar field $U_{\ell m}$,
transforms it into another function $U'_{\ell m}$ corresponding by value, that is
$U_{\ell m}(x, y, z) = U'_{\ell m}(x', y', z')$, which is equally harmonic because the Laplace
operator is rotation invariant. The rotation leaves the radius r invariant,
thus it also preserves the decomposition of the harmonic function U into homogeneous components of homogeneity degree $-\ell - 1$, that is of harmonic
degree ℓ. Thus the transformed $U'_{\ell m}$ can be expanded in spherical harmonics
by using only spherical harmonics of degree ℓ and any order $k = 0, \dots, \ell$.
We use the expansion in the form of eq (13.16)

$$
U'_{\ell m} = \frac{GM\, R_\oplus^\ell}{r^{\ell+1}} \sum_{k=0}^{\ell} \left[C'_{\ell k} Y_{\ell k1}(\theta', \lambda') + S'_{\ell k} Y_{\ell k0}(\theta', \lambda') \right]
$$

where the new coefficients $C'_{\ell k}, S'_{\ell k}$ are linear combinations of the old ones.
In the simple case $I = 0, \omega = 0$ we have $\theta = \theta'$ and $\lambda = \lambda' + \alpha$, thus the new
coefficients are obtained by

$$
\begin{bmatrix} C'_{\ell m} \\ S'_{\ell m} \end{bmatrix} = \begin{bmatrix} \cos(m\alpha) & -\sin(m\alpha) \\ \sin(m\alpha) & \cos(m\alpha) \end{bmatrix} \begin{bmatrix} C_{\ell m} \\ S_{\ell m} \end{bmatrix}
$$

with $C'_{\ell k} = S'_{\ell k} = 0$ for $k \neq m$. This results in an equation like (13.31).

The intermediate rotation around the ascending node axis has a more
complicated effect, mixing the spherical harmonics with the same order and
different degrees, thus it is expressed by a full $(2\ell + 1) \times (2\ell + 1)$ matrix
with coefficients function of I. There is a large literature on the computation of this matrix of conversion coefficients, with methods based either
on spherical trigonometry (Kaula 1966, Chap. 3) or on the theory of group
representations (Wigner 1959); they can be found in textbooks on quantum
mechanics such as (Edmonds 1957) as well as in papers about geophysics

---

[14] There are different types of Euler angles: this particular set, used in celestial mechanics, is of
type 3-1-3 in that the sequence of rotations is along the current z axis, the current x axis, the
current z axis.

(Sneeuw 1991, Jeffreys 1965). In the end, the component $U'_{\ell m}$ of the potential can be expanded with coefficients the **inclination functions** $F_{\ell m p}(I)$

$$U_{\ell m} = \frac{GM\,R_\oplus^\ell}{r^{\ell+1}} \sum_{p=0}^{\ell} F_{\ell m p}(I)\,[C_{\ell m}\,\cos(\psi_{\ell m p}) + S_{\ell m}\,\cos(\psi_{\ell m p})]$$

where the argument of the trigonometric function is

$$\psi_{\ell m p} = (\ell - 2p)(\omega + v) + m(\Omega - \phi) - \frac{\pi}{2}\,[(\ell - m)\,\mathrm{mod}\,2], \qquad (13.32)$$

the last term indicating that $C_{\ell m}, S_{\ell m}$ are replaced by $-S_{\ell m}, C_{\ell,m}$ when $\ell - m$ is odd. The inclination functions can be expressed as a trigonometric polynomial in $\sin I$ and $\cos I$ (Kaula 1966, eq. (3.62), p. 34)

$$
\begin{aligned}
F_{\ell m p}(I) &= \sum_{t=0}^{min(p,k)} \frac{(2\ell - 2t)!}{t!(\ell - t)!(\ell - m - 2t)!2^{2\ell-2t}} \sin^{\ell-m-2t} I \\
&\times \sum_{s=0}^{m} \binom{m}{s} \cos^s I \sum_{c} \binom{\ell - m - 2t + s}{c}\binom{m - s}{p - t - c}(-1)^{c-k}
\end{aligned}
$$

$$(13.33)$$

where k is the integer part of $(\ell - m)/2$ and c is summed over values making the binomial coefficients non-zero, that is with the lower index non-negative and not larger than the upper one. The inclination functions $F_{\ell m p}(I)$ with the indexes up to four are given by Kaula (1966, Table 1, pag. 34–35)[15].

The formula above, called the **Kaula expansion**, can be practically used, with some caution in the computation of the binomial coefficients, even for comparatively large ℓ, m. However, for near polar orbits, it is more convenient to use a formula based upon modified Jacobi polynomials introduced by Kinoshita *et al.* (1974); it contains only powers of $\cos I$ and therefore can be truncated for an approximate expansion near $I = 90°$. This formula has been converted to Kaula notation by Milani and Knežević (1995). For $k = \ell - 2p > 0$ the terms up to order 2 in $\cos I$ are

$$F_{\ell m p}(I) = \sum_{r=max(0,k-m)}^{min(\ell-m,\ell+k)} (-1)^t \frac{(2\ell - 2p - 1)!!(\ell + m)!(\ell - k)!}{2^{\ell+p}p!(m - k + r)!(\ell + k - r)!r!(\ell - m - r)!}$$

$$\times \left\{ 1 + (\ell - m - 2r + k)\cos I + \left[k(\ell - m - 2r) - r(\ell - m - r) \right. \right.$$

---

[15] However, Milani and Knežević (1995, Section 2.5) have found that the expression for $\ell m p = 420$ should have a factor $\sin^2 I$ instead of $\sin I$ as in the table, and the function for $\ell m p = 422$ should have $-(15/4)\sin^2 I$ instead of $+(15/4)\sin^2 I$.

$$+ \frac{k^2 - m + r(r-1) + (\ell - m - r)(\ell - m - r - 1)}{2}\Bigg] \cos^2 I \Bigg\} \qquad (13.34)$$

where t is the integer part of $(\ell - m + 1 + 2r)/2$. For $\ell - 2p < 0$ we set $k = 2p - \ell$ and use

$$F_{lmp}(I) = \sum_{r=max(0,k-m)}^{min(l-m,l+k)} (-1)^t \frac{(2l - 2p - 1)!!(l + m)!(l + k)!}{2^{l+p}p!(m - k + r)!(l + k - r)!r!(l - m - r)!}$$

$$\times \Bigg\{1 - (l - m - 2r + k)\cos I + \Bigg[k(l - m - 2r) - r(l - m - r)$$

$$+ \frac{k^2 - m + r(r-1) + (l - m - r)(l - m - r - 1)}{2}\Bigg] \cos^2 I \Bigg\}$$

$$(13.35)$$

where t is the integer part of $(3l - 3m + 1 + 2r)/2$.

When the expansions of the two-body problem are substituted in these expressions (in particular the expansion in powers of the eccentricity e), the dependence upon e is contained in **eccentricity functions** $G_{\ell pq}(e)$ and new arguments appear:

$$U_{\ell m} = \frac{GMR_\oplus^\ell}{a^{\ell+1}} \sum_{p=0}^{\ell} F_{\ell m p}(I) \sum_{q=-\infty}^{+\infty} G_{\ell pq}(e)\,[C_{\ell m}\,\cos(\psi_{\ell m p q}) + S_{\ell m}\,\cos(\psi_{\ell m p q})]$$

where the argument of the trigonometric function

$$\psi_{\ell m p q} = (\ell - 2p)\omega + (\ell - 2p + q)l + m(\Omega - \phi) - \frac{\pi}{2}\,mod(\ell - m, 2) \quad (13.36)$$

contains the mean anomaly l rather than the true anomaly. The eccentricity functions $G_{\ell pq}(e)$ are analytic in e and the lowest order term contains e^q. Their explicit computation is not simple; for the lowest order terms in the eccentricity functions with ℓpq up to 442 see (Kaula 1966, Table 2, p. 38).

13.5 Frequency analysis, ground track, and resonance

The most immediate consequence of the Kaula expansions for the geopotential perturbing function of the previous section is the possibility of listing all the frequencies which will appear in the first-order perturbations. By taking the time derivative of eq. (13.36)

$$\frac{d\psi_{\ell m p q}}{dt} = (\ell - 2p + q)\,n - m\,\dot\phi + \Big[(\ell - 2p)\,\dot\omega + m\,\dot\Omega\Big] = \nu_{lmpq} \quad (13.37)$$

where the dot stands for time derivative, and n is the mean motion. In a two-body approximation this is just a combination with integer coefficients of two constant frequencies, n and $\Omega_\oplus$. In a better approximation, the slow frequencies of precession of the elements ω, Ω, resulting from the zonal harmonics, also appear in the term between square brackets, thus in the frequency spectrum. The Kaula expansion allows us to compute this effect by averaging over the mean anomaly, that is by selecting the *secular terms* not containing it, with $\ell - 2p + q = 0$. For the simple case $e = 0$, or anyway to order zero in e: if $\ell = 2p$, that is for even order zonal harmonics,

$$\frac{1}{2\pi} \int_0^{2\pi} U_{\ell 0} \, dl = \frac{G M R_\oplus^\ell}{a^{\ell+1}} F_{\ell 0 p}(I) \, C_{\ell 0}.$$

The secular perturbation, that is the one generated by the secular terms, can be computed by the **Lagrange perturbative equations**, providing the perturbations in the elements to first-order in the small parameters, such as $C_{\ell 0}$. For the longitude of the node the Lagrange equation is

$$\frac{d\Omega}{dt} = \frac{1}{n \, a^2 \sqrt{1 - e^2} \, \sin I} \frac{\partial R}{\partial I},$$

where R is the perturbing function, i.e., the potential U without the monopole term. The secular perturbations on Ω result into a uniform precession with frequency

$$\overline{\frac{d\Omega}{dt}} = G M R_\oplus^\ell n \, a^3 \sin I \sum_{p=0}^{+\infty} \frac{R_\oplus^\ell}{a^\ell} F'_{2p \, 0 \, p}(I) \, C_{2p0} \left[1 + \mathcal{O}(e^2)\right].$$

If we use this secular value as $\dot{\Omega}$ in eq. (13.37) with $q = 0$ (for very low eccentricity), and also approximate $\dot{\omega}$ with a constant value (by a similar computation of the secular perturbation), for $q = 0$ the frequencies in the potential as a function of time, along the unperturbed two-body orbit, are

$$\nu_{\ell m p 0} = (\ell - 2p)(n + \dot{\omega}) + m(\dot{\Omega} - \dot{\Omega}_\oplus), \tag{13.38}$$

all integer combinations of two basic frequencies, although these are somewhat different from the two-body ones. The Lagrange equations contain only partials of the perturbing potential, thus the first-order perturbations can be obtained by the term-wise integration of a two-frequency Fourier series.

Resonance

Let us suppose the orbit of the satellite is exactly resonant with the rotation of the Earth, that is there are two integers j, k such that

$$\frac{n + \dot{\omega}}{j} = \frac{\dot{\Omega}_\oplus - \dot{\Omega}}{k} = \nu. \tag{13.39}$$

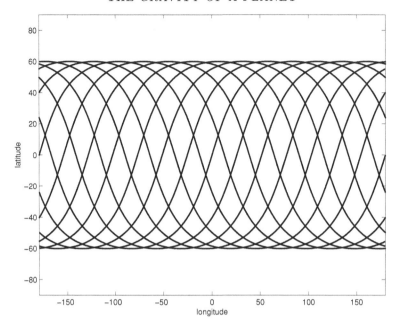

Fig. 13.1. The ground track of a circular orbit with $I = 60°$ and in a resonance $j/k = 14/1$.

In the two-body approximation, the period of the satellite would be exactly k/j sidereal days. Neglecting the $q \neq 0$ terms, the frequencies in the time series are multiples of the basic frequency ν

$$\nu_{\ell m p 0} = [(\ell - 2p)\,j - m\,k]\,\nu. \tag{13.40}$$

Thus the signal from all harmonics is periodic of period $2\pi/\nu$ and can be described as a Fourier series with arguments a multiple of $\nu\,t$. In a reference frame rotating with the Earth, the orbit is periodic and the **ground track**, the vertical projection of the orbit on a sphere of radius $R_\oplus$, returns on the same curve with a repeat cycle of j days (Figure 13.1).

The geometric properties of the periodic orbits are reflected in the analytical form of the first-order perturbations: if in eq. (13.40) the integer inside the square brackets is zero, there is a zero frequency in the perturbative equations and thus enhanced perturbations on some of the elements. For example, for the geosynchronous satellites $j = k = 1$, at order $\ell = 2$ there are zero frequencies for $p = 0, m = 2$ with coefficients containing C_{22}, S_{22} and for $p = 1, m = 0$ with C_{20}. In such cases the first-order solution obtained by simple quadrature is not a good approximation, and a different method needs to be used to identify the first approximation to be used, in the form of a pendulum-like equation (Kaula 1966, Section 3.6).

14

NON-GRAVITATIONAL PERTURBATIONS

The non-gravitational perturbations arise because outer space is not empty. First, planetary atmospheres extend to large altitudes, where they can be thin enough to allow for a satellite orbit but still generate a significant aerodynamic drag, given the high relative velocity of the spacecraft. As will be discussed in Chapter 16, there is interest in satellites orbiting as low as possible to determine high-order harmonics of the planetary gravity field. This may require propulsion to compensate for orbital decay and/or the use of on-board accelerometers to measure the amount of drag; that is, non-gravitational perturbations are critical in the design phase of the mission.

Second, outer space is pervaded by electromagnetic radiation: the light arriving directly from the Sun, reflected by the Earth, and by the other planets. The photons exchange momentum with spacecraft when they are absorbed and reflected; spacecraft themselves emit infrared radiation and electromagnetic waves carrying away some momentum. The resulting accelerations are small, but at the level of accuracy of current tracking systems they are not negligible, hence the need to model and/or measure them. Even small natural bodies, such as asteroids with diameters in the km range, have orbits affected by non-gravitational perturbations in a measurable way.

This chapter cannot be a full revision of the textbook by Milani *et al.* (1987), which should take into account all the new results accumulated in more than 20 years since its publication. It is just an update on the issues raised by non-gravitational perturbations in the orbit determination problem, taking into account state-of-the-art instrumentation. The conclusions, which will result from the orders of magnitudes and from an understanding of the complexity of the problem, will be the following: to model the non-gravitational perturbations is possible but they are anyway the main limitation to the accuracy of the orbit determination. For spacecraft, to measure non-gravitational accelerations directly with on-board instrumentation (see Section 16.1) allows for a much higher performance. The orbits of especially well observed asteroids, including some targets of space missions, may require special effort in non-gravitational force modeling.

14.1 Direct radiation pressure

Outer space is everywhere full of radiation, from different sources. We shall consider for now radiation, mostly visible light, coming from a single source, the Sun; we approximate the illumination from the Sun as if it were a point source. The photons of light carry energy and linear momentum: if the energy flux is measured by the intensity $\Phi_\odot$, the flux of momentum per unit cross-sectional area is $\Phi_\odot/c$, with c the speed of light. The total transfer of linear momentum upon impact of the photons from the Sun on the spacecraft surface generates the **direct radiation pressure**.

Interaction of radiation with the surface

To model what happens to the momentum carried by the radiation impacting the skin of the spacecraft, we can use a combination of three standard physical models: **absorption**, in which the photons are gobbled up by the spacecraft (acting as a black body), **reflection**, in which the photons bounce on a smooth surface following the laws of mirror reflection, and **diffusion** in which the photons are re-emitted with intensity following **Lambert's law**, that is proportionally to the cosine of the angle θ from the normal to the surface. The mix of these three phenomena is controlled by three positive constants α, ρ, δ (Milani *et al.* 1987, Chapter 4), with

$$\alpha + \rho + \delta = 1 \qquad (14.1)$$

expressing the fraction of the incident light that behaves according to the absorption, reflection, and diffusion law, respectively.

Under these hypotheses, the force applied by radiation pressure on an outer surface element dS of the spacecraft is directed in part along the normal $\hat{\mathbf{n}}$ to the surface, in part along the direction to the Sun $\hat{\mathbf{s}}$. The cross-sectional area of the surface element with respect to the radiation flux is $\cos\beta\, dS$ where $\cos\beta = \hat{\mathbf{s}} \cdot \hat{\mathbf{n}}$. For the surface to be illuminated, $\cos\beta > 0$ is necessary; the same condition is also sufficient if the shape is convex, otherwise there could be mutual shadowing between different parts.

The momentum of the absorbed photons is transferred to the spacecraft, thus the force acting on a surface element due to the absorption is

$$d\,\mathbf{F}_\alpha = -\,\frac{\Phi_\odot}{c}\,\alpha\,\cos\beta\,\hat{\mathbf{s}}\,dS.$$

The reflected photons transfer to the spacecraft the momentum they had upon arrival and the recoil momentum: the sum is directed along $\hat{\mathbf{n}}$

$$d\,\mathbf{F}_\rho = -\,\frac{\Phi_\odot}{c}\,2\rho\,\cos^2\beta\,\hat{\mathbf{n}}\,dS.$$

The photons of the fraction diffused are first absorbed, giving a force in the $-\hat{\mathbf{s}}$ direction equal to that of the absorption. Then they are re-emitted in different directions: for symmetry reasons, the resultant force is directed opposite to the surface normal $\hat{\mathbf{n}}$, and its intensity contains the integrals

$$\frac{\int_0^{2\pi} d\lambda \int_0^{\pi/2} \cos^2\theta\,\sin\theta\,d\theta}{\int_0^{2\pi} d\lambda \int_0^{\pi/2} \cos\theta\,\sin\theta\,d\theta} = \frac{2}{3},$$

thus the total force is

$$d\,\mathbf{F}_\delta = -\,\frac{\Phi_\odot}{c}\,\delta\,\cos\beta\left[\hat{\mathbf{s}} + \frac{2}{3}\hat{\mathbf{n}}\right]dS.$$

The effect on the spacecraft orbit is due to the resultant of the forces on all the surface elements: if S_I is the portion of the outer surface illuminated by the Sun,

$$\mathbf{F} = -\,\frac{\Phi_\odot}{c}\int_{S_I}\left[(1-\rho)\,\cos\beta\,\hat{\mathbf{s}} + \left(\frac{2}{3}\delta + 2\rho\,\cos\beta\right)\cos\beta\,\hat{\mathbf{n}}\right]dS, \quad (14.2)$$

where we have used (14.1), is the **radiation pressure force** acting on the spacecraft. Note that, unless the shape has some special symmetry, it is by no means guaranteed that the resultant force is applied at the center of mass of the spacecraft, thus radiation pressure will also affect the spacecraft rotation state, also called the **attitude**.

For some simple shapes the integral (14.2) can be computed analytically, e.g., for a sphere with constant α, ρ, δ the radiation pressure force is

$$\mathbf{F} = -\,\frac{\Phi_\odot\,\mathcal{A}}{c}\,\hat{\mathbf{s}} \qquad\qquad (14.3)$$

where the **effective cross-section** $\mathcal{A}$ is (Milani *et al.* 1987, p. 74–75)

$$\mathcal{A} = \left(\alpha + \rho + \frac{13}{9}\delta\right)\pi\,R^2 = \left(1 + \frac{4}{9}\delta\right)\pi\,R^2,$$

that is the geometric cross-sectional area times a coefficient depending upon

δ. For this the name "reflectivity coefficient" and the symbol C_R are used, but this is not logical, taking into account that this coefficient is always > 1.

The same formula (14.3) applies for a flat panel oriented orthogonally to $\hat{s}$, such as an optimally oriented solar array. For example, for $A = 16$ m^2 of solar panels, with large $\alpha \geq 0.8$ (thus $1 + 4/9\,\delta \simeq 1$ and $\mathcal{A} \simeq A$) and a total mass of the spacecraft $M = 500$ kg, the area-to-mass ratio is $A/M = 0.32$, the radiation pressure acceleration at 1 AU from the Sun, where $\Phi_\odot \simeq 1.38$ kW/m^2, is directed along $-\hat{s}$ with an intensity $\simeq 1.5 \times 10^{-5}$ cm/s^2. For a spacecraft with a comparatively simple shape, say a box shaped bus with $\simeq 4$ m^2 faces, the radiation pressure acceleration has a variable component smaller in size by a factor of 3 or 4, depending upon the bus attitude and surface properties.

For a realistic spacecraft model the radiation pressure force is a complicated function of the illumination direction $\hat{s}$ and of the spacecraft attitude. To compute it explicitly we need to know the exact shape, the attitude (including the state of the moving parts), and the three optical coefficients α, ρ, δ for each portion of the surface. For a spacecraft with a complex shape this can be difficult, unless special care is taken in the design phase in using a simple shape, surfaces with well known properties, and a simple operation mode. An additional difficulty is due to the fact that the optical coefficients α, ρ, δ change with time, as an effect of degradation of the spacecraft surface layers, damaged by charged particles from the Sun, and the magnetosphere of the relevant planet. Within a few years a white paint becomes brown and a mirror surface becomes irregular at a scale comparable to the wavelength of visible light; thus δ and ρ decrease, α increases.

For a natural body such as an asteroid, the a priori knowledge of both the optical properties and the shape of the surface is very poor (also the mass is poorly known). On the other hand, given the known shape of some real asteroids, to approximate their surface with a sphere in most cases is not a good approximation. However, if the purpose is not real-time orbit determination, but rather processing in batch at or near the end of an asteroid orbiter mission, the data from the entire mission can be used to build a direct radiation pressure model for the asteroid which could have a relative accuracy comparable to that of a spacecraft.

Secular perturbations

The relevance of radiation pressure as a source of perturbations on the orbit of both spacecraft and asteroid depends upon the way it accumulates with time. We shall use a model problem to develop the tools to discuss the orbital effects of small perturbations, such as the non-gravitational ones.

The model is just a two-body problem, with a satellite of mass M at $\vec{x}$ perturbed by a small force $\epsilon \mathbf{F}(t)$, where ϵ is a small parameter and $r = |\mathbf{x}|$ is the distance from the central body of mass $M_\oplus$:

$$\frac{d^2 \mathbf{x}}{dt^2} = -\frac{GM_\oplus}{r^3} \mathbf{x} + \epsilon \, \mathbf{F}/M.$$

In a mobile reference system defined by the orbit plane, orthogonal to the angular momentum vector $\mathbf{c} = \mathbf{x} \times d\mathbf{x}/dt$, we have

$$\hat{\mathbf{r}} = \frac{\mathbf{x}}{r}, \quad \hat{\mathbf{w}} = \frac{\mathbf{c}}{|\mathbf{c}|}, \quad \hat{\mathbf{t}} = \hat{\mathbf{w}} \times \hat{\mathbf{r}}$$

and the components of the non-gravitational acceleration are

$$R = \epsilon \, \mathbf{F} \cdot \hat{\mathbf{r}}/M, \quad T = \epsilon \, \mathbf{F} \cdot \hat{\mathbf{t}}/M, \quad W = \epsilon \, \mathbf{F} \cdot \hat{\mathbf{w}}/M$$

for the **radial**, **transversal**, and **out of plane** component, respectively. The total energy (per unit mass) E of the two-body approximation has a time derivative equal to the power of the perturbing force

$$\frac{dE}{dt} = \frac{\epsilon \, \mathbf{F}}{M} \cdot \frac{d\mathbf{x}}{dt} = R \, v_R + T \, v_T.$$

By the two-body formulae (similar to those of Section 4.2) the velocity components v_R and v_T (along the $\hat{\mathbf{r}}$ and $\hat{\mathbf{t}}$ directions, respectively) are

$$v_T = \frac{d\mathbf{x}}{dt} \cdot \hat{\mathbf{r}} = \frac{|\mathbf{c}|}{r} = \frac{G \, M_\oplus}{|\mathbf{c}|} (1 + e \, \cos v), \quad v_R = \frac{d\mathbf{x}}{dt} \cdot \hat{\mathbf{t}} = \frac{G \, M_\oplus}{|\mathbf{c}|} e \sin v,$$

with $|\mathbf{c}|$ the scalar angular momentum. This allows us to conclude how the orbital energy changes:

$$\frac{dE}{dt} = \frac{G \, M_\oplus}{|\mathbf{c}|} [T + e \, (R \, \sin v + T \, \cos v)]$$

and, by the relationship between energy and semimajor axis,

$$\frac{dE}{dt} = \frac{G \, M_\oplus}{2 \, a^2} \frac{da}{dt} \implies \frac{da}{dt} = \frac{2}{n \sqrt{1 - e^2}} [T + e \, (R \, \sin v + T \, \cos v)].$$

The main term for a low eccentricity orbit is

$$\frac{da}{dt} = \frac{2}{n} T + \mathcal{O}(e) \implies \frac{dn}{dt} = -\frac{3}{a} T + \mathcal{O}(e). \tag{14.4}$$

The main conceptual step in a perturbative approach is just to expand the solution to the complete equation of motion in a Taylor series with respect to the small parameter ϵ. For example, $a(t) = a_0(t) + \epsilon \, a_1(t) + \epsilon^2 \, a_2(t) + \cdots$

and the same expansion applies to the other five orbital elements. Then equations such as (14.4) can also be expanded in powers of ϵ and, by equating the terms of the same order in ϵ on both sides, we get a_0 constant and $\epsilon\, da_1/dt = 2T^{(0)}/n_0 + \mathcal{O}(e)$, where $T^{(0)}$ is T evaluated on the unperturbed orbit. That is, eq. (14.4) can be reinterpreted as a first perturbative order equation, providing the $\mathcal{O}(\epsilon)$ terms in the solution for $a(t)$, when the right-hand side is computed at the unperturbed orbit.

The corresponding **along-track effect** can be computed by using a set of orbital elements non-singular for $e = 0$ (Milani *et al.* 1987, Section 3.3), e.g., $\lambda = \omega + \ell$ for $e > 0$, where ℓ is the mean anomaly, while for $e = 0$ the element λ is just the angle on the circular orbit with origin at the ascending node.[1] Then the equations for the perturbed motion are

$$\frac{d\lambda}{dt} = n + \mathcal{O}(\epsilon), \quad a\,\frac{d^2\lambda}{dt^2} = -3\,T + \mathcal{O}(e) + \mathcal{P}\,\mathcal{O}(\epsilon) + \mathcal{O}(\epsilon^2) \qquad (14.5)$$

where $\mathcal{P}$ contains terms arising from the integration of $d\lambda/dt - n$. The non-trivial part of the above computation is to show that $\mathcal{P}$ contains only periodic terms with zero average. This implies that the effect of the R and W components does not accumulate quadratically with time in the along-track direction, at least not to order 1 in ϵ. A similar argument shows that there is no orbital effect accumulating quadratically with time in the other directions $\hat{\mathbf{r}}, \hat{\mathbf{w}}$. Thus for a nearly circular orbit the acceleration along-track is, to a good approximation, -3 times the perturbative transversal acceleration. If the transversal component T, as a function of time, can be decomposed into an average, or *secular*, part $\overline{T}$ and a short periodic part averaging out over one two-body orbital period $P = 2\pi/n$, starting from $t = t_0$,

$$T(t) = \overline{T} + T_{sp}(t), \quad \overline{T} = \frac{1}{P} \int_{t_0}^{t_0+P} T(t)\, dt$$

then the perturbation in the semimajor axis also decomposes into a **secular perturbation**, with linear growth in t, a **short periodic perturbation** averaging out over one period, and terms of higher order in ϵ

$$a(t) = a_0 + \frac{2\overline{T}}{n_0}t + \frac{2}{n_0}\int_0^t T_{sp}(s)\, ds + \mathcal{O}(e) + \mathcal{O}(\epsilon^2) \qquad (14.6)$$

where $a_0 = a(0)$ and $n_0 = n(a_0)$. The accumulated along-track effect is obtained by combining eqs. (14.5) and (14.6)

$$a(t)\,(\lambda(t) - \lambda_0 + n_0\,t) = -\frac{3\overline{T}}{2}t^2 + \mathcal{P}_1\mathcal{O}(\epsilon) + \mathcal{O}(e) + \mathcal{O}(\epsilon^2), \qquad (14.7)$$

---

[1] The variable defined in this way can be shown to be differentiable even for $e = 0$.

where $\mathcal{P}_1$, arising from the integral of $\mathcal{P}$ and the double integral of T_{sp}/ϵ, contains only periodic terms. In conclusion, for an orbit which is initially nearly circular, the only source of along-track effects quadratic in time is the averaged transversal acceleration $\overline{T}$, with the same coefficient -3 as the instantaneous transversal acceleration (14.5).

The above result has deep implications on the relevance of non-gravitational perturbations. For many sources of non-gravitational perturbations it is indeed the case that $\overline{T}$ is zero; e.g., if the orbit is circular and $\mathbf{F}$ is constant in time, $T(\lambda)$, when the unperturbed $\lambda = n_0\,t + \lambda_0$ is substituted, is a trigonometric function of time, averaging to zero.

We have outlined the argument to order zero in eccentricity, but in fact similar results can be proven to an arbitrary order. For example, let us assume that the vector $\mathbf{F}$ is constant in time, or even dependent upon the position $\mathbf{s}$ of the Sun, which in turn changes with time with frequencies much slower than n, assuming that the satellite-to-Sun vector can be approximated by the Earth-to-Sun vector. Then, as shown by Anselmo *et al.* (1983a) and Milani *et al.* (1987, Section 4.2), there is no secular perturbation in the semimajor axis, to first order in the ϵ and to all orders in e.

The same result applies also to many gravitational perturbations. For example, the gravitational perturbations from the Sun and from the other planets[2] have a zero $\overline{T}$, thus there is no secular perturbation in the semimajor axis to order one in the small parameters (which are those described in Section 4.5). This is a straightforward generalization of a classical result, going back to Lagrange. A simple comparison of the perturbative accelerations due to different causes (see Section 15.3) can be useful to discard from the dynamical model some exceedingly small effects. However, to decide which are the main effects we need to compute the secular along-track acceleration $\overline{T}$.

The above discussion assumes that the radiation pressure has a constant direction and intensity, or at least varying with a period much longer than the orbital period, e.g., a period of one year for an Earth satellite. If the area-to-mass ratio and the optical properties are not changing with time, this is enough to avoid quadratic accumulations with time of the perturbation on the orbit. On the contrary, if the radiation pressure acceleration undergoes changes with frequency equal to the orbital period, this does result in quadratic effects. Two examples of this are as follows: an Earth satellite with a constant attitude and an Earth pointing antenna may experience secular perturbations in the semimajor axis due to the antenna, not to the body (Anselmo *et al.* 1983a); an asteroid orbiting the Sun with a rotation axis not

---

[2] The same applies to the perturbations from the Moon on an Earth satellite, provided there is no low-order resonance between the orbital periods.

orthogonal to the orbit plane experiences a secular perturbation in the semi-major axis if the two hemispheres "north" and "south" have either a different shape or different optical properties (Vokrouhlický and Milani 2000).

The orbital elements e, ω can experience secular perturbations due to radiation pressure (Milani *et al.* 1987, Section 4.3): they appear as long periodic perturbations, with periods ≥ 1 year for an Earth satellite. For the elements I, Ω, if the radiation pressure $\mathbf{F}$ is constant the first-order secular effects are zero for $e = 0$. Anyway the perturbations on the elements e, ω, I, Ω result in changes in the spacecraft position with the orbital period. In conclusion, if $\overline{T} = 0$, the effects on the spacecraft position do not accumulate quadratically with time, thus they are in general a minor problem for the accurate orbit determination used both in satellite geodesy and in the control of active spacecraft. However, for large A/M space debris a long-term growth of eccentricity and inclination takes place: e, I can reach high values and this is a major problem, because it may result in a large relative velocity with respect to active satellites in the same region, especially in the geosynchronous belt (Valk *et al.* 2007).

14.2 Thermal emission

A passive celestial body exposed to solar radiation $\Phi_\odot A$, where A is the cross-sectional area, transforms the absorbed fraction $\alpha \, \Phi_\odot A$ into heat and reaches some thermal state. The surface temperature is not uniform and changes with time as a result of both the rotation and the orbital motion. Thus the entire surface re-emits thermal radiation anisotropically, carrying away some net linear momentum. This phenomenon of **thermal emission** results in a perturbative acceleration, affecting the orbit.

For each surface element dS the energy output due to thermal radiation is $\epsilon \, \sigma \, T^4 \, dS$, where T is the surface temperature, $\sigma = 5.67 \times 10^{-5}$ erg/cm^2 s K is the Stephan–Boltzmann constant, and ϵ is the emissivity coefficient ($\epsilon = 1$ for a black body). The thermal radiation is diffused according to Lambert's law, thus the flow of linear momentum results in a force

$$d\mathbf{F}_\epsilon = -\frac{2\,\epsilon\,\sigma\,T^4}{3\,c}\,\hat{\mathbf{n}}\,dS. \tag{14.8}$$

To model the surface temperature distribution and to compute the integral of $d\mathbf{F}_\epsilon$ is not simple: a full analytical solution exists only under very simple conditions. We shall briefly outline one such analytical solution, under the hypothesis that the surface is spherical, with radius R, and that the body rotates uniformly around a constant axis, which we use as the $\hat{\mathbf{z}}$ axis of the reference frame; let (r, θ, λ) be polar coordinates in that frame. We further assume that the vector to the Sun $\mathbf{s}$ is constant (an approximation applicable over a time span short with respect to the period of the heliocentric motion) and that the surface temperature differences are small with respect to the average temperature, allowing us to linearize the heat equation (Milani *et al.* 1987, Section 5.2).

The heat equation for a stationary state reduces to the Laplace equation $\Delta T = 0$. Thus, assuming that the rotation is fast enough to average out the temperature as a function of λ, the temperature T of the body should reach an equilibrium state expressed by zonal spherical harmonics

$$T(r, \theta) = T_0 + \sum_{i=1}^{+\infty} T_i \left(\frac{r}{R}\right)^i P_i(\sin \theta),$$

with T_0 the average surface temperature, T_i constants, and P_i the Legendre polynomials of (13.10). As a boundary condition, we have the balance between the outward heat flow caused by thermal conduction $-\chi \, \partial T / \partial r$ (where χ is the thermal conductivity of the body, assumed to be constant) and the net emission at the surface (the difference between the external irradiation from the Sun and the emission)

$$\epsilon \sigma T^4 - \alpha \, \overline{\hat{\mathbf{n}} \cdot \hat{\mathbf{s}} \, \Phi_\odot} = -\chi \frac{\partial T}{\partial r} \tag{14.9}$$

where the over-line indicates the average over λ

$$\overline{\hat{\mathbf{n}} \cdot \hat{\mathbf{s}}} = \frac{1}{2\pi} \int_0^{2\pi} g(\hat{\mathbf{n}} \cdot \hat{\mathbf{s}}) \, d\lambda = s(\theta)$$

with the function g equal to its argument if it is positive, zero otherwise (thus restricting the integral to the illuminated hemisphere). The function $s(\theta)$ can be computed analytically: if the latitude and longitude of the Sun are $(\xi, \pi/2)$ we get three formulae for the three latitude zones, the midnight Sun, the dark noon, and the sunrise–sunset zones, respectively

$$s(\theta) = \begin{cases} \sin \xi \sin \theta & \text{for } \frac{\pi}{2} - \xi \le \theta \\ 0 & \text{for } \theta \le \xi - \frac{\pi}{2} \\ \frac{1}{2\pi} \left[2 \cos \xi \, \cos \theta \, \cos \lambda_1 + (\sin \xi \sin \theta) \, (\pi - 2\lambda_1) \right] \\ \qquad \text{for } \xi - \frac{\pi}{2} \le \theta \le \frac{\pi}{2} - \xi \end{cases}$$

where λ_1 solves the terminator plane equation $\hat{\mathbf{n}} \cdot \hat{\mathbf{s}} = 0$, that is $\cos \lambda_1 =$

$\tan\theta \tan\xi$. With $s(\theta)$ expanded in Legendre polynomials

$$s(\theta) = s_0 + \sum_{i=1}^{+\infty} s_i\, P_i(\sin\theta)$$

equation (14.9), linearizing in the temperature harmonic coefficients T_i (assumed $\ll T_0$), gives a separate equation for each zonal harmonic coefficient

$$\epsilon\sigma\, T_0^4 = \alpha\, s_0\, \Phi_\odot, \qquad T_i = \frac{\alpha\, s_i\, \Phi_\odot}{4\epsilon\,\sigma\, T_0^3 + i\chi/R}, \qquad (14.10)$$

with the solutions of lowest degree: $s_0 = 1/4$ and $s_1 = \sin\xi/2$. By substituting in eq. (14.8) and integrating over the sphere, we get the net thermal emission force directed along the $\hat{\mathbf{z}}$ axis

$$\begin{aligned}
\mathbf{F}_\epsilon &= -\hat{\mathbf{z}}\,\frac{2\epsilon\sigma}{3c}\int_S \sin\theta\,[T_0^4 + 4T_0^3 T_1\, P_1(\sin\theta)]\,dS \\
&= -\hat{\mathbf{z}}\,\frac{4\pi\epsilon\sigma R^2}{3c}\int_{-\pi/2}^{+\pi/2}\sin\theta\,\cos\theta\,(T_0^4 + 4T_0^3 T_1\,\sin\theta)\,d\theta
\end{aligned}$$

where terms of higher degree have been neglected.[3] The average temperature gives an isotropic emission, and the degree 1 harmonic gives a net force

$$\mathbf{F}_\epsilon = -\hat{\mathbf{z}}\,\frac{4\,\pi\,\alpha\,\Phi_\odot\,R^2\,\sin\xi}{9\,c\,\beta} \qquad (14.11)$$

where the reduction factor $\beta = 1 + \chi\,T_0/\alpha\,R\,\Phi_\odot$ plays an important role. With $A = \pi R^2$ the cross-section, the acceleration is

$$\frac{\mathbf{F}_\epsilon}{M} = -\hat{\mathbf{z}}\,\frac{A\,\Phi_\odot}{M\,c}\,\frac{4\,\alpha\,\sin\xi}{9\,\beta}.$$

If we assume $\epsilon = \alpha$, and the distance from the Sun 1 AU, the average temperature is $T_0 \simeq 280$ K. The assumption $T_i \ll T_0$ used in the linearization of the equations implies $\beta \gg 1$, i.e., high conductivity. As an example we use the LAGEOS class satellites (see Section 15.2), whose body is an aluminum sphere with a radius of 30 cm and a mass of 400 kg. Then

$$\frac{\mathbf{F}_\epsilon}{M} = -\hat{\mathbf{z}}\,\frac{5.8 \times 10^{-8}}{\beta}\,\sin\xi \ \mathrm{cm/s}^2.$$

If it is modeled as a homogeneous body, given the conductivity of aluminum $\chi = 2.1 \times 10^7$ erg cm^{-1} s^{-1} K^{-1}, we have $\beta = 471$. If a modification of the above computation is used to take into account an insulating core of radius 25 cm and an outer aluminum shell, $\beta = 155$. In any case the thermal

---

[3] The degree 2 harmonic gives no contribution to the integral, T_i is assumed negligible for $i > 2$.

emission is one of the main sources of uncertainty in the dynamical model for spacecraft of the LAGEOS class (see Section 15.3).

For more complex shaped spacecraft, the explicit computation of a surface temperature model is a challenge, and the spherical approximation is too poor. The above computation can be used to give an order of magnitude, provided some estimate for the surface temperature excursion ΔT is available (to be used in place of the T_1).

Similar computations can be done for natural bodies, such as asteroids. The conductivity is expected to be much lower, and anyway it is essentially unknown: a reasonable guess would be for a range between 10 and 1000 erg/cm s K, depending upon the texture of the surface (e.g., regolith is a very good insulation, solid rock is more conductive). Moreover, neither the shape is close to spherical nor the conductivity is expected to be constant, given the very uneven distribution of regolith found in the few asteroids for which we have very close images (Eros, Itokawa). Thus to build a realistic thermal model of an asteroid is a challenge, and the thermal emission effects on the orbit cannot be accurately predicted.

The Yarkovsky effect

The same argument used in Section 14.1 applies to thermal emission: what matters is the fraction of the perturbing acceleration contributing to the secular change in the semimajor axis. For example, if the attitude in an inertial reference system and the thermal state were constant, at least on average, the acceleration induced by thermal emission would be a constant vector and its contribution to $\overline{T}$ would be zero. This is the case for an axially symmetric body rapidly spinning around a fixed axis $\hat{\mathbf{z}}$ orthogonal to the plane of a circular heliocentric orbit. The same applies to a planetocentric orbit, even an eccentric one, if we can assume that the surface temperature is not significantly affected by the thermal emission from the planet and there are no eclipses (Anselmo *et al.* 1983a). However, unlike direct radiation pressure, thermal emission can have secular effects in the semimajor axis for a heliocentric orbit, even for a spherical shape; this is called the **Yarkovsky effect**. A similar effect occurs for geocentric orbits, due to the uneven heating resulting from radiation emitted by both the Sun and the Earth.

It is important to realize that there is no *Yarkovsky force*, but just thermal emission forces, which under suitable circumstances have a comparatively small, but significant, mean transversal component $\overline{T}$. We shall discuss here the heliocentric case, in which there are two contributions to the Yarkovsky effect: the seasonal and the diurnal one.

Seasonal Yarkovsky effect

If a body with a fixed rotation axis were in a constant thermal state as it orbits around the Sun, then the thermal emission force of eq. (14.11) would be of constant size and direction, thus $\overline{T} = 0$. This condition can be violated for two reasons.[4] The first is when the obliquity ϵ, that is the angle between the spin axis and the orbital angular momentum, is not 0. Then the latitude ξ of the Sun in the body equatorial frame is not constant, and eq. (14.11) gives a thermal emission force changing with time, essentially with the frequency of the mean motion n. The second reason is that the illumination $\Phi_\odot$ is a function of the distance from the Sun, thus it changes for an eccentric orbit, mostly with the frequency n.

In both cases, the thermal emission force has an intensity which changes with a period equal to the orbital period (in a two-body approximation), and the same resonance effect mentioned for radiation pressure on asymmetric bodies can apply. This is called a **seasonal effect** because it depends on the fact that the heliocentric body has temperature variations depending upon the equatorial obliquity and upon the orbital eccentricity, similarly to the major planets. An explicit computation of the size of this effect is not simple, even for a spherical body (Vokrouhlický *et al.* 2000), and becomes very complicated for complex shaped bodies. Qualitatively, the secular drift is always towards the lower semimajor axis and its magnitude can be up to 15 m/y for asteroids with diameter in the 300–500 m range.

Diurnal Yarkovsky effect

The Yarkovsky **diurnal effect** arises because thermal inertia of the illuminated body results in a temperature maximum lagging some time after the maximum of illumination. This effect depends upon the conductivity χ, by no means linearly: for $\chi \to 0$ the thermal time lag goes to zero; for $\chi \to +\infty$ the surface temperature excursion goes to zero; for some intermediate value of χ there is a maximum effect, see (Vokrouhlický *et al.* 2000, Figure 1). Interestingly, in the realistic range of values of χ for a small asteroid, in many cases the dependence upon χ is not very strong, the effect changing by less than a factor of 2. Thus this effect is always of the same order of magnitude, once the mass is known; of course it depends upon the obliquity ϵ, with the semimajor axis secularly increasing for prograde rotation ($\epsilon < 90°$) and decreasing for retrograde ($\epsilon > 90°$). The magnitude of this effect can be larger than that of the seasonal effect, up to several tens of m/y.

---

[4] There is a third possible reason: the rotation axis could be changing with time, for an asteroid not in a simple rotation state but tumbling, either regularly or chaotically as for (4179) Toutatis. This case is too complicated to be discussed here.

When is the Yarkovsky effect relevant?

The Yarkovsky effect is very important as a source of secular perturbations to model the dynamical evolution of asteroids, e.g., it is relevant for the transport of meteorites and asteroids to the near-Earth region: 15 m/y $\simeq 10^{-4}$ AU/My accumulates to a large change over the age of the asteroids. From the point of view of orbit determination, there are only a few and so far exceptional cases where effects of this class are relevant to fit an orbit of an asteroid. This is because the secular perturbations are typically a few percent of the instantaneous thermal emission accelerations. Thus for an orbit determination with a data span shorter that an orbital period they are very small, and anyway less relevant than the short period perturbations due to both direct radiation pressure and thermal emission.

The exceptional cases are asteroids with a very long observed arc; very accurate observations may also be needed. As an example, the first asteroid for which the Yarkovsky effect has been measured by orbit determination is (6489) Golevka, which has been observed by radar during three separate close approaches to the Earth; the second case was (152563) 1992 BF, an asteroid with an exceptionally long arc due to attribution of precovery observations (Chesley *et al.* 2003, Chesley *et al.* 2008). With the accumulation of more data and also with the expected improvements in astrometric accuracies, such cases will become much more frequent.

14.3 Indirect radiation pressure

The case of a satellite orbiting around a planet (and the Moon) is more complicated because the planet is an additional source of radiation, ultimately coming from the Sun but either reflected, or diffused, or absorbed and re-emitted as infrared. Moreover, the planet casts a shadow which cuts off direct radiation pressure from the Sun and also results in thermal transient states of the spacecraft. This is a very complicated subject and our intent is not to explain in detail how to model the corresponding non-gravitational perturbation on the orbit, but just to list the different physical effects and give an idea of their relative importance for an accurate orbit determination.

Reflected radiation pressure

A planet illuminated by the Sun also shows a linear combination of absorption, reflection, and diffusion, with optical coefficients (14.1) which may change significantly with the position on the surface. For example, on the Earth the **planetary albedo** $1 - \alpha_\oplus$ can be ~ 0.8 for glaciers, fresh snow and clouds, ~ 0.2 for the ocean and intermediate for a continental area with clear sky, depending also on the vegetation cover. The ratio ρ/δ depends also upon the texture of the surface, with a smooth lake mirror reflecting more than a rough sea; there are phenomena intermediate between mirror reflection and diffusion, resulting in a concentration of reflected light near the perfect reflection direction, like in the *sword of the Sun* which can be observed at sunset from the seashore and from an airplane over the sea.

Visible light

To accurately model the radiation pressure on a spacecraft from the visible light reflected/diffused by a planet we would need a map of the optical coefficients α, ρ, δ values on the entire surface, in the case with an atmosphere with variable weather (as on the Earth and on Mars) including a full weather map giving at least the average cloud cover with good spatial and time resolution. Then the effect could be computed for each surface element, and some numerical approximation of a surface integral over the portion of surface visible from the spacecraft should be used. In practice, this has never been done, and although it may become technologically feasible in the future we have to question whether this would be useful.

To compute the orders of magnitude of the relevant effects, let us select the case of the geodetic satellite LAGEOS. Radiation reflected/diffused from the Earth has an instantaneous value of the order of 3×10^{-8} cm/s^2 (see Table 15.1 for comparison with other perturbations). The orbit of LAGEOS was found to be affected by a **mystery drag**, that is an unexpected secular decrease of the semimajor axis, corresponding to an average transversal deceleration $\overline{T} \approx -3 \times 10^{-10}$ cm/s^2; superimposed on this secular effect, there were long periodic terms corresponding to values of $\overline{T}$ of the order of 10^{-10} cm/s^2 and with periods up to three years. The secular term cannot be explained by radiation pressure (possibly by the Yarkovsky effect and/or drag), but the long period ones could be: this implies that a "brute force" model of the radiation pressure of the Earth would need to have a relative accuracy < 0.003 to roughly account for the long periodic terms of the mystery drag.

Anselmo *et al.* (1983b) have shown that the radiation from the Earth could well account for the long periodic perturbations to the semimajor axis of LAGEOS by a semiquantitative argument, in which what matters is the angle between the orbital plane of LAGEOS and the **terminator plane** through the Earth's center of mass and orthogonal to the Sun's direction. Neglecting topography, the intersection of the terminator plane with the Earth's surface is the line along which either sunrise or sunset takes place.

When the satellite orbit is crossing the terminator plane, e.g., with a ground track crossing from day to night, the radiation pressure from the illuminated portion of the Earth is pushing from behind the spacecraft and increasing the semimajor axis. After the true anomaly of the satellite has increased by $\simeq \pi$, there is another terminator plane crossing, this time with the ground track crossing from night to day, thus a push from the front and a decrease of the semimajor axis. If these two spikes of T were exactly equal and opposite, there would be no contribution to $\overline{T}$. However, one of the two terminator crossings is in the Northern hemisphere, the other in the Southern one with a much larger proportion of low albedo ocean area; if one of the two is with the ground track on a region experiencing summer, the other is in winter, with increased cloud cover. Thus there is unbalance and a long periodic perturbation, having as the main angular arguments the Sun mean longitude $\lambda_\odot$ and LAGEOS longitude of the node Ω, thus the main effects have periods between 156 and 1050 days, in qualitative agreement with the frequency spectrum of the mystery drag.

Infrared radiation

The thermal emission from the planet is regulated by the same heat equation (14.9), thus a solution for the thermal emission can be computed in the spherical surface approximation. However, unlike the asteroid case of Section 14.2, it is not possible to average over a revolution of the planet, because this period can be longer than the orbital period of the spacecraft; also the lag in the temperature maximum after the time of maximum illumination can be comparable to the satellite orbital period.

The average temperature T_0 results in isotropic emission which has the same effect of a change in the mass of the planet. The main contribution to the infrared radiation pressure perturbations arise from the first harmonic T_1; note that $T_1 \ll T$ is a reasonable approximation for a planet such as the Earth, but fails for the Moon and even more for Mercury. For an accurate model, the absorption α cannot be assumed to be constant; even on dark bodies like the Moon and Mercury there are comparatively bright surface features. Again the total effect of infrared radiation pressure needs to be computed as an integral over the portion of surface visible from the spacecraft.

The conclusion is that infrared radiation pressure can be modeled somewhat more easily than visible radiation pressure from the planet, because the surface behavior is less sharply variable, but still this can be a suitable model to simulate the effect, not an accurate model.

Eclipses

A planet, or anyway a large body, prevents sunlight from reaching a shadow cone: when the spacecraft is in the full shadow, it experiences an eclipse of the Sun, during which there is no direct radiation pressure from sunlight (Milani *et al.* 1987, Section 5.4). This effect is important because the average transversal component $\overline{T}$ of the direct radiation pressure force is not zero when the orbit plane is such that eclipses occur and the orbital eccentricity is not zero. It is possible to compute semianalytically the effect on the semimajor axis (Aksnes 1976).

The full shadow is surrounded by a region of penumbra; e.g., the diameter of the Sun is 2° as seen from Mercury, and a Mercury orbiter with an orbital period of a few hours experiences penumbra for a few tens of seconds just before and after the full shadow.

For orbits undergoing eclipses, the assumption of Section 1.1 that the right-hand side of the equations of motion is differentiable may fail. In practice, the direct radiation pressure acceleration may go from its full value to zero (and vice versa) in a very short time span, that of the penumbra phase. Since this time span for abrupt change can be shorter than, or comparable to, the step size of the numerical integrator used to propagate the orbit, numerical instabilities may occur. Indeed, they have been detected in numerical experiments with space debris.

For large A/M, of the kind which can occur in small space debris, the overall effect of eclipses, combined with other perturbations, can accumulate to very large values; the consequence is a very significant increase of the risk of impact on active satellites by high relative velocity debris (Valk and Lemaitre 2007).

14.4 Drag

Drag is caused by the direct interaction of the spacecraft with matter, assumed to be neutral (molecules). It is a resistive force, opposite in direction to the spacecraft velocity with respect to the average atmosphere

$$\mathbf{F}_v = -\frac{1}{2}\, C_D \, A \, \rho \, |\mathbf{v}|\, \mathbf{v} \qquad (14.12)$$

with ρ the density of the atmosphere, A the cross-sectional area orthogonal to $\mathbf{v}$, the velocity relative to the atmosphere, and C_D the adimensional **drag coefficient** (or **aerodynamic coefficient**), which is in general of the order of unity (Milani *et al.* 1987, Chapter 6).

It is already clear from eq. (14.12) that very accurate modeling of drag is not possible. The main unknown parameter is the density ρ, changing both in space and in time. The dependence upon the distance from the geocenter r can be described with moderate accuracy by an exponential model:

$$\rho(r) = \rho_0 \exp\left(\frac{r_0 - r}{\mathcal{H}}\right),$$

where $\rho_0 = r(r_0)$ and $\mathcal{H}$ is the **scale height** over which the density decreases by $1/\exp(1)$. This equation gives the solution for an isothermal column of gas in equilibrium with its own weight (Boltzmann law), and is a valid approximation when the temperature undergoes little change with height, as it happens in some high atmospheric layers (above 250 km). In practice, the scale height changes with height, and ρ_0 can experience variations by an order of magnitude or more as a result of solar and geomagnetic activity, on top of the changes driven by the solar illumination.

The computation of C_D is very complicated; when electromagnetic effects come into play, as a result of the charged particles in the ionosphere, the negative charging of the spacecraft surface may result in a coefficient C_D larger by one order of magnitude than the values typical of the neutral atmosphere (Milani *et al.* 1987, Section 6.3).

The atmosphere rotates with the Earth, more or less rigidly, thus the velocity $\mathbf{v}$ does not coincide with the inertial velocity but is closer to the velocity in a body fixed reference frame. Even the assumption, contained in eq. (14.12), that the drag force is along the direction $\hat{\mathbf{v}}$ is a simplification, because for some spacecraft shapes there could be a significant lift effect, e.g., for large panels oriented at an angle $\neq \frac{\pi}{2}$ with respect to $\hat{\mathbf{v}}$.

In conclusion, although drag forces have been the first non-gravitational ones to be included in models of satellite orbits (King-Hele 1964), in state-of-the-art satellite geodesy it is necessary to assume they are either measured by some on-board instrumentation (accelerometer), or compensated (drag free probe), or removed from the problem by using satellite navigation systems; see Chapter 16. They have to be taken into account when solving for the rapid orbital decay of low satellites and space debris, and in the planning of satellite geodesy missions whose lifetime is limited by orbital decay.

14.5 Active spacecraft effects

In an active spacecraft the energy from the Sun absorbed by the outer surface is processed in several ways before being re-emitted in different forms. Moreover, the internal temperature distribution is actively controlled by heaters, coolers, radiators, and by the dissipation of heat resulting from energy consumption. The thermal conditions (both the stationary states and the transients between them) are typically predicted with finite element algorithms, measured by on-board thermometers, and controlled by heat pipes and feedback loops activating heaters and variable surface radiators. All the clever methods used by aerospace engineers to maintain the on-board devices within their operational temperature range result in a more and more difficult task to model the thermal emission acceleration.

In practice, a model of the external surface temperatures with the required accuracy is never available.[5] This is even more the case when the external conditions are extreme, e.g., in the cold of the outer Solar System exploration probes and in the heat of the interior planet orbiters (see Chapter 17).

Radio wave beams

An active spacecraft needs to transmit to a ground station, by generating a directional radio wave beam; another possibility is a radio wave beam for radar, thus pointed to the planetary surface. Thus a fraction of the power $\alpha \, \Phi_\odot \, A$ absorbed through the surface is converted into electrical power, a fraction of which is used to generate radio waves. Both conversions having an efficiency less than unity, the power actually emitted as radio waves is a small fraction of the absorbed power, a few per cent. Nevertheless, the emitted beam has a direction different from the Sun and may contribute significantly to the secular along-track effect $\overline{T}$ (Milani *et al.* 1987, Section 5.3).

Possible solutions

If the accuracy of the orbit determination requires to take into account thermal emission and other subtle effects depending upon the spacecraft structure and activity, there are only two solutions. Either the non-gravitational perturbations are not modeled, but measured by accelerometers (see Chapter 16), or they are described by a set of empirical parameters to be solved with the orbit. The accelerometers are more suitable for the large and rapidly variable non-gravitational perturbations of the Mercury orbiters (see Chapter 17). In the cold and slowly variable conditions of the outer Solar System cruise phases, few parameters can describe the non-gravitational perturbations over a long arc. Two examples of the latter approach follow.

In (Bertotti *et al.* 2003a) the interplanetary orbit of the **Cassini mission** had to be modeled very accurately during a time span of a few weeks, during a superior conjunction, to determine the post-Newtonian parameter γ (see

---

[5] Engineers care much less about the surface temperature, provided the most extreme heat is avoided, especially in the solar cells.

Sections 6.6 and 17.5). The spacecraft was in a stationary state during the interplanetary cruise, with constant attitude and thermal state. Thus the non-gravitational acceleration (including both direct radiation pressure and thermal emission) over the observation arc time span used in the experiment could be modeled by a constant vector. The experiment was very successful (with an estimated $\mathrm{RMS}(\gamma) \simeq 2 \times 10^{-5}$) because of the extremely accurate tracking and of the simple operations mode of the spacecraft.

In (Olsen 2007) the orbit of the **Pioneer** spacecraft, while navigating beyond Saturn in an orbit escaping from the Solar System, has been solved assuming a constant perturbing acceleration directed towards the Sun. The interpretation of this "Pioneer anomaly" has been the subject of some controversy, but Olsen convincingly argues that a minor anisotropy ($\simeq 0.03$ of the isotropic term) in the thermal emission from the radioisotope power generator can account for the estimated 8×10^{-8} cm/s^2 acceleration. The time span of the data is not enough to discriminate between a constant acceleration and one decaying exponentially with the radioactive material (with a known half life of 87 years), but some indications of decay have been found.

The only way to prove that the effect is indeed due to non-gravitational perturbations, rather than to "anomalous gravity" of whatever origin, is to test the orbits of celestial objects with very different area-to-mass ratio and orbiting in the same region. Wallin *et al.* (2007) solved for the orbits of the best observed trans-neptunian objects adding a parameter to model the "Pioneer anomaly": if it were due to gravitational perturbations, by the equivalence principle it should also affect bodies of diameter > 100 km. They found a value for the unexplained radial acceleration an order of magnitude smaller than the Pioneer value, consistent with zero at the 1 RMS level, and inconsistent with the Pioneer value at the 5 RMS level. Thus the "Pioneer anomaly" is non-gravitational, and it should be due to thermal emission.

Maneuvers and leakages

It must not be forgotten that accurate orbit determination cannot be done on a spacecraft performing **maneuvers**, not only orbit control maneuvers, but also attitude control ones. Even when the attitude controlling torques are applied by using two thrusters acting in parallel and opposite directions, the amount of impulse from them cannot be balanced with very good accuracy; for the order of magnitude of this effect, see (Milani *et al.* 1987, Section 7.2). There are only two methods to control the degradation of orbit determination due to maneuvers, and they have to be chosen before, at the mission planning stage.

One method is to estimate, during the mission analysis study, how often the maneuvers have to be performed, and how much they affect the orbit. This is comparatively easy for the orbit maneuvers (Milani *et al.* 1987, Section 7.1), but it is not trivial for the attitude maneuvers. A space mission with a requirement for a very accurate orbit determination needs to use methods different from thruster activation to control attitude, such as reaction wheels, and the time span after which thruster activation is required for the unloading of the reaction wheels needs to be carefully predicted.

The second method is to agree, in the mission design phase, on a constraint on the time interval between maneuvers, and then use a multi-arc approach (see Chapter 15), with the arcs beginning and ending at the maneuver times. Still there is a requirement that the times are known.

Another similar problem is due to gas leaks from the thrusters. Even when the valves controlling the activation of the propulsion system are nominally shut, small leaks are difficult to avoid. The problem is that such leaks might be so small that the spacecraft designers do not worry about them, and still the impact on the orbit determination accuracy is significant. For example, a gas leak of hydrazine by an amount of 100 g per year might not be a problem from the point of view of fuel consumption, nevertheless with the gas at a temperature of $200\,\mathrm{K}$ on a spacecraft with mass $M = 500$ kg it would result in an acceleration of 4×10^{-7} cm/s^2.

14.6 Case study: asteroid orbiter

The main conclusion which should be drawn from the discussion of the intricacies of non-gravitational perturbations modeling is a simple advice: do not do it. If possible, a mission requiring very accurate orbit determination should be designed in such a way that its performance does not depend upon the accuracy, reliability, and stability with time of the non-gravitational perturbation model. However, there is a class of space missions for which this advice cannot be followed: the asteroid orbiters, whose purpose includes an extremely accurate orbit determination of the asteroid.

To understand why there could be the need to do this, please refer to Chapter 12. The basic idea is that we may need to deflect the orbit of an asteroid, which is predicted to impact our planet at some time in the foreseeable future, a few tens of years from now. Alternatively, we may wish to demonstrate that the technology to take such a defensive action is available, to be used if and when it may become necessary; this was the main goal of the space mission study **Don Quixote**, first performed on behalf of the European Space Agency (ESA) in 2002, and later updated both by internal ESA and industrial studies.[6]

Don Quixote was intended as a two-spacecraft mission, with one component, *Sancho*, orbiting around the target asteroid with the necessary complement of instruments to allow for an extremely accurate orbit determination, of both the spacecraft asteroid-centric orbit and the asteroid heliocentric orbit. The second component, *Hidalgo*, would arrive later and impact the asteroid at the largest possible relative velocity, thus transferring a significant amount of linear momentum and changing the heliocentric orbit by an amount which could be measured by Sancho. The purpose was to test this simple **kinetic method of deflection** and learn how effective is the transfer of linear momentum.[7]

Photo-gravitational symmetry

The Don Quixote method of deflection is attractive because it appears to be simple, not requiring new technologies (unlike other methods which have been proposed). However, there is a technology which needs to be demonstrated: non-gravitational perturbation modeling and/or determination. To understand this, we need to appreciate the orders of magnitude. Let us assume the target asteroid has a roughly spherical shape, with diameter $2R = 300$ m and density 1.3 g/cm^3, thus a mass $m \simeq 18 \times 10^6$ tons. If Sancho has a roughly circular orbit at a distance $r = 10\,R$, the acceleration due to gravity from the asteroid is $g \simeq 5.4 \times 10^{-5}$ cm/s^2, while the direct radiation pressure from the Sun is $f \simeq 1.8 \times 10^{-5}$ cm/s^2 (we are assuming $A = 20$ m^2 and $M = 500$ kg), thus the perturbation approach used in Section 14.1 is a rough approximation. To use an analytical formula as a tool to find the appropriate orders of magnitude, we need to find an exact solution to the *photo-gravitational problem* as a function of both f and g; we are neglecting the differential attraction from the Sun.

We assume that the spacecraft motion takes place in a plane orthogonal to the direction $\hat{s}$ to the Sun, that is parallel to the asteroid terminator plane. Let us assume only three accelerations are acting on the spacecraft in a rotating reference system: the gravitational monopole attraction **g**, the radiation pressure acceleration **f**, and the centrifugal acceleration $\omega^2 \mathbf{r}$, where ω is the angular rate around an axis parallel to $\hat{s}$ passing from the

---

[6] The Don Quixote project is not yet an approved mission, with a firm budget, thus it may or may not be implemented in the next decade.

[7] Because of the linear momentum carried away by the ejecta from the crater excavated by Hidalgo, the linear momentum transferred to the rest of the asteroid should be more than that carried by the impactor, but to estimate a priori how much more is very difficult.

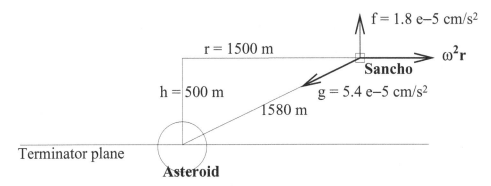

Fig. 14.1. For an asteroid with 300 m diameter, the simple two-body model of the photo-gravitational symmetry results in a very large displacement of the orbit plane with respect to the asteroid center of mass. The orbit plane does not even touch the asteroid!

asteroid center of mass, and $\mathbf{r}$ is the vector orthogonal to this axis pointing to the position of the spacecraft. Let $\mathbf{h}$ be the vector parallel to $\hat{\mathbf{s}}$ between the asteroid center of mass and the plane in which the spacecraft asteroid-centric orbit lies. We look for a relative equilibrium solution, stationary in the frame rotating with angular velocity $-\omega\,\hat{\mathbf{s}}$, such that $\omega^2\,\mathbf{r}+\mathbf{g}+\mathbf{f}=\mathbf{0}$. The components of the gravitational acceleration $\mathbf{g}$ along $\mathbf{r}$ and along $\mathbf{h}$ have to be balanced by centrifugal and radiation pressure accelerations (see Figure 14.1)

$$\mathbf{g}=-\frac{G\,m}{(r^2+h^2)^{3/2}}\,(\mathbf{r}+\mathbf{h}),\quad -\omega^2\,\mathbf{r}=-\frac{Gm}{(r^2+h^2)^{3/2}}\,\mathbf{r},\quad \mathbf{f}=\frac{Gm}{(r^2+h^2)^{3/2}}\,\mathbf{h}.$$

These equations have an exact solution for every h provided f, m change as

$$f(h)=\omega^2\,h;\quad m(h)=m(0)\left(1+h^2/r^2\right)^{-3/2},$$

where $m(0)$ is the mass corresponding to $h=f=0$.

In this simplified model, the circular solution of relative photo-gravitational equilibrium is valid for every value of r. In reality this solution does not exist for r smaller than the asteroid radius R because of the shadow, and becomes a poor approximation when r becomes comparable to the radius of the asteroid sphere of influence; then the differential gravitational attraction from the Sun becomes important.[8] For intermediate values of r, e.g., in the range $3\,R\le r\le 20\,R$, it is a useful first approximation.

The geometry of the orbit and its velocity are unchanged, since r and $\omega^2=G\,m(0)/r^3$ are independent of h. Thus the value of h is not observable by measuring the range and range-rate from Earth; the asteroid center of

---

[8] The photo-gravitational equilibrium point exists, as analytical continuation of the Euler collinear equilibrium in the restricted three-body problem (Simmons *et al.* 1985). The circular orbits above are very close to the Lyapounov periodic orbits around this equilibrium.

mass position (CoM), which is not observable directly, is shifted by $-\mathbf{h}$. The implication is that, in this approximation, if the list of parameters to be solved includes the asteroid CoM position, the quantity f, and the asteroid mass m, then the normal matrix is degenerate. These parameters cannot be solved at once, whatever the set of range and range-rate observations, and this is independent of the measurement accuracy. This exact symmetry is called the **photo-gravitational symmetry**. The only solution is to assign a priori constraints to f or h: either f is determined from a radiation pressure model, or h is measured by some local observation of the asteroid, e.g., from images of the asteroid taken from the spacecraft.

The implications for Don Quixote are apparent from a simple order of magnitude computation. For the nominal asteroid and spacecraft discussed before, $h = f/\omega^2 \simeq 500$ m. A relative error of 0.1 in the model for radiation pressure on the spacecraft would imply an error in estimating the position of the asteroid CoM of $\simeq 50$ m. Estimation of the position of the CoM of the asteroid from the images should give a smaller error, taking into account that such a small asteroid is expected to have a rather uniform density. Still an error of the order of $R/10 \simeq 15$ m would be unavoidable. What matters is that such an error would be much larger than the error in determining the position of Sancho with state-of-the-art tracking from the Earth, which could have submeter accuracy in range.

The above discussion is strongly simplified with respect to a realistic case, in which the shape of the asteroid is irregular, thus its gravity field contains significant low degree spherical harmonics, and the asteroid is in a heliocentric elliptic orbit, thus f is not constant. In reality the task of finding, from images of the asteroid, where the CoM is, would be far from trivial, but would need to be included in a complex orbit determination problem, in which the orbit of the asteroid, the orbit of Sancho, the rotation state of the asteroid, and some harmonics of its gravity field would appear as fit parameters. Nevertheless, the order of magnitude estimate is applicable, and there is indeed a limit at a few tens of meters in the possibility of determining accurately the asteroid-centric orbit, thus also the asteroid orbit from tracking of the Sancho orbiter.

Deflection by impact and its measure

Let us suppose the impactor Hidalgo has a mass of 400 kg and a relative velocity (with respect to the target asteroid) of 10 km/s. Then, even assuming that the linear momentum is transferred to the asteroid without gain due to ejecta, the asteroid changes its velocity by 0.02 cm/s. For an asteroid with semimajor axis $a = 0.9$ AU this implies a change in a, depending upon the angle θ between the direction of the velocity change and the heliocentric velocity, of $\simeq 1.8 \cos\theta$ km. This results in a change in the mean motion, accumulating an along-track drift of $-56 \cos\theta$ m/day. Thus the inaccuracy in the determination of the asteroid position, due to radiation pressure on Sancho, does not prevent a measurement with good relative accuracy, say 0.01 or better, of the deflection by continuing the tracking for weeks after the Hidalgo impact.

However, there is another element in the problem, namely non-gravitational perturbations on the orbit of the asteroid. The assumed asteroid has an area-to-mass ratio $A/M \simeq 4 \times 10^{-5}$, thus $\beta \simeq 2.7 \times 10^{-9}$ (see Section 4.6) and radiation pressure acceleration $\simeq 6 \times 10^{-10}$ cm s^{-2}. As a result, its semimajor axis changes by $\simeq 4 \cos\theta'$ m/day, where θ' is the angle between the direction $\hat{s}$ and the heliocentric velocity. Thermal emission is somewhat smaller, but more difficult to be modeled. Thus the main term of the error budget of the deflection measurement is neither the range measurement error nor the error in modeling/measuring the radiation pressure on Sancho, but the error in modeling the non-gravitational perturbations on the orbit of the asteroid. For the best performance of a Don Quixote class mission, to model accurately direct radiation pressure and thermal emission effects on the asteroid is a requirement.

The above is not necessarily an argument against the Don Quixote style deflection experiments. As shown by Chesley (2006), in the real case of the asteroid (99942) Apophis which could impact the Earth in 2036 (see Section 12.6), it might be possible neither to exclude, nor to predict with certainty that impact unless a much better model of the Yarkovsky effect is available for that asteroid. As pointed out above, there is no way to measure the Yarkovsky effect in a short time span; what is required is to follow an asteroid with a very accurate tracking, allowing us to fit the non-gravitational acceleration, represented as a time series (possibly as polynomial interpolations), for an entire orbital period of the asteroid, to directly determine the average along-track acceleration. This could be done with a Sancho-class spacecraft orbiting around Apophis for an orbital period of the asteroid. The conclusion of this phase of the mission could allow us to predict without uncertainty either the occurrence or the impossibility of an impact, allowing for a Hidalgo spacecraft to perform the deflection if necessary.

15

MULTI-ARC STRATEGY

One of the main assumptions used in Chapter 1 is that the dynamical model is deterministic. This assumption can be too optimistic for celestial bodies small enough to be significantly affected by complex non-gravitational interactions. Both drag and radiation pressure can be so poorly known that the errors in the dynamical model can affect the predictions by amounts exceeding, by orders of magnitude, the measurement accuracy.

When this is the case, there are three possible ways out, including the multi-arc strategy presented in this chapter. The others are the use of on-board accelerometers, see Chapters 16, 17, and the empirical parameterization of the unknown effects, see Section 14.5.

The multi-arc approach gives up the attempt to model the orbit of the spacecraft, over the entire time span of the observations, in a deterministic way with a single set of initial conditions. The time span of the observations is decomposed into shorter intervals and the set of observations belonging to each interval is called an observed arc, or just an **arc**. Each arc has its own set of initial conditions, as if there were a new spacecraft for each one of them. This results in over-parameterization, with the additional initial conditions absorbing the dynamical model uncertainties. Other parameters, e.g., in the dynamic model, can also be local to a single arc.

15.1 Local–global decomposition

The mathematics of the multi-arc method is a generalization of that of the marginal uncertainties discussed in Section 5.4. We use the notation $[a; b]$ to indicate the stacking of the two column vectors $\mathbf{a}$ and $\mathbf{b}$ to form a longer vector. The vector of all fit parameters $\mathbf{x} = [\mathbf{g}; \mathbf{h}]$ is split into a vector $\mathbf{g}$ of **global fit parameters** and a vector $\mathbf{h}$ of **local fit parameters**. The observations and the corresponding residuals are partitioned into n arcs by

311

some criterion, usually by time, in such a way that $\boldsymbol{\xi} = [\boldsymbol{\xi}_1; \boldsymbol{\xi}_2; \ldots; \boldsymbol{\xi}_n]$. The vector $\mathbf{h}$ is also split into vectors $\mathbf{h}_j$, one for each arc. Each subvector $\mathbf{h}_j$ is associated to the arc with the same index, in such a way that the residuals from one arc do not depend upon the local parameters of another arc

$$B_{\mathbf{g}}^{(j)} = \frac{\partial \boldsymbol{\xi}_j}{\partial \mathbf{g}}, \quad B_{\mathbf{h}_i}^{(j)} = \frac{\partial \boldsymbol{\xi}_j}{\partial \mathbf{h}_i} = \mathbf{0} \text{ for } i \neq j. \tag{15.1}$$

As a result the contributions of each arc to the overall normal equation are

$$C_{\mathbf{h}_i \mathbf{h}_j} = (B_{\mathbf{h}_i}^{(i)})^T B_{\mathbf{h}_j}^{(j)} = C_{\mathbf{h}_j \mathbf{h}_i}^T = \mathbf{0} \text{ for } i \neq j$$

$$C_{\mathbf{g}\mathbf{h}_i} = (B_{\mathbf{g}}^{(i)})^T B_{\mathbf{h}_i}^{(i)} = C_{\mathbf{h}_i\mathbf{g}}^T, \quad C_{\mathbf{g}\mathbf{g}} = \sum_{i=1}^{n} (B_{\mathbf{g}}^{(i)})^T B_{\mathbf{g}}^{(i)} = C_{\mathbf{g}\mathbf{g}}^T$$

giving to the normal matrix C an arrow-like structure (here we show the simplest case with two arcs only):

$$C = \begin{pmatrix} C_{\mathbf{g}\mathbf{g}} & C_{\mathbf{g}\mathbf{h}} \\ C_{\mathbf{h}\mathbf{g}} & C_{\mathbf{h}\mathbf{h}} \end{pmatrix} = \begin{pmatrix} C_{\mathbf{g}\mathbf{g}} & C_{\mathbf{g}\mathbf{h}_1} & C_{\mathbf{g}\mathbf{h}_2} \\ C_{\mathbf{h}_1\mathbf{g}} & C_{\mathbf{h}_1\mathbf{h}_1} & 0 \\ C_{\mathbf{h}_2\mathbf{g}} & 0 & C_{\mathbf{h}_2\mathbf{h}_2} \end{pmatrix}.$$

The contributions to the right-hand side D of the normal equation are

$$D = [D_{\mathbf{g}}; D_{\mathbf{h}}] = [D_{\mathbf{g}}; D_{\mathbf{h}_1}; D_{\mathbf{h}_2}; \ldots; D_{\mathbf{h}_n}],$$

with
$$D_{\mathbf{g}} = -\sum_{i=1}^{n} (B_{\mathbf{g}}^{(i)})^T \boldsymbol{\xi}_i, \quad D_{\mathbf{h}_i} = -(B_{\mathbf{h}_i}^{(i)})^T \boldsymbol{\xi}_i.$$

Then the normal equation can be written as a system of two vector equations:

$$\begin{cases} C_{\mathbf{g}\mathbf{g}} \Delta\mathbf{g} + C_{\mathbf{g}\mathbf{h}} \Delta\mathbf{h} = D_{\mathbf{g}} \\ C_{\mathbf{h}\mathbf{g}} \Delta\mathbf{g} + C_{\mathbf{h}\mathbf{h}} \Delta\mathbf{h} = D_{\mathbf{h}} \end{cases}$$

which can be solved as discussed in Section 5.4, by first solving for $\Delta\mathbf{h}$ in the second equation; the matrix $C_{\mathbf{h}\mathbf{h}}$ is block diagonal, thus it is possible to invert each diagonal block $C_{\mathbf{h}_j\mathbf{h}_j}$ separately

$$\Delta\mathbf{h}_j = C_{\mathbf{h}_j\mathbf{h}_j}^{-1} \left[D_{\mathbf{h}_j} - C_{\mathbf{h}_j\mathbf{g}} \Delta\mathbf{g} \right], \tag{15.2}$$

implying that the corrections to the local parameters $\mathbf{h}_j$ are not the same as would be obtained by ignoring the interaction with the global parameters, expressed by the submatrix $C_{\mathbf{h}_j\mathbf{g}}$. These expressions for $\Delta\mathbf{h}_j$ can be substituted into the first equation

$$[C_{\mathbf{gg}} - \sum_{j=1}^{n} C_{\mathbf{gh}_j} C_{\mathbf{h}_j \mathbf{h}_j}^{-1} C_{\mathbf{h}_j \mathbf{g}}] \, \Delta \mathbf{g} = C^{\mathbf{gg}} \, \Delta \mathbf{g} = D_{\mathbf{g}} - \sum_{j=1}^{u} C_{\mathbf{gh}_j} C_{\mathbf{h}_j \mathbf{h}_j}^{-1} D_{\mathbf{h}_j}$$

(15.3)

giving the solution for the global parameters

$$\Delta \mathbf{g} = \Gamma_{\mathbf{gg}} \, [D_{\mathbf{g}} - \sum_{j=1}^{n} C_{\mathbf{gh}_j} C_{\mathbf{h}_j \mathbf{h}_j}^{-1} D_{\mathbf{h}_j}], \qquad \Gamma_{\mathbf{gg}} = [C^{\mathbf{gg}}]^{-1} .$$

(15.4)

The corrections $\Delta \mathbf{g}$ and the covariance $\Gamma_{\mathbf{gg}}$ are in general not the same as in a separate global-only correction (that is, $C^{\mathbf{gg}} \neq C_{\mathbf{gg}}$ and $\Gamma_{\mathbf{gg}} \neq C_{\mathbf{gg}}^{-1}$).

The corrections to the local parameters are found by substituting $\Delta \mathbf{g}$ from (15.4) into (15.2):

$$\Delta \mathbf{h}_j = C_{\mathbf{h}_j \mathbf{h}_j}^{-1} \left[D_{\mathbf{h}_j} - C_{\mathbf{h}_j \mathbf{g}} \Gamma_{\mathbf{gg}} D_{\mathbf{g}} + C_{\mathbf{h}_j \mathbf{g}} \Gamma_{\mathbf{gg}} \sum_{k=1}^{n} C_{\mathbf{gh}_k} C_{\mathbf{h}_k \mathbf{h}_k}^{-1} D_{\mathbf{h}_k} \right] .$$

(15.5)

Their covariance can be deduced by comparing with the formula giving the correction by means of the full covariance matrix

$$\Delta \mathbf{h}_j = \Gamma_{\mathbf{h}_j \mathbf{h}_j} D_{\mathbf{h}_j} + \Gamma_{\mathbf{h}_j \mathbf{g}} D_{\mathbf{g}} + \sum_{k \neq j} \Gamma_{\mathbf{h}_j \mathbf{h}_k} D_{\mathbf{h}_k} ,$$

thus the covariance matrix of the local parameters $\mathbf{h}_j$ is

$$\Gamma_{\mathbf{h}_j \mathbf{h}_j} = C_{\mathbf{h}_j \mathbf{h}_j}^{-1} + C_{\mathbf{h}_j \mathbf{h}_j}^{-1} C_{\mathbf{h}_j \mathbf{g}} \Gamma_{\mathbf{gg}} C_{\mathbf{gh}_j} C_{\mathbf{h}_j \mathbf{h}_j}^{-1}$$

and the marginal uncertainty for $\mathbf{h}_j$ is larger than in a separate local solution. There are correlations between the local and the global parameters:

$$\Gamma_{\mathbf{h}_j \mathbf{g}} = -C_{\mathbf{h}_j \mathbf{h}_j}^{-1} C_{\mathbf{h}_j \mathbf{g}} \Gamma_{\mathbf{gg}}.$$

One advantage of the multi-arc decomposition is that the normal matrix has large portions of zeros, thus it is not necessary to store the full matrix in memory. However, the covariance matrix is in fact full, that is

$$\Gamma_{\mathbf{h}_j \mathbf{h}_k} = C_{\mathbf{h}_j \mathbf{h}_j}^{-1} C_{\mathbf{h}_j \mathbf{g}} \Gamma_{\mathbf{gg}} C_{\mathbf{gh}_k} C_{\mathbf{h}_k \mathbf{h}_k}^{-1}$$

for all $j \neq k$, is in general not zero (unless either $C_{\mathbf{h}_j \mathbf{g}}$ or $C_{\mathbf{gh}_k}$ is zero). In practice it may not be necessary to compute the correlations between the local parameters of different arcs, and anyway the full covariance matrix does not need to be stored.[1]

---

[1] Given the rapid increase in the RAM size, reducing memory usage is important only for very large problems; e.g., with 2008 technology, full matrices $20\,000 \times 20\,000$ can be stored in RAM.

Selection of the arc decomposition

Although the formalism of the previous section could be applied to an arbitrary decomposition of the observations into arcs, it is most useful when the decomposition is suggested by the distribution of the observations in time.

A satellite of the Earth is observable by each ground station only when it is above the local horizon (actually with an elevation $> 15-20°$ above the horizon, to avoid an increased error in the tropospheric correction). The time span over which this happens is called a **pass**. For an interplanetary probe, the rotation of the Earth controls the pass duration (see Section 17.2). If only a few stations can be used, the observational data are naturally concentrated in the time spans of the passes, with significant gaps in between.

Let us assume that the time span of a pass is of order dt, while the time span between two passes is on average Δt. If the uncertainty in the dynamic model is $\Delta \mathbf{F}$ and T is its along-track component (see Section 14.1), we use either eq. (14.5) or (14.7), depending upon the relationship between dt, Δt, and the orbital period P. Anyway the spacecraft position uncertainty is $\simeq 3/2 \, (dt/2)^2 \, T$ over one pass (assuming initial conditions at the center of the arc) and accumulates to $\simeq 3/2 \, (\Delta t)^2 \, T$ during the gap between observations. If $\Delta t \gg dt$ it may well be the case that the orbit can be modeled in a deterministic way during one pass, but not over an arc encompassing two or more passes. In this case the multi-arc approach can be effective.

15.2 Case study: satellite laser ranging

An important example of orbit determination where the local–global decomposition provides an effective strategy is **satellite laser ranging** (SLR). There are satellites, such as those of the **LAGEOS** class, including LAGEOS I (launched 1976) and LAGEOS II (1992), on high orbits well above the neutral atmosphere. These are passive spacecraft (without power) and covered with corner cubes to reflect back the laser pulses from the ground stations (see Figure 15.1).

The observations consist of ranges between some ground station, equipped with laser and timing equipment, and the satellite. The range is measured by the two-way light travel time divided by twice the speed of light c. It is corrected for the average distance between the reflecting corner cubes and the center of mass of the satellite, for the change in speed of light in the troposphere and for the finite time spans between transmission of the pulse, reflection, and reception at the ground station. The last correction is simpler than the one used for interplanetary tracking, see Section 17.2, because of the much shorter light times, allowing us to use just low-order corrections.

Fig. 15.1. The LAser GEOdynamics Satellite LAGEOS II, launched by a NASA/ASI collaboration from the Space Shuttle in 1992. It has 426 retroreflectors, each 3.8 cm in diameter.

A satellite on a high orbit, like the LAGEOS ones, has two to six passes per day on each station.[2] Modern SLR ground stations can generate thousands of pulses per pass (at a frequency $\simeq 10$ Hz or more), out of which a good fraction can result in a range measurement. The accuracy in range has been at the few cm level since the 1980s. The total data set collected over one year contains of the order of $m \simeq 10^5$ or more observations per station, with few tens of worldwide operational stations. Thus the accumulated data set is huge, and the ranges have a relative accuracy of the order of 1 part in 10^9.

With such a dataset it is possible to solve for a large set of parameters, including dynamical parameters (to be selected taking into account the orders of magnitude discussed in the next section), the initial conditions, and kinematical parameters. The latter include at least the station coordinates, but also their time derivatives, the Earth rotation parameters, and more. It is possible to solve for all the parameters at once, but such a fit would be subject to rank deficiency and to the effects of systematic errors.

15.3 Perturbation model

The LAGEOS class satellites were designed with a very low area-to-mass ratio $A/M = 0.007$ cm^2 g^{-1} and launched in a very high orbit ($\simeq 6000$ km above the Earth's surface), thus at the time of the launch of LAGEOS I

---

[2] Also depending on whether the station can operate in full daylight, or just at night.

it was believed that the non-gravitational perturbations would be a minor issue in the orbit determination. However, thanks to the extreme accuracy of the laser tracking, it was soon discovered that the first LAGEOS was experiencing a mystery drag, that is an **empirical acceleration** solved to allow a good fit to the data, with an average transversal component $\overline{T} \simeq -3 \times 10^{-10}$ cm/s^2. The cause of this deceleration is now interpreted as a combination of the Yarkovsky effect, charged particle drag, Earth's reflected radiation pressure, and eclipse effects (Bertotti and Iess 1991).

A useful exercise is to list the perturbations acting on the spacecraft orbit, in order of decreasing acceleration. For the LAGEOS class satellites, Table 15.1 lists the main perturbations down to $\simeq 1 \times 10^{-10}$ cm/s^2, from (Milani *et al.* 1987, Section 2.2) and (Bertotti *et al.* 2003b, Section 18.3). The non-gravitational perturbations contain the small quantity $A\Phi_\odot/(Mc) = F_{PR}$, those due to the shape of the Earth contain the coefficients $\overline{J}_{\ell m} = \sqrt{\overline{C}_{\ell m}^2 + \overline{S}_{\ell m}^2}$. The tidal terms contain the mass of the Moon $M_{\mathbb{C}}$, the Sun $M_\odot$, and the planets, the dynamic Love coefficient k_2, and the distances. Radiation pressure and drag have specific coefficients C_R and C_D.

Table 15.1. Accelerations in cm/s^2 acting on a LAGEOS class spacecraft.

Cause	Formula	Parameter	Uncertainty	Value
Earth monopole	$GM_\oplus/r^2 = F_0$	$GM_\oplus$	$2 \cdot 10^{-9}$	$2.8 \cdot 10^2$
Earth oblateness	$3\,F_0\,\overline{J}_{20}\,R_\oplus^2/r^2$	$\overline{J}_{20}$	$7 \cdot 10^{-8}$	$1.0 \cdot 10^{-1}$
Earth triaxiality	$3\,F_0\,\overline{J}_{22}\,R_\oplus^2/r^2$	$\overline{J}_{22}$	$2 \cdot 10^{-5}$	$6.0 \cdot 10^{-4}$
Moon tide	$2\,GM_{\mathbb{C}}\,r/r_{\mathbb{C}}^3$	$GM_{\mathbb{C}}$	$1 \cdot 10^{-7}$	$2.1 \cdot 10^{-4}$
Sun tide	$2\,GM_\odot\,r/r_\odot^3$	$GM_\odot$	$4 \cdot 10^{-10}$	$9.6 \cdot 10^{-5}$
Harmonic (6,6)	$F_0\,7\overline{J}_{66}\,R_\oplus^6/r^6$	$\overline{J}_{66}$	$5 \cdot 10^{-4}$	$8.8 \cdot 10^{-6}$
Solid tide	$3k_2\,GM_{\mathbb{C}}\,R_\oplus^5/(r_{\mathbb{C}}^3\,r^4)$	k_2	$2 \cdot 10^{-3}$	$3.7 \cdot 10^{-6}$
Radiation pressure	$C_R\,F_{PR}$	C_R	$2 \cdot 10^{-2}$	$3.2 \cdot 10^{-7}$
Relativistic Earth	$F_0\,GM_\oplus/(c^2\,r)$	$GM_\oplus$	$2 \cdot 10^{-9}$	$9.5 \cdot 10^{-8}$
Earth albedo	$C_R\,F_{PR}\,(1-\alpha_\oplus)\,R_\oplus^2/r^2$	$\alpha_\oplus, C_R$	0.2	$3.4 \cdot 10^{-8}$
Venus tide	$2\,F_0\,GM_{\mathbb{Q}}\,r/r_{\mathbb{Q}}^3$	$GM_{\mathbb{Q}}$	$3 \cdot 10^{-7}$	$1.3 \cdot 10^{-8}$
Indirect oblation	$3\overline{J}_{20}\,GM_{\mathbb{C}}\,R_\oplus^2/r_{\mathbb{C}}^4$	$GM_{\mathbb{C}}$	$1 \cdot 10^{-7}$	$1.4 \cdot 10^{-9}$
Thermal emission	$4/9\,F_{PR}\,\alpha\,\Delta T/T$	$\alpha, \Delta T$	0.5	$4 \cdot 10^{-10}$
Atmospheric drag	$C_D\,A\,\rho\,v^2/(2M)$	C_D, ρ	1	$1 \cdot 10^{-10}$

The list is long and contains exotic effects, like the main relativistic correction due to the mass of the Earth, and **indirect oblation**, the perturbation of the vector from the Earth–Moon center of mass to the center of the Earth, with respect to its two-body value, due to the oblateness of the Earth affecting the orbit of the Moon. However, what really matters is not the size of the accelerations (last column in the table), but the product of the acceleration size and its relative uncertainty (the column before the last). In this way we find that there are no significant uncertainties[3] down to a level of the order of $\Delta F \leq 10^{-8}$ cm/s^2. Below that level there is an accumulation of uncertain accelerations around a level of order 10^{-9} cm/s^2.

The analysis should focus on the perturbing accelerations with secular effects on the semimajor axis. The acceleration of the mystery drag appears as comparatively small fractions of thermal emission and Earth-reflected radiation pressure, and also as significant asymmetries of the thermal emission. An even smaller fraction of radiation pressure from the Sun may have effects along-track quadratic in time, but this requires either subtle properties of the eclipses or asymmetries in the optical properties of the two hemispheres of LAGEOS. A full discussion of the solution to the mystery drag problem is beyond the scope of this book: we shall just assume that there is a "modeling barrier" for the values of $\overline{T}$ of the order of $\Delta\overline{T} \simeq 10^{-10}$ cm/s^2.

Thus, over the time span of a pass ($dt \simeq 2 \times 10^3$ s) the errors in orbit propagations are $\leq 3/2\,(dt/2)^2\,\Delta F \simeq 0.015$ cm, while over the average interval between passes, say $\Delta t \simeq 2 \times 10^5$ s, the propagation error can be estimated by $3/2\,\Delta t^2\,\Delta\overline{T} \simeq 6$ cm. In practice, the propagation error is negligible with respect to the measurement error over one pass, and is always significant over the time span between two passes, not to speak of the effect over a longer arc. This suggests that the orbit of LAGEOS is a very suitable case for a multi-arc orbit determination.

15.4 Local geodesy

The simplest application of the multi-arc method to LAGEOS is obtained by assuming that the goal is to solve for the station positions (and possibly motions due, e.g., to continental drift). If the arcs are shorter than the orbital period ($\simeq 3 + 1/2$ hours) and the measurement accuracy is of the order of 1 cm, there is no need to solve for any dynamical parameter.

Thus the only global parameters **g** are kinematical, namely the coordinates of the vector positions $\mathbf{s}_i, i = 1, y$, of the y stations contributing ranges to the satellite. These coordinates are in a body-fixed frame, rotating with the Earth in a rigid body approximation.[4] This is appropriate for an orbit

---

[3] The uncertainty in the zero degree coefficient $G\,M_\oplus$ is mostly a problem of scale definition; the uncertainty within a consistent solution is much less.

[4] The deformations of the Earth due to tides have to be accounted for, with the Love coefficients h_2 and l_2 describing the elastic response of the Earth and the lag angle due to tidal dissipation. An even more complicated effect is the **oceanic loading**, inducing displacements (mostly vertical) due to the oceanic tide for stations located near some oceanic shore.

determination using only data from a limited time span, such as one year
or less. For a solution based on data from a longer time span, the station
velocities $\mathbf{v}_i$ need to be added to the global parameters, accounting for both
local motions and continental drift effects.

The only local parameters $\mathbf{h}_j$, with $j = 1, n$, to be solved for each of the n
arcs, are the initial conditions vectors $\mathbf{z}_j$, of dimension 6, containing some set
of orbital elements; we shall assume Cartesian coordinates are used, that is
$\mathbf{h}_j = [\mathbf{p}_j(t_j); \dot{\mathbf{p}}_j(t_j)]$ consists of three position and three velocity coordinates
at some epoch t_j (chosen at the center of the time span of the j-th arc).

Selection of the passes and data preparation

To reduce the observed arcs to a short time span, we need to use a local
network of SLR stations. Milani *et al.* (1995) present an experiment based
on the European SLR stations ranging to LAGEOS I. The European stations
are so close that an arc containing all the passes included in the same orbit
lasts only about half an hour. When the arc contains only 1–2 passes,
the information is not enough to solve for the initial conditions; the matrix
$C_{\mathbf{h}_j \mathbf{h}_j}$ is not invertible, or is very badly conditioned.

Thus we select as arcs only the intervals spanning passes over three European stations in the same orbit, but there are still enough data: $\simeq 1000$
arcs with 4.3 million ranges from seven European stations to LAGEOS I in
1985–1991. In these arcs with ≥ 3 passes the network of ground stations
behaves like a rigid body, providing a reference system for the orbit.

When the observations are closely spaced in time, the outliers can be identified by fitting all the data to smooth functions of time, such as polynomials
with a degree selected to capture the useful information. The outliers can be
identified by the size of the residuals, without the need for corrections (see
Figure 5.4). After a number of outlier removals, the value of the kurtosis
can be used as control; outlier removal should stop when the kurtosis is ≤ 3.
A reliable procedure for a large data set can be found in (Milani *et al.* 1995,
Section 4); the fraction of outliers was 2.4%.

After achieving a satisfactory polynomial fit, it is expedient to compress
the data by generating **normal points**. They are predictions based on the
polynomial model, at times selected to represent the useful information. For
example, if the raw data points are uniformly spaced at equal intervals of
time of size Δt, the normal points can be computed by a new sampling at
intervals $k \Delta t$, with the k such that $\nu = 2\pi/k \Delta t$ is less than the highest
frequency contained in the signal generated by the parameters being solved.
If the sampling of the raw data is non-uniform, the distribution of the times

of the normal points needs to account for the data gaps. Thus the observations were compressed in $\simeq 46\,000$ normal points, which have to be used as observables, but they are correlated, that is they have a full covariance matrix. Following Section 5.3, the normal matrix of the normal points has to be computed as the inverse of the covariance and to be used as weight matrix W. It is often necessary to add a component to account for the errors which are systematic, or at least with much lower frequencies.

15.5 Symmetries and rank deficiencies

A difficulty in this approach is rank deficiency. We shall use the method of Chapter 6: we find an exact symmetry in an approximate problem, then we show that an approximate symmetry remains in the full problem. We will also use a different approach looking directly for approximate symmetries.

To find an exact symmetry, we use as approximate equation of motion the geocentric two-body dynamics for LAGEOS; it has the group of symmetry $SO(3)$ of the rotations around the Earth's center of mass. That is, for each initial conditions $\mathbf{h}_j = [\mathbf{p}_j(t_j); \dot{\mathbf{p}}_j(t_j)]$ and each matrix $R \in SO(3)$ the solution $[\mathbf{p}'_j(t); \dot{\mathbf{p}}'_j(t)]$ with initial conditions $\mathbf{h}'_j = [R\,\mathbf{p}_j(t_j); R\,\dot{\mathbf{p}}_j(t_j)]$ can be obtained by rotating the original solution, that is considering $[\mathbf{p}'_j(t); \dot{\mathbf{p}}'_j(t)] = [R\,\mathbf{p}_j(t); R\,\dot{\mathbf{p}}_j(t)]$. However, when the observation equation is included, the $SO(3)$ symmetry is broken. The observables are the distances r_i between the satellite and the i-th station on the rotating Earth; if $S(t)$ is the rotation matrix between the reference system body fixed with the Earth and the inertial system in which the orbit of the satellite is computed, then $r_i(t) = |\mathbf{p}_j(t) - S(t)\,\mathbf{s}_i|$. If all the stations are also rotated by R, that is $\mathbf{s}'_i = R\,\mathbf{s}_i$, then the distance

$$r_i(t) = |\mathbf{p}_j(t) - S(t)\,\mathbf{s}_i| = |R\,\mathbf{p}_j(t) - S(t)\,R\,\mathbf{s}_i|$$

is exactly invariant if and only if $S(t)\,R = R\,S(t)$.

Let us assume that the rotation of the Earth is uniform and with a fixed axis: $S(t) = R_{\mathbf{\Omega}_\oplus t}$, where the angular velocity $\mathbf{\Omega}_\oplus$ is a constant angular velocity. Then R commutes with $S(t)$, for every t, if and only if R is a rotation around the same axis $\hat{\mathbf{\Omega}}_\oplus$. Note that this exact symmetry applies also to an Earth with non-spherical shape, provided it is axially symmetric (zonal harmonics only) with respect to the rotation axis. By inspecting Table 15.1 we find that this symmetry is very accurate even in a realistic case, the largest perturbation (due to the Earth's equatorial ellipticity) having a relative size of a few parts in 10^{-6} of the monopole. Moreover, by rotating the orbit by an angle ϵ in longitude ($\lambda \to \lambda + \epsilon$) the C_{22}, S_{22} perturbations change only by a fraction 2ϵ, thus the orbit difference with respect to the exact symmetry is much smaller than 1 cm, even for rotation angles affecting the station positions by a length $\epsilon R_\oplus$ of hundreds of meters.[5]

We conclude that at least one constraint needs to be applied to avoid an approximate rank deficiency: it can be obtained by fixing the longitude of one station, or better the longitude of the barycenter of the local network. In fact, numerical experiments on short-arc orbit determination with distances only (Milani and Melchioni 1989) show that there is an approximate rank deficiency of order four, that is three more than what is explained by the symmetry discussed above.

There is no way to find other exact symmetries, in particular to recover the $SO(3)$ symmetry group, unless we assume not only a spherical Earth but also that the Earth-fixed network of stations is non-rotating ($S(t)$ is constant in time). However, this is by no means an approximation, since the distances have relative changes of order unity with respect to the realistic case. Nevertheless the rotation of the Earth, although it affects the observables $r_i(t)$, has a reduced effect over one arc which is short enough. This occurs because the time span over which the orbit has to be propagated within a single arc ($dt/2 \simeq 1\,000$ s) is short with respect to one day, that is the Earth rotation angle over the time $dt/2$ is just $\eta = 0.073$ radians.

We look at the orbit determination problem in a body-fixed reference system,[6] in which the ground stations are not moving (besides the tidal and continental drift effects). Then the equation of motion contains the apparent **Coriolis** and **centrifugal** accelerations (see Section 16.1):

$$\mathbf{F}_{app}(\mathbf{p}_j(t), \dot{\mathbf{p}}_j(t)) = -2\,\boldsymbol{\Omega}_\oplus(t) \times \dot{\mathbf{p}}_j(t) - \boldsymbol{\Omega}_\oplus(t) \times [\boldsymbol{\Omega}_\oplus(t) \times \mathbf{p}_j(t)]\,.$$

Let R be a small rotation, that is $R = I + Z + \cdots$, where Z is the infinitesimal rotation by small angles $\mathcal{O}(\epsilon)$, represented by an antisymmetric matrix ($Z^T = -Z$), and the dots stand for $\mathcal{O}(\epsilon^2)$ terms. In the spherical Earth approximation, the equation of motion changes only by

$$\begin{aligned}\mathbf{F}_{app}(R\,\mathbf{p}_j(t), R\,\dot{\mathbf{p}}_j(t)) &- \mathbf{F}_{app}(\mathbf{p}_j(t), \dot{\mathbf{p}}_j(t)) \\ &= -2\boldsymbol{\Omega}_\oplus(t) \times Z\,\dot{\mathbf{p}}_j(t) - \boldsymbol{\Omega}_\oplus(t) \times [\boldsymbol{\Omega}_\oplus(t) \times Z\,\mathbf{p}_j(t)] + \mathcal{O}(\epsilon^2).\end{aligned}$$

The main centrifugal term can have a significant along-track component. It can be estimated in size as $\leq \Omega_\oplus^2\, p_j\, \epsilon$, thus its orbital effect over $dt/2$ is

---

[5] There are other perturbations due to the non-uniform rotation of the Earth, but they can be shown to be even smaller, see (Milani *et al.* 1987, Table 2.4).

[6] There is an alternative argument, by computing explicitly the commutators $RS - SR$ of the rotations group, but the following approach is simpler.

estimated by $\leq 3/2\,a\,\eta^2\,\epsilon$, where a is the semimajor axis of LAGEOS, to be compared with station displacements $\epsilon\,R_\oplus$. With $a \simeq 2\,R_\oplus$ we find that the ratio (orbit change)/(stations displacement) is estimated as $\leq 3\,\eta^2 \simeq 0.015$. For example, for $\epsilon \simeq 10^{-7}$ the stations are displaced by 60 cm and still the rotated orbit is distorted by no more than 1 cm.

The Coriolis term for a circular orbit has zero along-track component, thus it affects much less the orbit of LAGEOS; it can be estimated as $\leq 2\,\Omega_\oplus\,\dot{p}_j\,\epsilon$, and the orbital effect over $dt/2$ cannot exceed $3\,e\,\eta\,n\,dt/2\,\epsilon\,a$, with e, n the eccentricity and mean motion of LAGEOS, which is even smaller than for the centrifugal term. Another acceleration breaking the symmetry for rotations displacing the pole is due to the Earth's oblateness. It changes by a relative amount $\leq 2\,\epsilon$, thus for $\epsilon = 10^{-7}$ the acceleration difference is $\simeq 2 \times 10^{-8}$ cm/s^2, again not important over 1000 s. Thus we have identified approximate symmetries, although the corresponding exact symmetries (for $\eta = 0$) are not an approximation of a realistic problem.

The last approximate symmetry is well known to specialists of geodesy (both satellite and ground based): the geodetic network formed by the SLR stations can experience a *lift*, i.e., a translation away from the center of the Earth, by an amount d, provided the initial positions $\mathbf{p}_j(t_0)$ of the satellite are also translated in the same direction (for all j). Then the change in the two-body acceleration on the satellite is $\simeq 2\,(G\,M_\oplus/a^2)\,(d/a)$. For example, for a 1 m lift, $d/a \simeq 10^{-7}$ implies $|\Delta F_0| \simeq 4 \times 10^{-5}$ cm/s^2. Taking into account that the monopole gives a small contribution to the along-track perturbation T, the change in the orbit over 1000 s is much less than 1 cm.

Constraints and rigidity of the network

The four constraints required to avoid approximate rank deficiencies could be described as fixing to their initial value the three coordinates of the barycenter of the stations network, and inhibiting the rotations around the geocenter–barycenter axis.[7] They can be obtained with the a priori observation formalism of Section 6.1. Let $\mathbf{s}_i^{(0)}, i = 1, y$, be the initial values of the station coordinates; the barycenter coordinates are observed with uncertainty σ

$$\sum_{i=1}^{y} \frac{\mathbf{s}_i - \mathbf{s}_i^{(0)}}{y} = N(\mathbf{0}, \sigma^2 I),$$

that is, the probability density of the deviations from this a priori constraint is a Gaussian distribution with zero mean and covariance matrix $\sigma^2 I$ (I is

---

[7] In ground based geodesy, where the full $SO(3)$ symmetry applies, a traditional method is to fix two coordinates of one station and the longitude of another one.

a 3×3 identity matrix). The corresponding normal equation (6.3)

$$\sum_{i=1}^{y} \frac{\mathbf{s}_i}{y\,\sigma^2} = \sum_{i=1}^{y} \frac{\mathbf{s}_i^{(0)}}{y\,\sigma^2}$$

is added to the normal equation from the real observations. To inhibit rotations around the barycenter $\mathbf{b} = \sum \mathbf{s}_i^{(0)}/y$ we use an a priori constraint

$$\frac{1}{K} \sum_{i=1}^{y} \mathbf{b} \times (\mathbf{s}_i^{(0)} - \mathbf{b}) \cdot (\mathbf{s}_i - \mathbf{b}) = 0$$

with RMS $= \sigma$ and $K = \sum_{i=1}^{y} |\mathbf{b} \times (\mathbf{s}_i^{(0)} - \mathbf{b})|$; the normal equation is

$$\frac{1}{K} \sum_{i=1}^{y} \mathbf{b} \times (\mathbf{s}_i^{(0)} - \mathbf{b}) \cdot \mathbf{s}_i = \frac{1}{K} \sum_{i=1}^{y} \mathbf{b} \times (\mathbf{s}_i^{(0)} - \mathbf{b}) \cdot \mathbf{b}.$$

To impose a tight constraint, a small value of σ is used, e.g., $\sigma = 0.1$ for coordinates expressed in cm, thus the constraints have to be satisfied at the mm level. To assess the relevance of the approximate rank deficiency, the constraint is weakened to the meter level, with $\sigma = 100$.

Stability test

In (Milani *et al.* 1995) the stability of the solution is tested by splitting the data set (ranges from seven stations during one year) into two halves, formed by the odd and even arcs, numbered in order of time. With $\sigma = 0.1$, the RMS of the differences between the station coordinates in the two "half" solutions is 1.63 cm. The component of rigid motion contained in the differences has RMS $= 0.39$ cm; this is due to the two remaining degrees of freedom in the group of rigid motions, the *tilt* along axes passing from the barycenter $\mathbf{b}$. The instabilities which are deformations of the network have RMS $= 1.58$ cm, thus the constraints are effective. If the constraints are relaxed with $\sigma = 100$, the RMS of the coordinate differences grows to 72.5 cm.

The accuracy of such a local geodetic network, obtained many years ago, was such that in a solution with seven years of data (1985–1991) the same "two halves" stability test gave a RMS of the station position differences of 0.58 cm, with only 0.19 of RMS tilt. The differences in the stations' velocities had RMS $= 0.32$ cm/year, of which only 0.06 of tilt. In a solution including station velocities as global parameters, the number of constraints has to be increased to eight, inhibiting a translation velocity of the barycenter and a uniform rotation around the geocenter–barycenter axis.

16

SATELLITE GRAVIMETRY

In this chapter we deal with the problem of solving for the gravity field of a planet without degradation of the results due to non-gravitational perturbations. This problem is severe for low orbits around a planet with an atmosphere, like the Earth, because atmospheric drag sharply increases when the orbit altitude decreases. Even around planets without atmosphere, like Mercury, low orbits are much more affected by reflected and infrared radiation pressure from the planet's surface (see Section 17.3). Thus we need to estimate how low the spacecraft needs to orbit, to be sensitive to the portion of the gravity field we wish to measure. Let us suppose the spacecraft is in a nearly circular orbit at an altitude h above the surface of a planet with equatorial radius R. The potential due to a spherical harmonic of degree ℓ and order m is

$$\frac{GM}{r} \left(\frac{R}{r}\right)^{\ell} \overline{Y}_{\ell m i}$$

times the coefficient $\overline{C}_{\ell m}$ for $i = 1$, $\overline{S}_{\ell m}$ for $i = 0$ (see Chapter 13). Since $r = R + h$, if we assume that h coincides with the **spatial scale** of the harmonic, that is half of the smallest spatial wavelength, from the Kaula expansion (13.32) we have $h = \pi R/\ell$ and

$$\left(\frac{R}{r}\right)^{\ell} = \left(1 + \frac{h}{R}\right)^{\ell} = \left(1 + \frac{\pi}{\ell}\right)^{-\ell}, \quad \lim_{\ell \to +\infty} \left(1 + \frac{\pi}{\ell}\right)^{-\ell} = \frac{1}{\exp(\pi)} \simeq \frac{1}{23.14},$$

that is, for h equal to the spatial scale, the ratio of the monopole potential to the harmonics with high degree ℓ is essentially independent of ℓ and is close to the irrational number $\exp(\pi)$. For higher orbits, say $h = k\,\pi R/\ell$, the ratio becomes $\exp(\pi)^k$ and the gravity signal sharply decreases as k grows.

If we assume the gravity field is measured from the potential, good sensitivity can be obtained by keeping the altitude of the order of the spatial scale. One way to increase sensitivity is to measure derivatives of the potential: for the second radial derivative the sensitivity is increased by a factor $(\ell + 1)(\ell + 2)/2$; however, for $\ell = 100$ this only corresponds to $k \simeq 2.7$.

In conclusion, if a gravimetry mission targets a short spatial scale, just a few times the scale height of the atmosphere, drag is a critical problem. For example, for the Earth $\ell \simeq 200$ (scale $\simeq 100$ km) can be reached only with equipment to neutralize the effect of drag. Such equipment is described, from the point of view of its impact on the orbit determination, in the next section.

16.1 On-board instrumentation

We list the instruments which could be used to neutralize the effect of non-gravitational perturbations, with their advantages and problems.

Navigation systems

On the surface of the Earth, navigation instruments provide accurate positioning, in the reference system defined by a satellite constellation, such as GPS, GLONASS, Galileo. The versions used on satellites in low Earth orbit provide information equivalent to a position every few seconds.

The position of the low satellite is not measured instantaneously: the phases received from the navigation satellites can be fitted to a **reduced dynamics orbit**, the solution of a simplified equation of motion, containing as free parameters initial conditions and some empirical acceleration, to absorb both the non-gravitational effects and the inaccuracies in the gravity field model. That is, spacecraft positioning is an orbit determination problem with an overwhelming amount of data, such that very short arcs (a few minutes) can be used and still tightly constrain the orbit. In practice, with state-of-the-art space-borne GPS navigators,[1] we can assume that the orbit of a low satellite is known to a few centimeters at all times.

There are different ways to include the navigation data in the orbit determination. One way is a *brute force* method: a least squares fit with all the observables from navigation and the other instruments, and the equation of motion containing all the parameters (geopotential coefficients, initial conditions, and more). This can be used, but the same accuracy may be achieved by using methods with lower computational complexity, see Section 16.3.

More efficient is a **kinematic method**, in which the orbit determination

---

[1] The European navigation system Galileo should be even more accurate, but is not yet operational.

is split into two steps. In the first step, the **precision orbit determination** (POD), the time series of spacecraft positions is determined by using the navigation data only. Since the non-gravitational perturbations are absorbed by the empirical accelerations, in the POD there is no need to use the data from on-board accelerometers. In the second step the other parameters are determined by using as observables the spacecraft positions and/or the measurements of the other instruments.[2] The uncertainty of the positions from the POD has to be taken into account, but this is not the main source of error. Anyway, it is possible to iterate, by using the improved gravity field and the calibrated accelerometer data (see below) from the second step for the POD, replacing the empirical accelerations.

Accelerometers

An **accelerometer** measures the relative acceleration of a sensitive element with respect to the instrument rigid frame. The sensitive element position needs to be controlled by a feedback loop, in such a way that it does not undergo any large-scale motion with respect to the frame; the amount of these corrections, actuated by electrostatic forces, is the actual measurement.

Two technologies are currently used for accelerometers. In the *electrostatic accelerometer* the sensitive element is a conductive mass levitating inside a cavity of capacitors. One sensitive element is enough to measure a vector acceleration. The main limitation is that the levitation is hard to be maintained with the same equipment on the ground and in space, thus testing on the ground is limited and expensive space-borne tests are needed.[3]

The alternative is a *spring accelerometer*, in which the sensitive element is free to move only along an axis, with a spring as restoring force and electrostatic forces as controls. Three separate units, with orthogonal axes, measure an acceleration vector, but this complexity is somewhat compensated by the ease of testing on the ground: each unit can be tested in a micro-gravity environment by orienting it normal to the local gravity field.

In fact, there are two main limitations to the use of accelerometers in a very accurate orbit determination. First, an accelerometer anyway provides only a relative measurement; the electrical quantities corresponding to a zero acceleration are simply not known, and ground tests would provide values

---

[2] It could be argued that the second step is not orbit determination, since it solves for everything but the orbit, still it complies with the definition of the problem as given in Chapter 1.
[3] For example, CHAMP, the first scientific mission with a promising electrostatic accelerometer, was plagued by a partial failure.

different from those applicable in space.[4] Thus accelerometers measure only changes, over limited time spans, of the accelerations. Second, there is no way to build an accelerometer which does not act also as a thermometer, that is the reading is a function of both acceleration and temperature.

In conclusion, spring accelerometers are robust and reliable, but they do not reach the same accuracy as the electrostatic ones because they are generally more temperature sensitive. Thus at present the electrostatic accelerometers are used for the top accuracy satellite gravimetry missions around the Earth, while the spring ones are used for the gravimetry of other planets.

Apparent accelerations

The purpose of orbit determination for satellite gravimetry is not to determine the position of any specific point on the spacecraft, but to locate some point for which we can write an equation of motion, containing the dynamic parameters: the equation for the spacecraft center of mass (CoM) $\mathbf{x}$ is

$$\ddot{\mathbf{x}} = \nabla U(\mathbf{x}) + \mathbf{a}_{ng}, \tag{16.1}$$

where U is the gravitational potential and $\mathbf{a}_{ng}$ the non-gravitational acceleration. The non-gravitational forces act on the external surface of the spacecraft, and assuming the accelerometer frame is attached to a rigid spacecraft structure including the surface, the accelerometer cage is accelerated by $\mathbf{a}_{ng}$, the sensitive element does not feel this acceleration, and the instrument measures the **apparent acceleration** $-\mathbf{a}_{ng}$.

However, the accelerometer cannot be placed exactly at $\mathbf{x}$, but at some position displaced by the vector $\mathbf{Y}$ from the CoM in a spacecraft fixed reference system; let $\mathbf{y} = R\,\mathbf{Y}$ be the same displacement in the inertial system, with R a time-dependent rotation. The accelerometer velocity in an inertial frame is

$$\dot{\mathbf{y}} = \dot{R}\,R^T\,\mathbf{y} + R\,\dot{\mathbf{Y}} = \boldsymbol{\omega} \times \mathbf{y} + R\,\dot{\mathbf{Y}}$$

with $\boldsymbol{\omega}$ the **angular velocity** (Arnold 1976). The inertial acceleration is

$$\ddot{\mathbf{y}} = [\boldsymbol{\omega} \times (\boldsymbol{\omega} \times \mathbf{y}) + \dot{\boldsymbol{\omega}} \times \mathbf{y}] + 2\boldsymbol{\omega} \times R\,\dot{\mathbf{Y}} + R\,\ddot{\mathbf{Y}} = \mathbf{a}_{rot} + \mathbf{a}_{\mathbf{Y}} \tag{16.2}$$

where the part inside square brackets is the **rotation acceleration** $\mathbf{a}_{rot}$ of the accelerometer, and $\mathbf{a}_{\mathbf{Y}}$ is the acceleration due to a possible drift of the CoM in the spacecraft frame, due to either movable parts or fuel consumption. Both are applied by solid state forces on the accelerometer cage.

---

[4] An accelerometer is not an *inertial guidance system*; it can also act as inertial guidance, but only with errors larger by orders of magnitude.

Moreover, the accelerometer sensitive element is accelerated by the gravity field $\nabla U(\mathbf{x} + \mathbf{y})$, while the cage is accelerated by the gravity field at the CoM: thus the accelerometer also measures a **gravity gradient acceleration**. This acceleration can be computed, neglecting $\mathcal{O}(|\mathbf{y}|^2)$ terms, from the matrix of second derivatives of the gravitational potential U:

$$\mathbf{a}_{gg}(\mathbf{y}) = \frac{\partial^2 U}{\partial \mathbf{x}^2}(\mathbf{x})\, \mathbf{y}.$$

Thus there are differential accelerations, functions of $\mathbf{y}$, while $\mathbf{a}_{ng}$ does not depend upon $\mathbf{y}$, and the accelerometer measures the combination

$$\mathbf{a}_{acc}(\mathbf{y}) = -\mathbf{a}_{ng} - \mathbf{a}_{rot}(\mathbf{y}) - \mathbf{a}_{\mathbf{Y}} + \mathbf{a}_{gg}(\mathbf{y}) \qquad (16.3)$$

where the minus sign applies to accelerations acting on the cage, plus when acting on the sensing element directly. The equation of motion of the CoM using the accelerometer is obtained by substituting (16.3) into (16.1):

$$\ddot{\mathbf{x}} = \nabla U(\mathbf{x}) - \mathbf{a}_{acc} - \mathbf{a}_{rot}(\mathbf{y}) - \mathbf{a}_{\mathbf{Y}} + \mathbf{a}_{gg}(\mathbf{y}). \qquad (16.4)$$

It is possible to compute the equation of motion for the accelerometer $\mathbf{x} + \mathbf{y}$ by adding eq. (16.4) to (16.2), with cancellation of $\mathbf{a}_{rot}, \mathbf{a}_{\mathbf{Y}}$

$$\ddot{\mathbf{x}} + \ddot{\mathbf{y}} = \nabla U(\mathbf{x}) - \mathbf{a}_{acc} + \nabla(\nabla U)(\mathbf{x})\,\mathbf{y} = \nabla U(\mathbf{x} + \mathbf{y}) - \mathbf{a}_{acc} \qquad (16.5)$$

with the surprising result that, when using the data from the accelerometer as a term in the equation of motion, the equations of motion for the accelerometer are simpler than those for the CoM.[5]

The above presentation is somewhat simplified: care needs to be taken of three points. First, the tracking instruments are neither in $\mathbf{x}$ nor in $\mathbf{x}+\mathbf{y}$, but have some other reference point (e.g., the antenna phase center) displaced by a vector $\mathbf{Z}$ in spacecraft axes from the CoM. If the tracking data are spacecraft positions, they refer to $\mathbf{x} + \mathbf{z}$ (where $\mathbf{z} = R\mathbf{Z}$) and have to be corrected by subtracting $\mathbf{z} - \mathbf{y}$; on the other hand, $\mathbf{Z} - \mathbf{Y}$ is better known than $\mathbf{Z}$, since the position of the CoM inside the spacecraft structure may depend upon the poorly known content of fuel in the tanks. If the tracking data are range and/or range-rate, corrections containing $\mathbf{z} - \mathbf{y}$ and $\dot{\mathbf{z}} - \dot{\mathbf{y}}$, respectively, have to be applied, and the requirements on the knowledge of $\mathbf{Z} - \mathbf{Y}$ and $\dot{\mathbf{Z}} - \dot{\mathbf{Y}}$ are severe for state-of-the-art tracking systems (see Chapter 17).

Second, for an electrostatic accelerometer there is really a single sensitive reference point, the CoM of the levitating mass. For a spring accelerometer, there are three separate sensitive points $\mathbf{Y}_i$, $i = 1, 2, 3$, with mutual

---

[5] This method was suggested by H.-R. Schulte of EADS-Astrium in 2007.

distances of several cm which cannot be ignored. The solution is to select a conventional reference point $\mathbf{Y}$ in the accelerometer structure, then correct the readings of the three channels for the displacements $R(\mathbf{Y}_i - \mathbf{Y})$, with $\mathbf{Y}_i - \mathbf{Y}$ well known (and presumably constant).

Third, this discussion assumes that the rotation state, not just R but also $\boldsymbol{\omega}$ and $\dot{\boldsymbol{\omega}}$, are well known. In reality there will be a contribution from the knowledge of these quantities in the error budget. Experience shows that the corresponding requirements on the attitude control subsystem need to be clearly specified at an early stage in the design of the space mission.

Calibration

The relationship between the measured electrical quantity q and the actual acceleration a_{acc} for each accelerometer sensitive element is of the form

$$a_{acc} = a\,q + b\,T + c + \cdots \qquad (16.6)$$

where a is the **scale calibration**, T the local temperature, b the **thermal sensitivity**, c the **absolute calibration**, and the dots stand for the nonlinear effects, typically negligible provided the accelerometer dynamic range is not exceeded. The above formula is for an instrument with a single sensitive axis; for a three-axis accelerometer there are three such formulae.[6]

The scale calibration can be measured by a known acceleration: internal calibration with an ad hoc electrostatic force and external calibration with, e.g., the planet monopole gravity gradient, or the apparent force from spacecraft rotation. The thermal sensitivity can be measured on the ground.

The most critical calibration is the absolute one, for which a dedicated calibration device and/or procedure would be very difficult. Thus one key issue of this chapter will be the **a posteriori calibration**, that is the determination of the values of c as part of the orbit determination problem. The only information we have on c is that it changes slowly with time, but how slowly we do not know yet: every ground based laboratory is affected by too much acceleration noise, thus tests in space are the only possibility.[7] We conclude that c needs to be determined as three constants for each observed arc, possibly even as coefficients of interpolation models for each arc, depending upon the arc time span, see Section 17.6 and 16.3, respectively.

---

[6] In this discussion we are neglecting the problem of the alignment of the sensitive axes of the accelerometer; this error source has to be taken into account in a complete error budget.

[7] At the time this book is being written, these space-borne tests have not been completed yet. However, the Fourier analysis of long-term ground tests on spring accelerometers indicates that the stability time-scales could exceed one day.

Drag free missions

A **drag free** spacecraft uses an accelerometer coupled to the orbit control subsystem with a feedback loop, such that the measured accelerations $\mathbf{a}_{acc}$ are controlled to zero by thrusting. Ideally, the sensitive element of an electrostatic accelerometer should follow a purely gravitational orbit. The discussions above, on the measurements not being at the CoM and about calibration, make clear that this ideal condition can be realized only roughly. In particular it is difficult to obtain the absolute calibration in real time.

Nevertheless, an approximate drag free system may be necessary. A high spatial resolution gravimetry mission around the Earth needs to orbit where drag is significant, and two problems arise. The first is the mission duration, which could be cut short by orbital decay unless the drag is compensated by thrust: a solution is to have an orbit which is drag free, but only on average, e.g., impulsive thrust used when necessary.

The second problem is the accelerometer saturation: to ensure linearity of the calibration (16.6), the acceleration measured needs to be controlled, and this is a requirement applicable at all time, implying a drag free system with continuous thrust. The measured $\mathbf{a}_{acc}$ does not need to be controlled to the accelerometer sensitivity, but below the saturation level, at least 10^3 times larger. For a low satellite experiencing significant drag, an effective strategy is to have continuous thrust along a single direction, opposite to the velocity, controlling the low-frequency portion of the deceleration. Lift and radiation pressure accelerations (acting in different directions) do not need to be controlled, and the motion of the accelerometer reference point follows eq. (16.5), with $\mathbf{a}_{acc}$ including the apparent force due to thrust.

Gradiometers

A very powerful instrument for gravimetry missions is a **gradiometer**, directly measuring some second derivatives of the potential. It contains a number of accelerometers, electrically coupled in such a way that the differences of the accelerations are directly measured. A gradiometer is particularly effective when coupled to a navigation system; then the orbit can be considered known, and the differences in acceleration are linearly related to the gravity gradient matrix $\partial^2 U / \partial \mathbf{x}^2$, e.g.,

$$\frac{d^2 U}{dr^2} = \frac{\frac{\partial U}{\partial r}(\mathbf{x} + d/2\,\mathbf{r}) - \frac{\partial U}{\partial r}(\mathbf{x} - d/2\,\mathbf{r})}{d} + \mathcal{O}(d/r), \qquad (16.7)$$

where the $\mathcal{O}(d/r)$ term is typically below the instrument sensitivity.

The calibration of gradiometers is not very different from that of accelerometers, with one equation (16.6) for each differential measurement, where the temperature T is replaced by the difference in temperature between the two units. Thus the temperature needs to be controlled only in a relative sense: if all the accelerometers are contained in a well-insulated container, the thermal correction can be much smaller. The gradiometers are used to increase sensitivity to short spatial scales; the time-scale over which the signal changes can be short, e.g., for $\ell \simeq 200$ the shortest period is < 30 s, but the longest period from the $\ell - 2p = 2$ term is > 2700 s, see (13.38). Thus the critical issue is the a posteriori calibration, discussed in Section 16.3.

Apparent accelerations

The question is how many accelerometers, and how many independent measurements of components of $\partial^2 U/\partial \mathbf{x}^2$, are to be used. The apparent acceleration $\mathbf{a}_{rot}$ is linear in the positions $\mathbf{y}_i$ of the accelerometers, and so is $\mathbf{a}_{gg}$. Thus no matter where the CoM is, the differential apparent accelerations can be computed by using the relative positions $\mathbf{y}_i - \mathbf{y}_j$. If the gravimeter can be considered as a rigid body, there is no Coriolis term. The centrifugal apparent acceleration can be obtained from a potential W:

$$-\boldsymbol{\omega} \times [\boldsymbol{\omega} \times (\mathbf{y}_i - \mathbf{y}_j)] = \nabla \left[\frac{1}{2} |\boldsymbol{\omega}|^2 \, |\Pi(\mathbf{y}_i - \mathbf{y}_j)|^2 \right] = \nabla W$$

where Π is the projection on the plane orthogonal to the angular velocity $\boldsymbol{\omega}$. Then the matrix of second derivatives measured by the gradiometer is in fact $\partial^2 (U + W)/\partial \mathbf{x}^2$. U satisfies the Laplace equation, thus

$$\Delta(U + W) = \Delta W = 2|\boldsymbol{\omega}|^2. \tag{16.8}$$

If the knowledge of the rotation state (from the attitude control subsystem) is good enough, each differential measurement can be corrected for the centrifugal term.[8] However, the gradiometer could be better in measuring rotation than the attitude control; then the accuracy would be limited by the errors in the centrifugal term. A solution is to measure all three diagonal components of $\partial^2 (U + W)/\partial \mathbf{x}^2$ and compute $|\boldsymbol{\omega}|^2$ from the formula above: this requires two accelerometers along each of the three orthogonal axes, and leaves only two diagonal components independently measured.

---

[8] There is also a term $-\dot{\boldsymbol{\omega}} \times (\mathbf{y}_i - \mathbf{y}_j)$, requiring knowledge of the angular acceleration.

16.2 Accelerometer missions

An *accelerometer mission* for gravimetry could have just two instruments: an accelerometer and a navigation system. The orbit determination can be decomposed into two steps. The first is the POD, using the phases from the navigation satellites and solving for the satellite positions (with empirical accelerations absorbing the dynamic model errors).

There are several ways to define the second step. We shall use as an example the method used in a simulation of one such mission, the Italian Space Agency project SAGE (Albertella and Migliaccio 1998). In this approach, the second step uses as observables the three coordinates of the satellite, and has to solve for a large set of parameters, including the harmonic coefficients and the initial conditions for each arc. The latter are also determined by the navigation, but with an accuracy insufficient to solve for the gravity field.

A simple formula to estimate the maximum arc time span uses the assumption that the accelerometer measurements only contain noise, with known constant spectral density ΔA. If x_T is the coordinate of the satellite position $\mathbf{x}$ in the transverse direction at the extremes of an arc time span P

$$\mathrm{RMS}(x_T) = 2\,\Delta A\,(P/2)^{3/2}.$$

For example, if $\Delta A = 3 \times 10^{-6}$ cm s^{-2} Hz$^{1/2}$ and the positions from navigation have an RMS > 1 cm, the random error due to accelerometer noise is less, even for arcs covering an orbital period (5560 s for $h = 400$ km). This is an optimistic estimate of the orbit errors due to the accelerometer (calibration errors have a larger effect), giving just an upper bound for the arc time span. Still, this estimate has severe implications; e.g., for a six months simulation, with arcs of $1/2$ orbital period, there are $\simeq 5500$ arcs, thus $\simeq 33\,000$ initial conditions. If the target is to solve for the 8281 coefficients up to degree and order 90, the dimension of the normal/covariance matrix is $\simeq 41\,300$. This strongly suggests the decomposition of the normal equation.

Local–global decomposition

The local–global decomposition of Section 15.1 can be used as a solution to the problem of a too large normal matrix. If we can neglect the a priori correlation between initial conditions of different arcs, this method allows us to invert $\simeq 5500$ matrices $C_{\mathbf{h}_i\mathbf{h}_j}$ which are just 6×6 matrices if the local parameters are just the initial conditions, somewhat larger, e.g., 9×9, if accelerometer calibration parameters are also estimated for each arc. Then the solution for the harmonic coefficients $\mathbf{g}$ are obtained by inverting a matrix 8281×8281, by using eq. (15.4). Also the equations for $\mathbf{g}$ can be decomposed by resonant decomposition, see Section 13.5.

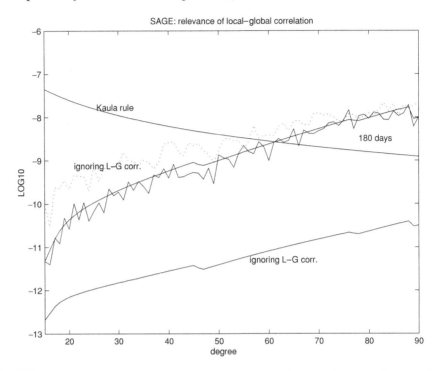

Fig. 16.1. A simulation of the SAGE gravimetry mission. The smooth curves plotted as RMS values ($\log_{10}$ scale) for each harmonic degree ℓ, from top to bottom on the left, are: the expected values of the coefficients according to the Kaula rule; the formal RMS of the results taking into account local–global correlations; the same RMS neglecting them. The jagged curve is the actual error with local–global correlations, the dotted one the actual error neglecting them.

The SAGE simulations provided a numerical experiment on the quantitative aspects of the local–global decomposition. The covariance matrix of the global parameters $\Gamma_{\mathbf{gg}}$ is not the inverse of the normal matrix $C_{\mathbf{gg}}$, but of another matrix $C^{\mathbf{gg}}$ defined by eq. (15.3), which is obtained by subtracting from $C_{\mathbf{gg}}$ a matrix with non-negative eigenvalues. This implies that $C^{\mathbf{gg}} < C_{\mathbf{gg}}$, by which we mean that for each vector $\Delta\mathbf{g}$

$$\Delta\mathbf{g} \cdot C^{\mathbf{gg}} \, \Delta\mathbf{g} \leq \Delta\mathbf{g} \cdot C_{\mathbf{gg}} \, \Delta\mathbf{g}.$$

Now $\Delta\mathbf{g} \cdot C^{\mathbf{gg}} \, \Delta\mathbf{g} = \sigma^2$ is a confidence ellipsoid for the global parameters $\mathbf{g}$, and the above inequality implies that the confidence ellipsoids with matrix $C_{\mathbf{gg}}$ would be smaller (for the same σ). Ignoring the local–global correlations, and replacing the marginal covariance matrix of $\mathbf{g}$ (for whatever local parameters $\mathbf{h}$) with the conditional covariance matrix of $\mathbf{g}$ (for $\mathbf{h}$ kept fixed) gives an optimistic assessment of the accuracy of the results (see Section 5.4).

Figure 16.1 shows the results of a simulation, for a six-month mission, with and without the local–global terms $C_{\mathbf{h}_j\mathbf{g}}$. The difference between the two cases is dramatic: neglecting the local–global terms gives an illusory accuracy estimate, by more than two orders of magnitude in the RMS values. In this example, using the conditional covariance of $\mathbf{g}$ would give the impression that the coefficients up to degree and order 90 can be estimated with a S/N ratio > 10, while in fact S/N > 1 is possible only up to degree $\simeq 60$. Moreover, the actual error (difference between the ground truth value used in the simulation and the obtained solution) is larger.

The SAGE simulations were simplified (the initial conditions were the only local parameters), but the conclusions on the need of a full solution for all parameters still stand. With more local parameters (for accelerometer calibration) the local–global terms would have an even bigger impact.

16.3 Gradiometer missions

A *gradiometer mission* could have just two instruments, a navigation system and a gradiometer, but the latter is large and complex; as shown in Section 16.1, six accelerometers are required. Since at present electrostatic three-axis accelerometers are used (ESA 1999), there are 18 accelerometer channels, and the measurement procedure has a complexity which we are not going to discuss. To allow for a self-contained discussion of the gravity field solution with a gradiometer we use two simplifying assumptions. First, the spacecraft position is determined by the navigation, in a POD independent from the accelerometer data. Second, we assume that the differential measurements from the accelerometer channels are summarized in independent measurements of two out of three diagonal terms of $\partial^2 U/\partial \mathbf{x}^2$, the effect of apparent accelerations having been removed by pre-processing (the errors in this procedure contribute to the error model). The gradiometer calibrations are summarized as two slowly varying functions of time.

A basic choice is the organization of the gravimetry observations in one of three possible ways. In a **spacewise method** the data points are ordered by their position in space, in a reference frame rotating with the Earth; this is a sampling of gravity around a geocentric sphere (a thin shell if the orbit is almost circular). The advantage is that most of the gravitational signal to be determined is also organized spatially, although there are time-dependent signals due to tides and other deformations. In a **timewise method** the data points are considered as a discretized time series. This is a less natural way of looking at the static gravity: indeed, each spherical harmonic appears as a sum of signals with different frequencies, by eq. (13.36). The advantage of the timewise methods is in the treatment of the gradiometer calibrations. In a **frequencywise method** the time series of the gradiometer data are Fourier transformed into the frequency domain: the effect of each spherical harmonic is represented by a Fourier polynomial and the fit can be directly performed in the frequency space. If each harmonic is solved independently, the computational cost is low: this method is used in the mission design phase, to convert requirements in the error spectrum of the gradiometer into an error spectrum of the recovered gravity. However, the correlations between the harmonics have to be taken into account sooner or later, see Section 16.5.

These three approaches are equivalent for a perfect distribution of data. With a spatial distribution uniform on a sphere and a time distribution uniform over an unlimited time span there is a well conditioned one-to-one correspondence between spherical harmonics, a linear subspace of the signal as function of time, and its discrete spectra. However, such uniformities are impossible in real missions, and there are superpositions between the frequencies of the gravitational signal and the one of the calibrations.

We will thus follow the timewise approach for a second step of orbit determination, in which the positions are assumed as given, that is with a kinematic method. A significant advantage is that it is not necessary to consider the positions from the navigation as observables, and is not necessary to solve for the initial conditions, that is to recompute a more refined orbit (as in the accelerometer mission case). The gradiometer measures directly the gravity potential, through some of its second derivatives, without the intermediary of the orbit. The position time series is needed only to assign the measurement taken at some time to a specific point in space.

The accuracy requirements for this positioning can be assessed by estimating the third derivative of the potential; e.g., for the radial derivatives and the monopole term, with a radial displacement by $\Delta r \simeq -10$ cm

$$\frac{\partial^3 U_0}{\partial r^3} = -\frac{3}{r}\frac{\partial^2 U_0}{\partial r^2} \implies \Delta \frac{\partial^2 U_0}{\partial r^2} \simeq \frac{\partial^3 U_0}{\partial r^3}\,\Delta r = -3\frac{\Delta r}{r}\frac{\partial^2 U_0}{\partial r^2} \simeq 5\times 10^{-8}\frac{\partial^2 U_0}{\partial r^2},$$

for an orbit at altitude $h = 250$ km. With a monopole gradient $\simeq 7\times 10^{-7}$ s^{-2}, the change in the radial gradient is $\simeq 3\times 10^{-14}$ s^{-2}, or 3×10^{-5} E (in Eötvös unit: $1\ E = 10^{-9}\ s^{-2}$). This is well below the sensitivity of the current gradiometers, implying that, in the context of gradiometer missions, the positioning provided by state-of-the-art navigation systems is accurate enough to be used neglecting entirely its uncertainty.

Gradiometer error models

To decide how to model the calibration functions, we need to follow three principles of good practice. First, calibration models should be based on some understanding of the measurement physics. Second, the error models used in the simulations must contain non-random terms, to describe the unavoidable systematics. Third, the solutions should try to compensate for the systematic errors, without using information on the specific functional expression of the systematics introduced in the simulation. We shall elaborate on these three principles, by using a case study, the European Space Agency mission GOCE, to be launched very soon,[9] with the goal of determining the Earth's gravity field up to very high harmonic degree (ESA 1999).

The performance of gradiometers (also of accelerometers and satellite-to-satellite tracking systems) is generally represented as a noise spectral density S, a function of the frequency f. The **measurement band** is a frequency interval $[f_m, f_s]$ in which the noise spectral density S is minimum, e.g., for GOCE $f_s = 1/10$ Hz and $f_m = 1/200$ Hz, with a requirement $S \leq 4 \times 10^{-3}$ E Hz$^{-1/2}$ in the measurement band. The noise increases for frequencies $> f_s$, that is for short integration time; typically, the observations are taken with sampling interval dt of the order of $1/f_s$. The **Nyquist frequency** $f_N = 1/(2\,dt)$ needs to be $\leq f_s$ to avoid aliasing from noise at higher frequencies. $S(f)$ increases again for frequencies $< f_m$: a typical assumption is that the increase is like $1/f$. Thus the shape of $S(f)$ is trapezoidal in a log-log plot.

However, this is a model for noise, and the most important error terms are not noise. For example, the thermal signals are low frequency (the accelerometers are thermally insulated) and have as forcing frequencies integer combinations of the orbit mean motion n and of the Earth's rotation frequency $\Omega_\oplus$. Unfortunately some of these appear also in the gravitational signal, see eq. (13.32). An active thermal control might change the dominating frequencies, but might also introduce some systematic signal with another frequency.

Another source of systematics is the attitude control system: errors in the estimation of $R, \omega, \dot{\omega}$ appear as spurious signals in the measured gravity gradient. Since the torques acting on the spacecraft attitude contain signals with frequencies n and $n_\oplus$, the rotation matrix R contains the same frequencies, and some of these would alias with the gravity signal. The active attitude control might suppress some of these frequencies, and again introduce new ones, which should be included in the error model.

We use as example the simulation published by Milani *et al.* (2005d). The random error component was simulated as uncorrelated Gaussian noise with

---

[9] Added in proofs: launched on 17 March 2009.

RMS $= 0.004$ E (Eötvös units, 10^{-9} s^{-2}). This noise term was used to define
the formal covariance, that is the gravity gradient residuals are weighted
dividing by 0.004 E. The data simulation did include some systematic errors,
represented by a finite number of harmonics, with an amplitude growing as
$1/f$: a daily term (with amplitude of 1.73 E), "once per rev" and "twice per
rev" terms (with amplitude of 0.1 and 0.055 E), supposedly accounting for
thermal changes. A very long term drift (supposedly due to seasonal thermal
effects) had a period of one year and an amplitude of 18.6 E, decreased, with
respect to the f^{-1} law, by a factor 0.03 (supposedly an a priori calibration
by means of a temperature measurement accurate to $\simeq 2 \times 10^{-3}$ kelvin).
Moreover, a term with period 1000 s and amplitude of 0.02 E, five times
the RMS of the noise component, was added to investigate the effects of
systematic errors due to other causes, such as the attitude control.

A posteriori gradiometer calibration

The simulation of Milani *et al.* (2005d) adopted a *kinematic* timewise
method, that is only the gravity harmonic coefficients $\mathbf{g}$ (e.g., $201^2 - 4$
coefficients for $\ell_{max} = 200$) and the gradiometer a posteriori calibration
parameters $\mathbf{h}$ were fit to the gradiometer observations. The main issue is
the dimension of $\mathbf{h}$: e.g., if two calibration parameters had to be solved for
each interval of $\simeq 200$ s, then for a nominal eight month mission $\mathbf{h}$ is a
vector with $\approx 2 \times 10^5$ components. Note that, if the parameters $\mathbf{h}$ appear
linearly in the calibrations as functions of time, the least squares problem is
linear, although a very large one. To solve for $> 100\,000$ parameters would
be a problem, not only for the computational load, but also for the bad
conditioning of the normal matrix.

   The challenge was how to neutralize the systematics (by absorbing them in
the calibrations) with acceptable computational load and numerical stability,
and this without cheating; e.g., to solve for a finite number of sinusoids would
mean using the information on the systematics introduced in the simulation.
This would produce illusory results, not reproducible with real data, where
the systematic errors cannot have such a simple spectrum.

   The time-dependent absolute calibration of the gradiometer, for each component, can be represented as a linear combination of N suitable base functions $c(t) = \sum c_i\, b_i(t)$. This is applicable only to the time span Δt of an
arc, which cannot be too long, both to keep low the number N of base functions solved at once (to avoid computational complexity and instability) and
not to impose one specific functional representation of the absolute calibration. On the other hand, Δt cannot be too short: the calibration to be
removed has to be the low-frequency component, with frequencies below a

calibration band upper limit f_c well separated from f_m. As an example, the tests of Milani *et al.* (2005d) have used a calibration band with frequencies below $f_c = 1/2000$ Hz. Thus Δt, N, and the base functions b_i have to be selected to model an arbitrary signal in the calibration band $f < f_c$; spurious signals in the intermediate band with $f_c < f < f_m$ are not removed.

The base functions b_i must be such that the normal equation for the calibration parameters $\mathbf{h}_i$ of each arc i is well conditioned. After testing different choices, including Fourier expansions and Chebichev polynomials, Milani *et al.* (2005d, Section 2.3) concluded that a good choice was a base containing a constant, a linear function of time, sine and cosine terms with periods $\Delta t/k$ for $k = 1, \ldots, K$. To limit the removal of signal to the calibration band, $K \simeq \Delta t \cdot f_c$. For example, $\Delta t = 10\,000$ s, $K = 5$ are acceptable choices. The total number of local parameters $\mathbf{h}$ for a simulation over eight months, with $N_{arc} = 2000$ arcs, was $2\,(2K + 2) \times N_{arc} = 48\,000$.

Local–global correlations

Given the number of parameters to be solved, e.g., 40 397 global and 48 000 local, some decomposition of the normal equation is necessary. The first one is the local–global decomposition of Section 15.1. The impact of the local–global correlations on the results and their estimated uncertainty, discussed in Section 16.2, also occurs in gradiometer missions. Only some of the global coefficients, with low degree and order ℓ, m, are severely affected.

The uncertainty estimated without the local–global terms is significantly smaller than in the complete computation, especially for $\ell = 2, 3, 6, 7, 8$; the difference is not significant for $\ell \geq 25$ (Milani *et al.* 2005d, Figure 1). For the Fourier component with $k = 4$ (frequency 1/2500 Hz) and the C_{20} harmonic (main frequency 1/2680 Hz), the local–global correlations (from $\Gamma_{\ell g}$) are $\simeq 0.2$ (a significant effect does not require a very high correlation). The actual error is consistent with the higher formal uncertainty, not with the lower one.

The spherical harmonics with $\ell < 25$ generate gradiometer signals with frequencies outside the measurement band, thus GOCE should not be used to solve for them. The simplest solution is to include in $\mathbf{g}$ only the coefficients of degree $\ell \geq 25$. In this way the formal RMS is somewhat underestimated, due to the correlations between the harmonics with $\ell \geq 25$ and those with $2 \leq \ell \leq 24$; this becomes negligible for $\ell > 75$. A better solution is to use a **collocation method**, e.g., by adding a priori observations of the harmonics with $2 \leq l \leq 24$ weighted with the inverse of their covariance matrix, as resulting from previous gravimetry missions.

16.4 Resonant decomposition

To obtain the scientifically interesting results $\mathbf{g}$, we need to compute $\Gamma_{\mathbf{gg}} = (C^{\mathbf{gg}})^{-1}$, see eq. (15.4). To solve for coefficients up to a large degree, e.g., $\ell = 200$, the matrix is still very large. Thus we want to decompose the problem into smaller ones; the solution of the normal equation (15.3) has to be found as the limit of a sequence of independent differential corrections for subsets of the $\mathbf{g}$ parameters. The simplest solution is solving for each coefficient independently of the others, that is approximating $C^{\mathbf{gg}}$ with a diagonal matrix: this would work if the observations were uniformly distributed on a sphere, so that the different harmonics would be orthogonal, see Section 13.3. This is not the case for any possible satellite orbit, as it is already clear from Figure 13.1: if the orbit has an inclination $I < 90°$ with respect to the equator, the **polar caps** with latitude $> I$ and $< -I$ are never overflown.[10] It can be shown that indeed the spherical harmonics are not orthogonal over a latitude band (Albertella *et al.* 1999, Pail *et al.* 2001).

Colombo (1989) proposed that the normal matrix should be decomposed by the value of the harmonic order m: the harmonics with the same m share the same frequencies (13.37), thus they are more correlated. This could work, but something better is possible, by using the Kaula expansion of the geopotential perturbing function (Section 13.4) as a function of orbital elements. Let us suppose that the orbit of the gravimetric satellite is exactly resonant, in the sense of eq. (13.39), and that e is small, thus we can neglect the $q \neq 0$ terms. Then the only frequencies in $U_{\ell m}$, as a function of time along the orbit, are: $\nu_{\ell m p 0} = [(\ell - 2p)\, j - m\, k]\, \mu$. The same frequencies appear in all the partial derivatives of the potential, including the gradiometer observables; this follows from (13.24) and (13.25). Two harmonics (ℓ, m) and (ℓ', m') share some frequencies if and only if

$$k\, m' = \pm k\, m\, (\mathrm{mod}\ j), \tag{16.9}$$

a significant constraint, since for a low satellite j is larger than k by a factor > 10. The signals from two harmonics with disjoint sets of frequencies are orthogonal over an infinite time span. Thus we can define a **resonant decomposition**: the harmonics are reordered by remainder class $r = \pm k\, m\, (\mathrm{mod}\ j)$, with $0 \leq r \leq j/2$, and the linear system (15.3) is

$$(M - N)\, \Delta\mathbf{g} = D_{\mathbf{g}} \tag{16.10}$$

where M is the block diagonal part of $C^{\mathbf{gg}}$, with off-diagonal terms only

---

[10] For retrograde orbits with $I > 90°$, the polar caps include latitudes $> 180° - I$ and $< I - 180°$. Even for an exactly polar circular orbit, the sampling would not be uniform, and anyway other rank deficiencies would occur, see Section 16.6.

when row and column correspond to harmonics in the same remainder class. The blocks are larger than in the decomposition by m, and still M^{-1} can be computed block by block with acceptable computational load. In the assumptions (13.39), $e = 0$, and an infinite observation time span, $N = \mathbf{0}$. In practice, N is not zero, because none of these assumptions applies exactly to a realistic case. If N is smaller than M, to solve the linear system (16.10) we can transform it into a fixed point problem (Bini *et al.* 1988, Section 5.2)

$$M \, \Delta\mathbf{g} = N \, \Delta\mathbf{g} + D_{\mathbf{g}} \Leftrightarrow \Delta\mathbf{g} = M^{-1} N \, \Delta\mathbf{g} + M^{-1} D_{\mathbf{g}} = P \, \Delta\mathbf{g} + Q_{\mathbf{g}}$$

which can be solved by iteration

$$\Delta^{(1)}\mathbf{g} = P \, \Delta^{(0)}\mathbf{g} + Q_{\mathbf{g}}, \quad \Delta^{(2)}\mathbf{g} = P \, \Delta^{(1)}\mathbf{g} + Q_{\mathbf{g}}, \quad \ldots$$

starting from an arbitrary initial guess $\Delta^{(0)}\mathbf{g}$; if, in some matrix norm, $||P|| < 1$, the sequence $\Delta^{(k)}\mathbf{g}$ converges to $\Delta^*\mathbf{g}$ solving the complete normal equation. For an initial guess $\Delta^{(0)}\mathbf{g} = \mathbf{0}$, $\Delta^{(1)}\mathbf{g} = Q_{\mathbf{g}} = M^{-1} D_{\mathbf{g}}$ is the solution with approximate inversion, the first step of a convergent iteration.

For GOCE, the exact value of the semimajor axis was not yet known at the time of the simulation (ESA 1999, Section 6.2.4); it has been decided later, by selecting an optimal ground track repeat period > 60 days, thus a different resonant decomposition would be better. Anyway, a decomposition corresponding to a resonance not too close to the nominal one, that is $j = 31$, $k = 2$, has been used to test the robustness of the method. The upper part of Figure 16.2 shows the results for the gravity coefficients, and their covariance, for the remainder class $r = 1$, which includes $m = 15, 16, \ldots$.

The norm of P is small because the off-block correlations are small, but there is no quantitative estimate, thus we cannot predict how many iterations would be necessary for satisfactory convergence. In (Milani *et al.* 2005d) there is a numerical test of the second iteration $\Delta^{(2)}g$: this is possible only after a complete first iteration, 16 separate differential corrections with $r = 0, 1, \ldots, 15$, by performing a second iteration for some remainder class. The upper plot of Figure 16.2 shows both the result of the first and the second iteration, and the two curves are so close that most points are superimposed. Albertella and Migliaccio (1998, Section 3.5.2) report another numerical test computing explicitly, for the accelerometer mission case, the off-block correlations and showing that they are < 0.01.

16.5 Polar gaps

In a solution for the geopotential coefficients up to a large degree ℓ_{max}, after controlling the local–global correlations (with an appropriate minimum

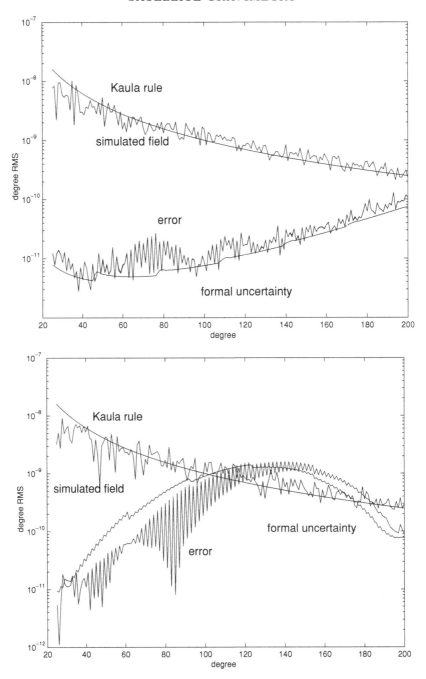

Fig. 16.2. We plot as a function of the degree ℓ: the gravity field signal (Lemoine *et al.* 1998), the approximating Kaula rule, the formal uncertainty, and the actual error (estimate minus "true" value used in the data simulation). Above: for the remainder class $r = 1$; the actual error is shown for the first and the second iteration; the two curves are almost superimposed. Below: for the $r = 0$ class, including the zonals, there is a large bulge such that both the formal and the actual errors exceed the signal for degree $\ell > 100$. Reproduced with permission of Springer from Milani *et al.* (2005d).

degree ℓ_{min}), and after decomposing the normal matrix of the global parameters $C^{\mathbf{gg}}$ (see the previous section), some diagonal blocks are still very badly conditioned. As an example, in Figure 16.2, lower panel, we show the results for the remainder class $r = 0$ in a solution for $25 \leq \ell \leq 200$. The bulge in both the formal uncertainty, and in the actual error corresponds to very high correlations among the zonal spherical harmonics. The error/signal ratio reaches $\simeq 1$ around $\ell = 100$, unlike the $r = 1$ case of the upper plot. The analogous figure for $\ell_{max} = 90$ would show a much lower bulge.

Large RMS (and actual errors) occur in this remainder class only for the zonals with $m = 0$; the other harmonics, with $m = 31, 62, \ldots$, are determined with error/signal ratio well below 1 for all ℓ. The correlations are significant only among the zonals, e.g., the $\ell = 150$ zonal has significant correlations with all the even zonals, correlation 0.9999 with the $\ell = 148$, and the $\ell = 152$ one. This is clear from the frequency analysis: for $m = 0$, low e, $\nu_{l0p0} = (\ell - 2p)\,(n + \dot{w})$, hence the highly correlated zonals share the same frequencies, consecutive even (or odd) ℓ have only one frequency not in common; in any case zonal harmonics with the same parity have some frequencies in common, and are correlated.

Similar results are obtained for the remainder classes containing harmonics with low m: e.g., for $r = 2$, that is for $m = 1, 29, 33, \ldots$ the error-to-signal ratio reaches 1 around $\ell = 90$. For $m = 7$ the formal and actual error curves still show the bulge at intermediate degrees, but the error-to-signal ratio is below 1. These difficulties with the low m coefficients in GOCE were known, by the designers of the mission (Aguirre-Martinez and Sneeuw 2003, Figure 4), to be due to the polar gaps; the non-polar orbit of GOCE has ground tracks with latitude neither above $\simeq 83.5°$ nor below $-83.5°$.

Principal components analysis

Given the covariance matrix $\Gamma_{\mathbf{gg}}$, or at least one of the blocks $\Gamma^r_{\mathbf{gg}}$ of the resonant decomposition, we can perform a principal components analysis. Let $\lambda_1 > \lambda_2 > \cdots > \lambda_s$ be the square roots of the eigenvalues of $\Gamma^r_{\mathbf{gg}}$.

The corresponding unit eigenvectors $V_j, j = 1, \ldots, s$, contain harmonic coefficients and can be interpreted as gravity anomalies. For the $r = 0$ block λ_1, λ_2 are not very different and significantly larger than λ_3; in V_1, V_2 the components corresponding to harmonics with $m \neq 0$ are very small $(< 10^{-4})$, thus the harmonic functions with coefficients $\lambda_1 V_1$ and $\lambda_2 V_2$ are essentially zonal harmonics, functions of latitude, e.g., in Figure 16.3 they are represented as anomalies of the geoid. The anomalies are concentrated on the two polar caps: $\lambda_1 V_1$ more pronounced on the South pole, $\lambda_2 V_2$ on

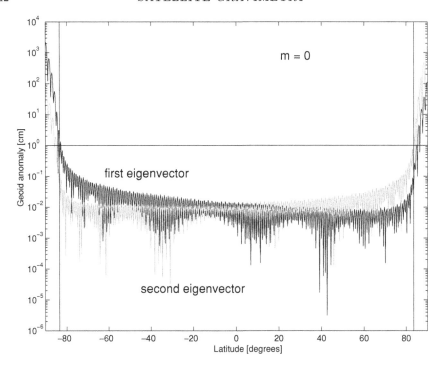

Fig. 16.3. The gravity anomalies corresponding to the two principal axes of the confidence ellipsoid (for $\sigma = 1$) represented by the corresponding geoid anomalies, as a function of latitude. The vertical lines bound the latitude band covered by GOCE, the horizontal line is at 1 cm. Reproduced with permission of Springer from Milani *et al.* (2005d).

the North pole. The size of the anomalies is huge with respect to the target accuracy of GOCE, up to $\simeq$ 20 m on the poles; to understand this, it is enough to compute one of the GOCE observables, e.g., the radial component of the gravity gradient, for the anomalies $\lambda_1 V_1$ and $\lambda_2 V_2$. Although the undetermined anomalies are not only on the polar caps, the signal for latitudes between $-83.5°$ and $+83.5°$ is well below the noise of the GOCE measurements.

Symmetry and degeneration

Why are the undetermined anomalies as large as shown in Figure 16.3, and why does the situation become worse as the solution is pushed to higher ℓ?

Typically, the origin of an approximate rank deficiency is due to a "nearby" exact rank deficiency, in turn connected to a number of exact symmetries, see Section 6.3. When the symmetries are broken, approximate symmetries remain, and they result in an approximate rank deficiency.

The case of a gradiometer mission is somewhat more complicated, because the symmetry group is an infinite dimensional subspace of harmonic functions. Its existence can be proven by selecting, on the sphere of radius $R_\oplus + h$, an arbitrary smooth (C^∞) function Φ with support in the polar caps (e.g., the function is exactly zero for $-83.5° <$ latitude $< 83.5°$). By the solution of the exterior Dirichlet problem given in Section 13.3, there is a harmonic function for geocenter distance $> R_\oplus + h$ and coinciding with Φ on the sphere of radius $R_\oplus + h$. Such a function may not exist on the sphere of radius $R_\oplus$ because the downward continuation may well be divergent; that is, it does not need to be a "realistic" gravity anomaly.

If we take the approximation that the satellite flies at an exactly constant altitude, then the gradiometer is measuring the second derivatives where Φ is zero. In this approximation there would be an exact symmetry. This symmetry is broken for two reasons: first, the altitude is not constant, although the eccentricity is small ($e < 0.0045$ for GOCE). Second, the harmonic function we are trying to fit to the gradiometer data is the sum of only a finite number of harmonics, with limited degree $\ell \le \ell_{max}$. The cap function Φ cannot have a finite spherical harmonic expansion, because it is not an analytic function. If Φ is expanded in a series of spherical harmonics, the series is convergent on the sphere. When this series is truncated to degree ℓ_{max}, the remainder is small on the latitude band where Φ is zero, with maximum $\to 0$ for $\ell_{max} \to +\infty$. This is what is shown in Figure 16.3: the observable signal is not zero but very small. Thus as ℓ_{max} increases the observable signal becomes smaller and smaller, and the bulge becomes more pronounced. At the limit for $\ell_{max} \to +\infty$ the undetermined geoid anomaly on the sphere of radius $R_\oplus$ might be arbitrarily large.

Outside the polar caps

On a block of harmonics not including low m ones the polar gaps have little effect: e.g., for $r = 1$, Figure 16.2, above. The actual error is larger than the formal one, but the ratio does not exceed 4 (and is ≤ 2 for most orders ℓ). Both the formal and actual error are well below the signal up to $\ell_{max} = 200$, and it is possible to solve harmonics with signal-to-ratio ≥ 1 up to $\ell_{max} = 220$–230. Figure 16.3 suggests a positive interpretation of the polar gaps effects: the undetermined gravity anomalies are concentrated on the polar caps, e.g., the geoid anomalies are at the mm level in the "overflown region", the latitude band of the ground tracks. The harmonics up to $\ell_{max} = 220$ cannot all be determined, but the gravity in the overflown region can be determined down to a spatial scale of $\pi R_\oplus/220 \simeq 91$ km.

Even in the overflown region the solution is reliable only provided there is no aliasing between spurious signals and the gravity signal. To detect such an

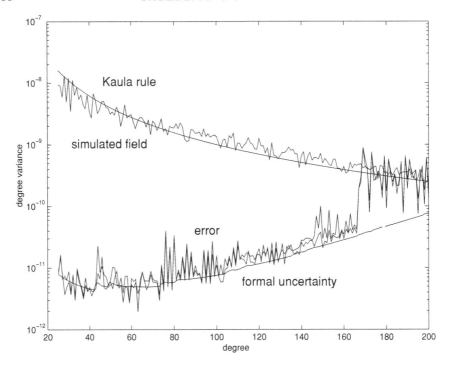

Fig. 16.4. For the $r = 7$ remainder class the formal and actual errors are compared with the signal. The harmonics with $\ell > 160$ cannot be determined with a signal/error ratio > 1. Reproduced with permission of Springer from Milani *et al.* (2005d).

aliasing, we analyze the specific harmonics where the largest actual/formal error ratio holds. For example, for the remainder class $r = 7$ (Figure 16.4) the actual error grows significantly for $\ell > 160$. Among the harmonics in this block, those with $m = 167$ show a particularly high actual error. The spurious signal with frequency $f = 1/1000$ Hz introduced in the simulation, and not removed by the calibration fit because $f > f_c$, is responsible for this behavior. Looking for gravity field signals with similar frequencies, we find, e.g., that for $l - 2p = -5, m = 167$ the frequency ν_{lmp0} corresponds to a period of 990 s; for $l - 2p = -19, m = 167$ the frequency is negative, with a period of 957 s.

Limitations of gravimetry missions

The main general conclusion we can draw from this case study is that each gravimetry mission has some limitations, due to the performance of the instruments and to the orbit: in the case of GOCE, these limitations could be summarized as follows. First, if the orbit does not cover the entire surface with the ground track, then it is not possible to accurately solve for all the spherical harmonics up to some high degree. Nevertheless, in the overflown region the results can be accurate down to a short spatial scale.

Second, given the noise model, the formal error deduced from the covariance matrix $\Gamma_{\mathbf{gg}}$ sets an upper bound for the degree ℓ to be determined, where it exceeds the signal: for GOCE this occurs for $\ell \simeq 230$. Since this value is obtained in the optimistic assumption that the systematic errors are zero, this upper bound does not depend upon the orbit determination method used, but only on the specifications for the measurement noise.

Third, if in the intermediate frequency band there are spurious systematic signals, they can alias with gravity signals. The period 1000 s for the spurious signal is arbitrary, but any period in that band would affect some harmonics. Sometimes a numerical simulation has to be used to convince the skeptics, even though a back of the envelope computation could be enough. The frequencies $\nu_{\ell m p 0}$ for $\ell \leq 200$ are $\simeq 40\,000$; to invent some spurious signal without frequencies close to some harmonic would be a difficult task. Thus each spurious signal generates a wrong value for some harmonics, with an actual error significantly larger than the formal one, an illusory result limited to a few coefficients.

16.6 Satellite-to-satellite tracking

A **satellite-to-satellite tracking** mission uses two spacecraft, in low Earth orbits, and relative tracking data between the two, such as range and/or range-rate. The main choice in mission design is to decide the distance d between the two, and how tightly to control it. For a target minimum spatial scale $L = \pi R_{\oplus}/\ell_{max}$, at short distances $d \ll L$ the measurement has the same information content as a gradiometer (16.7), although the data processing is somewhat more complex (see below). The short distance allows us to use a relative tracking method sensitive to distance and insensitive to the atmospheric propagation disturbances, but requires a drag free system.

If $d \simeq L$ the orbit control requirement is weaker: this has been the choice of the NASA mission GRACE (Tapley *et al.* 2005), launched in 2002. The two spacecraft are allowed to decay because of drag, sporadic orbit maneuvers controlling the distance between $\simeq 120$ and $\simeq 270$ km. The difficulty is to model with the required accuracy the propagation of radio waves: GRACE has a complex multifrequency link between the two spacecraft.

In any case, for an Earth gravimetry mission, an accelerometer is required on both spacecraft: apparent forces and calibration are a problem as they are for an accelerometer mission. Around the Moon this is not necessarily the case. The MORO lunar mission study of the European Space Agency (Coradini *et al.* 1996) proposed to measure the range-rate to a simple lunar subsatellite released from the main polar orbiter (Milani *et al.* 1996). The sensitivity to the lunar gravity field can be good, down to a scale $\simeq 100$ km, with orbital height $h \simeq 100$ km, which can be maintained even without propulsion (Knežević and Milani, 1998). The problem of the sectorial weakness (see below) was discussed in the MORO study. The Japanese lunar mission KAGUYA, launched in 2007, has implemented another version of the subsatellite concept, using VLBI differential measurements from Earth.

Laser Doppler interferometry for gravimetry

We shall use as an example of satellite-to-satellite tracking the LDIM study commissioned by the European Space Agency to Thales Alenia Spazio, for a long-duration Earth gravimetry mission (Cesare *et al.* 2005, Cesare *et al.* 2006). This study selects laser interferometry to generate extremely accurate measurements of changes in the distance between reference points on the two spacecraft, at a distance $d \simeq 10$ km. This distance gives good sensitivity to harmonics with degree up to $\ell_{max} \simeq 200$, because $d/L \simeq 1/10$, where $L \simeq 100$ km is the spatial scale. An accelerometer with a measurement band between $1/1000$ Hz and $1/100$ Hz and noise spectral density $S(f)$ comparable to the units on GOCE is assumed; for the accelerometer and interferometers performances in measuring differences of accelerations see (Cesare *et al.* 2006, Figure 2). The selected altitude $h = 325$ km is maintained with an ion thruster, providing drag free control and relative distance control. The two spacecraft maintain a constant attitude in the $(\hat{\mathbf{r}}, \hat{\mathbf{t}}, \hat{\mathbf{w}})$ orbit frame of Section 14.1, and are separated in the $\hat{\mathbf{t}}$ direction.

Given the preliminary nature of such a study, only a simplified orbit determination simulation was included,[11] for a six month mission segment, to assess how small changes in the geopotential, with periods ≥ 1 year, can be measured. The observations were simulated as gradiometer measurements, an approximation with small parameter $d/r \simeq 1.5 \times 10^{-3}$.

The orbit inclination was constrained by the requirement of a Sun synchronous orbit (with period of the node equal to one year, to simplify the thermal control) to $I = 96.8°$. As a consequence the polar gaps effects were comparable to those found for GOCE and discussed in Section 16.5: see (Cesare *et al.* 2005, Figures 4.2–5, 4.2–6). To summarize the results as a function of degree ℓ we have separated the harmonics with $m \leq 15$, for

---

[11] Gravity field solutions obtained by the uncorrelated frequencywise method were used to perform a trade-off, e.g, to select the value of h.

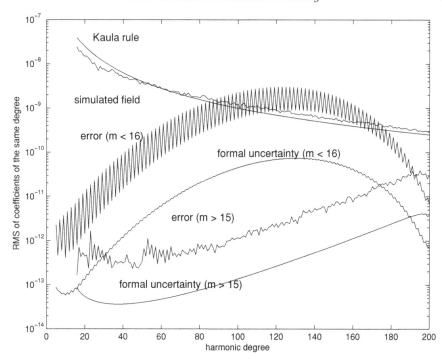

Fig. 16.5. Signal and different estimates of the error as a function of the harmonic degree ℓ. From top to bottom on the left: Kaula rule and actual signal (Lemoine *et al.* 1998); actual and formal error for $m \leq 15$, actual and formal error for $m > 15$.

which there was a significant bulge, in Figure 16.5, which looks like the superposition of the two plots of Figure 16.2. This was the LDIM goal, to achieve results comparable to GOCE from a higher orbit, with two smaller spacecraft, which could avoid orbit decay for more than five years.

However, there is an interesting difference. If we show the results for each individual harmonic coefficient $C_{\ell m}$, $S_{\ell m}$ in a representation with the low m near the axes and low $\ell - m$ near the diagonal as in Figure 16.6, the two "ears" on the sides show the effects of the polar gaps, the ridge in the middle indicates the presence of another *weakness* of the solution, although not quite a rank deficiency; the increase in the noise for the sectorial spherical harmonics, with $\ell = m$, is just by a factor $\simeq 10$.

Still it is possible to find a nearby exact symmetry, because the only component of $\partial^2 U / \partial \mathbf{x}^2$ which can be measured is the one in the direction $\hat{\mathbf{t}}$. If the orbits were exactly polar, all the derivatives of the sectorial harmonics with respect to the latitude would be zero: $P_{\ell\ell} = const$. Since $\pi/2 - I \simeq 1/10$ rad, the weakness is not important in the results. It is important to remind us that there is no ideal orbit allowing a perfect gravity field solution; if the

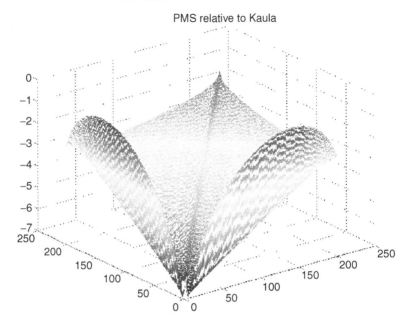

Fig. 16.6. For each harmonic ℓ, m the smooth noise/signal computed as formal RMS divided by the value suggested by the Kaula rule. In the horizontal plane, the two coordinates are (ℓ, m) for $\overline{C}_{\ell m}$ and (m, ℓ) for $\overline{S}_{\ell m}$.

orbit was closer to polar, the polar gap would be smaller and the bulge less pronounced, but the sectorial weakness could become a problem.

The numerical precision requirements

There is a specific additional problem in the satellite-to-satellite tracking missions, not fully addressed by the first LDIM study, namely the numerical precision of the orbit. With laser interferometry, the sensitivity Δd to changes in the distance d between the two spacecraft could be as small as 10^{-7} cm. With $r \simeq 6.7 \times 10^{8}$ cm, we have $\Delta d/r \simeq 1.5 \times 10^{-16}$. If the two orbits computed separately are $\mathbf{x}_1$ and $\mathbf{x}_2$, in computing $d = |\mathbf{x}_1 - \mathbf{x}_2|$ the components of the difference vector are differences between two large numbers. This implies that **rounding off** is a problem with current computers, supporting in hardware only a mantissa of 53 bits, thus with relative rounding off error $2^{-52} = 2.2 \times 10^{-16}$. Extended precision is supported in software, but this significantly increases the computational load. There are ways to use extended precision only for some of the computations, not including the spherical harmonic expansion of the gravity fields and its derivatives.

17

ORBITERS AROUND OTHER PLANETS

BepiColombo is a European Space Agency mission to be launched in 2014, with the goal of an in depth exploration of the planet **Mercury**. This chapter contains a case study: the orbit determination for this mission, including the techniques to be used for both the simulations and the real data processing. Some tools not yet described in this book will be developed, such as the light time and range-rate implicit equations, the equations for the rotation of the planet, etc. This chapter is based on our papers (Milani *et al.* 2001b, Milani *et al.* 2002) and on ongoing research.

17.1 Science goals for an orbiter around Mercury

In a first exploration of a planet, every piece of information which can be gathered is valuable, starting from the first close up images. Mercury was visited by the NASA **Mariner 10** probe, with three flybys[1] in 1974–1975. With these short visits, the orbit determination of Mariner 10 provided estimates of the harmonic coefficients $\overline{C}_{20} \simeq -2.68 \times 10^{-5}$, $\overline{C}_{22} \simeq 1.55 \times 10^{-5}$ (Anderson *et al.* 1987) and gave evidence of a bipolar magnetic field.

Further exploration of a planet needs to follow a very different logic. Scientists would like to define a set of **science goals** and then obtain from the space agencies the resources to achieve them, but this is not really possible. The science goals have to be limited to what is affordable, taking into account how expensive are the resources, such as the mass, the power, the data rate and the thermal control, when deployed around another planet. The experiments are limited by the available technology, also taking into account the need for extreme miniaturization.

---

[1] The technique used to obtain three close approaches to the planet without consuming much fuel was invented by Giuseppe (Bepi) Colombo: it is a resonant, multiple gravity assist, which is strictly related to the resonant return described in Section 12.4.

A window of opportunity was opened for the BepiColombo mission with the proposal (Iess and Boscagli 2001) of including a two-frequency radio system, with transponders for both the X and the Ka band. With a five-way link between the spacecraft and the ground antenna it is possible to eliminate almost completely the uncertainty on the speed of the radio waves, which is lower than the speed of light because of the plasma content along their path. This is possible even for radio wave paths passing comparatively close to the Sun, down to a few Sun radii.

Then the question has to be reversed: which scientific goals can be achieved with such good quality of tracking data, and which are the additional requirements to fully exploit the extraordinary accuracy possible in the measurements of both range and range-rate to a **Mercury orbiter**. The possible fields of scientific interest are two. First, geophysics, with the study of the internal structure of the planet, including the core, the mantle, and the source of the bipolar magnetic field. Second, the theory of gravitation, since Mercury is a probe going into the gravity potential well of the Sun deeper than any other body large enough not to be affected significantly by non-gravitational perturbations (see Section 4.6).

Geophysics of Mercury

It is possible to determine the gravity field of Mercury from the orbit determination of a satellite of that planet; the question is how far we can go with the degree and order of the spherical harmonics. The problem, as discussed in Section 13.2, is that the gravity field outside a planet, as represented in (exterior) spherical harmonics, does not uniquely constrain the internal mass distribution. In particular the degree 2 harmonics are five, and there is a linear relationship with the six coefficients of the inertia quadratic form, resulting in one free parameter which may be expressed as the concentration coefficient $C/(MR^2)$, where C is the maximum moment of inertia. This makes it difficult to constrain the size of the core of Mercury by gravimetry.

Moreover, the static gravity field alone does not constrain the physical state of the interior layers, such as the outer core. On the Earth, the outer core is liquid (as established with seismic wave analysis) and it is believed to be the source of the Earth's bipolar magnetic field, through a dynamo effect. If Mercury had no liquid layer the dynamo theory could be challenged.

Both the presence of a liquid layer and the size of the core can be investigated by constraining the rotation state of the planet. Also the time-dependent part of the gravity field of Mercury, that is the response of the planet's gravity to the tides raised by the Sun, can be used as a constraint.

The rotation state of the planet can be measured by comparing images of the same area on the planet taken at different times, see Section 17.4. The question is how accurately it is possible to measure the harmonic coefficients (as a function of the degree ℓ), the tidal response of the planet, and the rotation state, given the technology practically available around Mercury.

Theory of gravitation

The parametric post-Newtonian approach, outlined in Section 6.6, allows us to describe a number of theoretically interesting violations of the current theory of gravitation, namely general relativity, and other interesting astrophysical phenomena by a set of parameters such as $\gamma, \beta, \eta, \alpha_1, \alpha_2, \zeta, J_{2\odot}$. Of course these parameters can be determined only within an orbit determination procedure solving also for the initial conditions of Mercury and of the Earth. The question is how accurate can the solution be for the post-Newtonian parameters, given the accuracy and time distribution of the tracking data. If the constraints on the possible violations of general relativity can be significantly improved with respect to all the other available experimental constraints, this will become a very important experiment.[2]

17.2 Interplanetary tracking

The observables are the distance r between the ground antenna and the spacecraft, and its time derivative $\dot{r}$. They can be computed from solutions for the motion of five different state vectors

$$r = |(\mathbf{x}_{sat} + \mathbf{x}_M) - (\mathbf{x}_{EM} + \mathbf{x}_E + \mathbf{x}_{ant})| + S(\gamma) \qquad (17.1)$$

where $\mathbf{x}_{sat}$ is the Mercury-centric position of the orbiter, $\mathbf{x}_M$ is the Solar System barycentric position of the center of mass of Mercury, $\mathbf{x}_{EM}$ is the position of the Earth–Moon center of mass in the same reference system, $\mathbf{x}_E$ is the vector from the Earth–Moon barycenter to the center of mass of the Earth, and $\mathbf{x}_{ant}$ is the position of the ground antenna center of phase with respect to the center of mass of the Earth.

$S(\gamma)$ is the **Shapiro effect**, the difference between distance in a flat space and the geodesic length in curved space-time, depending upon the post-Newtonian parameter γ. Thus the distinction between dynamical and kinematical parameters, introduced in Chapter 1, is not sharp, because

---

[2] The results would appear more interesting if they were to prove a violation rather than just confirming general relativity to a better accuracy, but the effort done in the experiment and in particular in the orbit determination is exactly the same in the two cases.

γ appears in the observation equation (17.1) and also in the relativistic dynamics, see Section 6.6. If the distance is measured by light-time, the flat-space distance of two heliocentric vectors $\mathbf{r}_1, \mathbf{r}_2$ has to be corrected by (Moyer 2003)

$$S(\gamma) = \frac{(1+\gamma)\, G\, m_0}{c^2} \log\left[\frac{r_1 + r_2 + r_{12}}{r_1 + r_2 - r_{12}}\right],$$

where the vectors $\mathbf{r}_1$ and $\mathbf{r}_2$ correspond to the vectors $\mathbf{x}_{EM} + \mathbf{x}_E + \mathbf{x}_{ant}$ and $\mathbf{x}_{sat} + \mathbf{x}_M$, but they have to be converted to a heliocentric frame, which is moving with the velocity $\dot{\mathbf{x}}_\odot$ of the Sun, thus a relativistic coordinate transformation needs to be applied. The length r_{12} is similarly converted from the r of eq. (17.1). This conversion introduces a number of small terms of post-Newtonian order > 1 which may be observable, given the very high signal-to-noise in the range. Other terms relevant for the level of accuracy of this experiment appear in the denominator inside the logarithm: when the radio waves are passing near the Sun, at just a few solar radii, even corrections of the order of $G\, M_\odot/c^2 \simeq 1.5$ km have to be computed, although they introduce a correction to $S(\gamma)$ which is of post-Newtonian order 2.

The five vectors of (17.1) have to be computed at the epoch of different events, e.g., $\mathbf{x}_{ant}$, $\mathbf{x}_{EM}$, and $\mathbf{x}_E$ are to be considered at both the antenna **transmit time** t_t and the **receive time** t_r of the signal. $\mathbf{x}_M$ and $\mathbf{x}_{sat}$ are computed at the **bounce time** t_b, when the signal has arrived at the orbiter and is sent back, with corrections for the delay of the transponder. Thus there are two different light-times, the up-leg $\Delta t_{up} = t_b - t_t$ for the signal from the antenna to the orbiter, and the down-leg $\Delta t_{do} = t_r - t_b$ for the return signal. Given the down-leg and up-leg distances

$$\begin{aligned} \mathbf{r}_{do}(t_r) &= \mathbf{x}_{sat}(t_b) + \mathbf{x}_M(t_b) - \mathbf{x}_{EM}(t_r) - \mathbf{x}_E(t_r) - \mathbf{x}_{ant}(t_r) \\ r_{do}(t_r) &= |\mathbf{r}_{do}(t_r)| + S_{do}(\gamma) \end{aligned} \quad (17.2)$$

$$\begin{aligned} \mathbf{r}_{up}(t_r) &= \mathbf{x}_{sat}(t_b) + \mathbf{x}_M(t_b) - \mathbf{x}_{EM}(t_t) - \mathbf{x}_E(t_t) - \mathbf{x}_{ant}(t_t) \\ r_{up}(t_r) &= |\mathbf{r}_{up}(t_r)| + S_{up}(\gamma) \end{aligned} \quad (17.3)$$

with somewhat different Shapiro effects S_{do}, S_{up}; by definition of distance in a relativistic space-time the light-times are $\Delta t_{do} = r_{do}/c$ and $\Delta t_{up} = r_{up}/c$ respectively. If the measurement is labeled with the receive time t_r, the iterative procedure needs to start from eq. (17.2) by computing the states $\mathbf{x}_{EM}$, $\mathbf{x}_E$, and $\mathbf{x}_{ant}$ at epoch t_r, then selecting a rough guess t_b^0 for the bounce time.[3] Then the states $\mathbf{x}_{sat}$ and $\mathbf{x}_M$ are computed at t_b^0 and a first guess

---

[3] In fact, $t_b^0 = t_r$ is good enough, although it is possible to do better.

r_{do}^0 is given by (17.2). This allows a better estimate $t_b^1 = t_r - r_{do}^0/c$. This is repeated computing r_{do}^1, and so on until convergence, that is, until $r_{do}^k - r_{do}^{k-1}$ is smaller than the required accuracy.

After accepting the last value of t_b and r_{do} we start with the states $\mathbf{x}_{sat}$ and $\mathbf{x}_M$ at t_b and with a rough guess t_t^0 for the transmit time.[4] Then $\mathbf{x}_{EM}$, $\mathbf{x}_E$, and $\mathbf{x}_{ant}$ are computed at epoch t_t^0 and r_{up}^0 is given by eq. (17.3); then $t_t^1 = t_b - r_{up}^0/c$ and the same procedure is iterated to convergence, that is to achieve a small enough $r_{up}^k - r_{up}^{k-1}$. Then the two-way range is just $r_{up} + r_{do}$; a one-way range can be conventionally defined as $r(t_r) = (r_{up} + r_{do})/2$.

The iterative procedure above is also used for **planetary radar** to a natural body, such as an asteroid (Yeomans *et al.* 1992), in which case state-of-the-art accuracies can be $\simeq 50$ m in range and $\simeq 4$ mm/s in range-rate. With an active transponder, and using higher frequencies, the accuracies can now be > 100 times better, and this implies that also post-Newtonian corrections of order 1 need to be taken into account. Thus we need to add to (17.2) and (17.3) relativistic corrective terms Δ_{do}, Δ_{up} accounting for the different time coordinates; see an example in the next section.

The instantaneous range-rate is computed with the unit vectors $\hat{\mathbf{r}}_{up}$ and $\hat{\mathbf{r}}_{do}$, e.g. down-leg

$$\dot{r}_{do}(t_r) = \hat{\mathbf{r}}_{do} \cdot \dot{\mathbf{r}}_{do} + \dot{S}_{do}(\gamma). \tag{17.4}$$

The problem is the computation of $\dot{\mathbf{r}}_{do}$. A first approximation can use the velocities for each of the five position vectors, at the same times t_r and t_b, t_t obtained at convergence of the two light-time iterations

$$\dot{\mathbf{r}}_{do} = (\dot{\mathbf{x}}_{sat} + \dot{\mathbf{x}}_M) - (\dot{\mathbf{x}}_{EM} + \dot{\mathbf{x}}_E + \dot{\mathbf{x}}_{ant}).$$

However, this neglects the fact that t_b, t_t depend on t_r also through r_{do}, r_{up}

$$\frac{dt_b}{dt_r} = 1 - \frac{\dot{r}_{do}}{c} + \frac{d\Delta_{do}}{dt_b}, \quad \frac{dt_t}{dt_r} = 1 - \frac{\dot{r}_{do}}{c} - \frac{\dot{r}_{up}}{c} + \frac{d\Delta_{do}}{dt_b} + \frac{d\Delta_{up}}{dt_b}$$

and the corresponding corrections to $\dot{r}_{do}$

$$\dot{\mathbf{r}}_{do} = (\dot{\mathbf{x}}_{sat} + \dot{\mathbf{x}}_M)\left(1 - \frac{\dot{r}_{do}}{c} + \frac{d\Delta_{do}}{dt_b}\right) - (\dot{\mathbf{x}}_{EM} + \dot{\mathbf{x}}_E + \dot{\mathbf{x}}_{ant}) \tag{17.5}$$

are large with respect to the Doppler measurement accuracy, the first term being $\mathcal{O}(\dot{r}/c)$; the one due to $\Delta_{do}(tc)$ is smaller, but still significant. Thus the improved value of $\dot{\mathbf{r}}_{do}$ has to be inserted in eq. (17.4), the correction (17.5) recomputed, and so on until convergence of the value $\dot{r}_{do}$. Similarly, an iteration loop is necessary for $\dot{r}_{up}(t_r)$. Note that also the computation of $\dot{S}_{do}(\gamma), \dot{S}_{up}(\gamma)$ requires corrections $\mathcal{O}(\dot{r}/c)$.

---

[4] $t_t^0 = t_b - (t_r - t_b)$ is good enough.

Conventionally, $\dot{r}(t_r) = (\dot{r}_{up}(t_r) + \dot{r}_{do}(t_r))/2$ is the instantaneous value. However, the measurement is not instantaneous: an accurate measure of the Doppler effect requires us to fit the difference in phase between carrier waves, the one generated at the station and the one returned from space, accumulated over some **integration time** Δ, typically between 10 and 1000 s. Thus the observable $\dot{r}$ is really obtained from a difference of ranges

$$\frac{r(t_b + \Delta/2) - r(t_b - \Delta/2)}{\Delta} = \frac{1}{\Delta} \int_{t_b - \Delta/2}^{t_b + \Delta/2} \dot{r}(s) \, ds \qquad (17.6)$$

or, equivalently, an averaged value of range-rate over the integration interval, which can be computed with a quadrature formula (see Appendix B).

The computation of the observables, as presented in this section, is already complex, but still the list of subtle technicalities is not complete. To understand the computational difficulty we need to take into account also the orders of magnitude. For state-of-the-art tracking systems, such as those using a multifrequency link in the X and Ka bands, the accuracy of the range measurements can be $\simeq 10$ cm and that of the range-rate 3×10^{-4} cm/s (over an integration time of 1000 s). Let us take an integration time $\Delta = 30$ s, which is adequate for measuring the gravity field of Mercury.[5]

The accuracy over 30 s of the range-rate measurement can be, by Gaussian statistics, $\simeq 3 \times 10^{-4} \sqrt{1000/30} \simeq 17 \times 10^{-4}$ cm/s, and the required accuracy in the computation of the difference $r(t_b + \Delta/2) - r(t_b - \Delta/2)$ is $\simeq 0.05$ cm. The distances can be as large as $\simeq 2 \times 10^{13}$ cm, thus the relative accuracy in the difference needs to be 2.5×10^{-15}. This implies that rounding off is a problem with current computers, with relative rounding off error of $\varepsilon = 2^{-52} = 2.2 \times 10^{-16}$; extended precision is supported in software, but it has many limitations. The practical consequences are that the computer program processing the tracking observables, at this level of precision and over interplanetary distances, needs to be a mixture of ordinary and extended precision variables. Any imperfection may result in "banding", that is residuals showing a discrete set of values, implying that some information corresponding to the real accuracy of the measurements has been lost in the digital processing. As an alternative, the use of a quadrature formula for the integral in eq. (17.6) can provide a numerically more stable result, because the S/N of the range-rate measurement is $\ll 1/\varepsilon$.

---

[5] If the orbital period is $\simeq 8000$ s, the harmonics of order $m = 26$ have periods as short as $\simeq 150$ s, see Section 13.5.

Time-scales and science goals

Of the five state vectors used in eq. (17.1), $\mathbf{x}_{ant}$ and $\mathbf{x}_E$ can be assumed known, that is their current knowledge (at the cm level) cannot be improved by ranging to a Mercury orbiter. To observe the orbit of the Moon it is more effective to measure the range to a point on the surface of the Moon, as it is done with lunar laser ranging. Both tracking navigation satellites and using very long baseline interferometry give, by far, more information on the position of the antenna and on the rotation of the Earth.

On the contrary, $\mathbf{x}_{sat}$ contains information on the gravity field of Mercury, $\mathbf{x}_M$, $\mathbf{x}_{EM}$, and $S(\gamma)$ on the orbits of the planets and on the theory of gravitation. Of the underlying dynamics, that of $\mathbf{x}_{sat}$ has orbital periods of $\simeq 8000$ s, the planetary orbits have periods starting at $\simeq 7 \times 10^6$ s for $\mathbf{x}_M$. The Shapiro effect $S(\gamma)$ during **superior conjunction**, when the Sun is close to the path of radio waves from the Earth to the spacecraft, changes over an intermediate time-scale of $\simeq 3 \times 10^5$ s.

The distribution in time of the observations is tightly constrained by visibility conditions, e.g., Mercury has to be well above the horizon of the ground station, thus a pass observable by a given station lasts about eight hours, with seasonal variations (longer in summer, shorter in winter). The spacecraft must not be behind Mercury, thus for some relative orientations of the Mercury-centric orbit plane and the direction to the Earth the passes are interrupted by occultations. The radio waves must not meet the Sun in their path to the spacecraft, including not just the photosphere (the visible Sun) but also the inner solar corona, where the radio waves are too much disturbed by plasma turbulence. Overall, for a Mercury polar orbiter, tracking from a single station is possible only about 1/4 of the time.

For the state-of-the-art tracking systems discussed above, the accuracy of the range-rate measurements is better than that of the range, when integrated over a time span $< 33\,000$ s; the one in range is more accurate over longer times. It follows that over one pass the measurements of $\dot{r}$ provide the most accurate constraints on $\mathbf{x}_{sat}$. On the contrary, the constraints to the planetary orbits $\mathbf{x}_M$ and $\mathbf{x}_{EM}$ are essentially from r measurements. The determination of γ from the Shapiro effect is possible by using $\dot{r}$ to constrain $\dot{S}(\gamma)$ during a superior conjunction experiment (Bertotti *et al.* 2003a). Still, if r measurements are available, they constrain the value of γ even during a superior conjunction with an accuracy improved by about an order of magnitude with respect to the $\dot{r}$ measurements alone.

Thus it is possible to separate conceptually (although not in the data processing) a *gravimetry experiment* and a *relativity experiment*.

17.3 The gravimetry experiment

The orbit $\mathbf{x}_{sat}(t)$ depends upon the gravity field of Mercury and is a function of the mass of Mercury, of the harmonic coefficients $\overline{C}_{\ell m}, \overline{S}_{\ell m}$ of its static field (in a frame rotating with Mercury), and of the coefficients of the tidal deformations affecting the potential (mostly the Love number k_2).

The orbit also depends upon the coefficients describing the rotation of Mercury, including the obliquity ϵ_1 and the amplitude of the libration in longitude ϵ_2 (see the next section). However, the response of the orbit $\mathbf{x}_{sat}(t)$ to ϵ_1 contains the coefficient $\overline{C}_{20}$ and the response to ϵ_2 contains $\overline{C}_{22}$ (Cicalò 2007), of course there would be no effect of the rotation on the gravity field if the planet were spherically symmetric. Thus the sensitivity to ϵ_1, ϵ_2 is too weak for a robust orbit determination based on the orbit only.

Table 17.1. Accelerations acting on a spacecraft in orbit around Mercury, in a planetocentric reference frame, with $a = 3000$ km, $A/M = 0.05$ cm^2/g.

Cause	Formula	Parameters	Value cm/s^2
Mercury monopole	$GM_{\text{☿}}/r^2 = F_0$	$GM_{\text{☿}}$	$2.4 \cdot 10^2$
Mercury oblateness	$3 F_0 \overline{C}_{20} R_{\text{☿}}^2/r^2$	$\overline{C}_{20}$	$1.3 \cdot 10^{-2}$
Mercury triaxiality	$3 F_0 \overline{C}_{22} R_{\text{☿}}^2/r^2$	$\overline{C}_{22}$	$7.8 \cdot 10^{-3}$
Radiation pressure	$C_R F_{PR}$	C_R	$6.8 \cdot 10^{-5}$
Thermal emission	$4/9\, F_{PR}\, \alpha_{\text{☿}}\, \Delta T/T$	$\alpha_{\text{☿}}, \Delta T$	$3 \cdot 10^{-5}$
Sun tide	$2\, GM_{\odot}\, r/r_{\odot}^3$	$GM_{\odot}$	$2.3 \cdot 10^{-5}$
Effect of ϵ_1	$(9/2)\, \epsilon_1 F_0 \overline{C}_{20} R_{\text{☿}}^2/r^2$	$\epsilon_1 \overline{C}_{20}$	$1.9 \cdot 10^{-5}$
Effect of ϵ_2	$(9/2)\, \epsilon_2 F_0 \overline{C}_{22} R_{\text{☿}}^2/r^2$	$\epsilon_2 \overline{C}_{22}$	$3.3 \cdot 10^{-6}$
Solid tide	$3k_2\, GM_{\odot} R_{\text{☿}}^5/r_{\odot}^3 r^4$	k_2	$2.8 \cdot 10^{-6}$
Mercury albedo	$C_R F_{PR}\, (1 - \alpha_{\text{☿}})\, R_{\text{☿}}^2/(2 r^2)$	$\alpha_{\text{☿}}, C_R$	$2.7 \cdot 10^{-6}$
Venus tide	$2\, GM_{\text{♀}}\, r/r_{\text{♀☿}}^3$	$GM_{\text{♀}}$	$4 \cdot 10^{-8}$
Relativistic Mercury	$F_0\, G M_{\text{☿}}/(c^2 r)$	$G M_{\text{☿}}$	$1.9 \cdot 10^{-8}$

A list of the perturbations acting on the orbit $\mathbf{x}_{sat}(t)$ with their orders of magnitude (Table 17.1) easily shows that non-gravitational perturbations, in the fiery radiation environment around Mercury, are large enough to mask the effect of Mercury's gravity field, besides the lowest degree harmonics, and even these could not be determined with the needed accuracy. The direct radiation pressure from the Sun generates a perturbing acceleration of the order of 0.01 times the $\overline{C}_{22}$ effect; note that the thermal emission from the planet is of the same order, also because the planetary albedo of Mercury

$1 - \alpha_{\mancury}$ is low. Thus the requirement for an accelerometer: it is in fact available on BepiColombo as the ISA instrument, a spring accelerometer (Iafolla and Nozzoli 2001, Lucchesi and Iafolla 2006).

Accelerometer observables

As discussed in Section 16.1, an accelerometer directly measures the acceleration resulting from non-gravitational perturbations acting on the outer surface of the spacecraft, with a minus sign, see eq. (16.3). Apparent forces due to the displacement of the sensing heads with respect to the spacecraft center of mass (CoM) are also included in the measurement, but this does not introduce an error in the acceleration measurements if the reference point, for which an orbit is computed, is rigidly attached to the accelerometer rather than being the CoM, see eq. (16.5).

A more serious problem is the thermal perturbations on the accelerometer, resulting in time-variable accelerometer calibration parameters; if these variations were slow enough, the determination of constant calibration parameters separately for each observed arc would solve the problem, but if the temperature variations are large this fails. For an accelerometer without active thermal control the temperature changes are large. For example, over a time span of 22 days (1/4 of Mercury orbit) the temperature could change by $\simeq 10°$, and the resulting spurious signal would degrade the solution.

Full size simulations of the BepiColombo gravimetry experiment have been performed (Milani *et al.* 2001b, Milani *et al.* 2003) to assess the feasible accuracy in the determination of the Mercury gravity coefficients, static and tidal, and to define the requirements on the accelerometer a priori calibration. The results are summarized in Figure 17.1, showing the effects of temperature changes on the accelerometer body, assuming the temperature changes are either uncontrolled or reduced to 10% or to 1% of the value applicable to the spacecraft frame. In all cases the true error, the difference between the ground truth value used in the simulation and the nominal solution, is significantly larger than the formal accuracy as deduced from the covariance matrix. However, with temperature control reducing by 1–2 orders of magnitude the changes occurring along an orbit of Mercury around the Sun, the gravitational signal is still well above the error up to order and degree 25. This resulted in the definition of a requirement for the accelerometer unit, which was adopted in the design of the instrument.[6]

---

[6] Note that it does not matter whether this temperature calibration is obtained by control rather than by measurement of the temperature, provided the thermal sensitivity coefficient b of eq. (16.6) is well known.

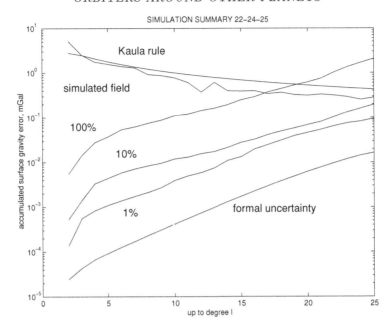

Fig. 17.1. Signal and error for three simulations with 100%, 10%, and 1% of spurious accelerometer signals resulting from thermal sensitivity. For each degree ℓ the results are expressed in terms of the gravity anomalies at the surface, in milliGal (10^{-3} cm/s^2). The errors for each degree are accumulated as the RMS sum of the errors for all the degrees up to ℓ. The top curves are the simulated signal and Kaula rule, the bottom one is the formal uncertainty from the covariance matrix.

The best simulation, with temperature control at the 1% level, allowed us to solve for the **Love number** k_2, giving the elastic response of Mercury's gravity to the degree 2 tidal potential of the Sun, with a true error of 0.004, the formal accuracy being one order of magnitude less.

Relativity in Mercury-centric orbit

The Mercury-centric orbit of the spacecraft is coupled to the orbit of the planet, mostly through the difference between the acceleration from the Sun on the probe and that on the planet (the Sun's tidal term). This coupling is weak because the Sun's tide is just 10^{-7} of the monopole acceleration from Mercury. The relativistic perturbations containing the mass of Mercury are small, as shown in the table, to the point that they are not measurable, being easily absorbed by the much larger accelerometer calibrations. Should we conclude that general relativity does not matter in the computation of the Mercury-centric orbit? The answer is negative, but the main relativistic effect does not appear in the equation of motion for $\mathbf{x}_{sat}$.

In the equations for the range-rate observable (17.5) the term $d\Delta_{do}/dt_b$

accounts for the difference in **time coordinates**. In fact, there are three
different time coordinates to be considered. The dynamics of the planets,
as described by the Lagrangian (6.18), is the solution of differential equations with as independent variable a time belonging to a space-time reference frame with origin in the Solar System barycenter (6.16). There can
be different realizations of such a time coordinate; the currently published
planetary ephemerides are provided in a time called TDB (for **dynamic
barycentric time**).[7] The observations are based on averages of clocks and
frequency scales located on the Earth's surface; this corresponds to another
time coordinate called TT (for **terrestrial time**). Thus for each observation the times t_t, t_r need to be converted from TT to TDB to find the
corresponding positions of the planets, e.g., the Earth and the Moon, by
combining information from the precomputed ephemerides and the output
of the numerical integration for Mercury and the Earth–Moon barycenter.
This time conversion step is necessary for the accurate processing of each set
of interplanetary tracking data; the main term in the TT-TDB difference is
periodic, with period 1 year and amplitude $\simeq 1.6 \times 10^{-3}$ s, while there is
essentially no linear trend, as a result of a suitable definition of the TDB.

The equation of motion of a Mercury-centric satellite can be approximated, to the required level of accuracy, by a Newtonian equation of motion
provided the independent variable of the spacecraft equation of motion is
the proper time of Mercury. Thus, for the BepiColombo radioscience experiment, it is necessary to define a new time coordinate TDM (**dynamic
Mercury time**) containing terms of post-Newtonian order 1 depending
mostly upon the distance from the Sun r_{10} and velocity v_1 of Mercury. The
relationship with the TDB scale, truncated to post-Newtonian order 1, is
given by the differential equation

$$\frac{dt_{TDM}}{dt_{TDB}} = 1 - \frac{v_1^2}{2\,c^2} - \sum_{k \neq 1} \frac{G\,m_k}{c^2\,r_{1k}}$$

which can be solved by a quadrature formula (see Appendix B) once the
orbits of Mercury, the Sun, and the other planets are known. Figure 17.2
plots the output of such a computation, showing a drift due to the non-zero
average of the post-Newtonian term. The periodic term, with the period of
Mercury's orbit, is almost an order of magnitude larger than the difference
TT-TDB. The time derivative of the periodic correction is $\simeq 10^{-8}$; in (17.5)
it is multiplied by the velocity of Mercury $\simeq 50$ km/s, resulting in a change in

---

[7] They are available from NASA Jet Propulsion Laboratory, Pasadena, California. Different time
coordinates have been proposed and should be adopted in the near future for the computation
of planetary ephemerides, see Appendix C.

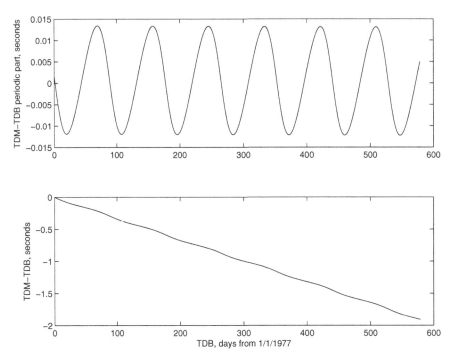

Fig. 17.2. Bottom: difference between the time coordinate TDM, in which the Mercury-centric orbit of the spacecraft is computed, and TDB, in which the planetary orbits are computed, as a function of TDB. Top: the same difference after removing a linear trend, showing the dependence upon the distance between Mercury and the Sun.

range-rate by up to 0.05 cm/s, $\simeq 30$ times larger than the accuracy of the range-rate with an integration time of 30 s. The linear drift is relevant, although it could be removed by a change of scale, see Section 4.1, of the dynamic time and of the mass of Mercury. Caution must be used also in the space portion of the space-time coordinate change, used to refer the spacecraft orbit to the center of Mercury, see Appendix C.

17.4 The rotation experiment

The **rotation experiment** will use correlation of images from the high-resolution camera of BepiColombo to directly constrain the rotation state of the planet. The theory of the rotation of Mercury was established based on the early planetary radar data, which allowed us to measure the planet rotation frequency $\nu \simeq 3/2\, n_1$, where n_1 is the mean motion of the orbit of Mercury. Because of the significant eccentricity $e_1 \simeq 0.2$ of the orbit, the torque applied by the Sun's tidal attraction on the long axis of the equator of Mercury forces a periodic perturbation of the uniform rotation. In a simplified model with Mercury in a Keplerian orbit and the rotation axis orthogonal to the orbit plane, coincident with the axis of maximum moment of inertia, the equation of motion for the rotation phase ϕ is (Colombo and Shapiro 1966)

$$C\frac{d^2\phi}{dt^2} = \frac{3}{2}\frac{G\,M_\odot\,(B-A)}{r^3}\,\sin(v_1-\phi) \qquad (17.7)$$

where A, B, C are the principal moments of inertia (eigenvalues of the inertia quadratic form), r is the distance to the Sun, and v_1 the true anomaly of Mercury. From this equation it is possible to derive a first approximation solution

$$\phi = \frac{3}{2}l_1 + \frac{3}{2}\frac{B-A}{C}\sin l_1 + \cdots, \qquad (17.8)$$

where l_1 is the mean anomaly of Mercury, showing a libration of the rotation angle around the uniform rotation rate $\nu = 3/2\,n_1$. The phase of this **libration in longitude** is 0 for $v_1 = 0$, that is with Mercury at perihelion. However, if we assume that Mercury has a liquid layer separating a rigid outer shell from an inner core, the moment of inertia C appearing in eqs. (17.7) and (17.8) should be replaced by C_m, the moment of inertia of the outer shell (Peale 1972). Then the amplitude of the libration in longitude would be larger by a factor C/C_m, which could be as large as $\simeq 2$. The first goal of the rotation experiment is to estimate the amplitude ϵ_2 of this libration, thus constraining $(B-A)/C_m$.

A complete theory of the rotation of Mercury, taking into account the secular perturbations to the orbit, should also take into account the **Cassini laws**, by which the rotation axis belongs to the plane spanned by the normal to the orbit plane and the axis of the orbital plane precession of Mercury (Colombo 1966). The **obliquity**, the angle ϵ_1 between the normal to the orbit plane and the rotation axis, is proportional to the concentration coefficient $C/(MR^2)$ where R is the mean radius of Mercury. Then (Peale 1988)

$$\frac{C_m}{C} = \frac{C_m}{B-A}\frac{MR^2}{C}\frac{B-A}{MR^2}, \qquad (17.9)$$

where the first two factors are measured by ϵ_1, ϵ_2 and the third by the harmonic coefficient C_{22}, see eq. (13.23). Thus it is possible, by measuring the rotation state of Mercury, in particular the parameters ϵ_1, ϵ_2, and the gravity field, in particular C_{22}, to draw some conclusions about both the physical state and the size of the core of Mercury.

A more complete model of the rotation of Mercury would include planetary short periodic perturbations on the orbit, indirectly affecting the rotation state (Dufey *et al.* 2008). The main terms introduced in this way in the longitude libration (17.8) contain the anomalies of the planets, thus have periods of a few years. The largest term is due to Jupiter: it has period 11.86 y and amplitude $\simeq$ 13 arcsec. This implies that the phase of the complete libration in longitude is not 0 at perihelion. If the duration of the mission in orbit around Mercury is small compared to the periods of the main planetary perturbation terms, the latter are approximately a constant shift in the rotation phase at perihelion. We assume that the expansion in spherical harmonics is done with spherical polar coordinates (r, θ, λ) with the origin of longitudes at the meridian of Mercury facing the Sun at some perihelion. Then the presence of a non-zero rotation phase implies that the axis of minimum inertia is at some angle δ_{22} from the reference meridian. Then there is a non-zero S_{22} coefficient, which can be solved in the orbit determination and used to compute δ_{22} by the equation $C_{22} \cos(2\lambda) + S_{22} \sin(2\lambda) = J_{22} \cos(2\lambda + 2\,\delta_{22})$. In conclusion the planetary perturbations generate effects which should be observable by the BepiColombo radioscience experiment.

The observing conditions

The peculiar rotation of Mercury in a 3/2 resonance with the orbital motion results in sharp constraints on the possibility of observing multiple times the same portion of the surface from an orbiting spacecraft. A **Mercury solar day** is a time span ($\simeq$ 176 Earth days) in which the planet completes three sidereal rotations (the rotation phase ϕ is the curve F in Figure 17.3) and the Sun as seen from Mercury's orbit revolves twice in the sky (the mean anomaly l_1 of Mercury is the curve M). Thus the Sun, as seen from a point on the surface, makes a full revolution on the celestial sphere in one Mercury solar day. The solar time on Mercury is given by $\phi - v_1$ (curve T); for a given meridian on Mercury's surface the times in which there is sunlight are the continuous part of the curve, the dotted part corresponds to darkness. The phase of the torque from the solar tide acting on the C_{22} harmonic is $2\phi - 2v_1 = 3\,l_1 - 2v_1 + \mathcal{O}(\epsilon_1)$ (curve C), thus the forced libration in the longitudinal main term has the period of the orbit of Mercury, see (17.8).

For a polar orbit there are only six times per Mercury solar day when the spacecraft ground track passes in a given area (of the size of a high-resolution image), as shown in Figure 17.3 by the intersections of the horizontal lines and the rotation phase, and three of these are in darkness, on average. Moreover, the illumination conditions and the spacecraft altitude are bound to be different in the two images to be compared. If the mission lasts one Earth year, the same cycle of observing conditions repeats twice, but the same longitude is observed again at the same value of the anomaly of

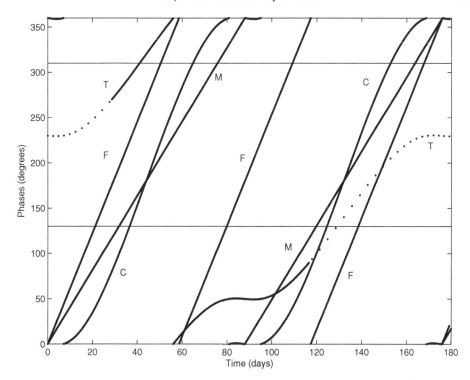

Fig. 17.3. The angles involved in the libration in longitude of Mercury over half a year. M is the mean anomaly of the orbit of Mercury. F is the phase of the planet spin. T is the local solar time in degrees; note that the Sun can go back in the sky as seen from Mercury near perihelion. C is the phase of the torque applied by the Sun on Mercury's equatorial bulge: its period is the orbital period. The two horizontal lines represent one choice for the orbital plane of the spacecraft.

Mercury, thus the libration in longitude also has the same phase. As a result of these observational constraints, it is essential to take all the opportunities to record images of the same area, including times when the spacecraft is not being tracked by a ground station. It has been shown by the gravimetry experiment simulations that, with reduction to 1% of the accelerometer bias, the orbit $\mathbf{x}_{sat}(t)$ is accurate enough to perform the rotation experiment by comparing images taken at arbitrary times during the mission. However, this implies constraints on the maneuvers, even attitude ones, performed between two tracking passes.

The measurements are direction angles to reference points on the surface, as seen from the spacecraft and referred to an inertial reference frame by using the knowledge of the attitude (from star mappers) and of the alignment of the camera with respect to the spacecraft-fixed reference frame. Thus the error budget for these measurements includes the spacecraft orbit error, the attitude error, the thermo-mechanical stability of the camera and star mapper alignments, and the error in correlating two images to find the relative displacement (Milani *et al.* 2001b).

17.5 The relativity experiment

To test the theory of gravitation, the orbit of Mercury has to be determined with unprecedented accuracy; some improvement is necessary also for the orbit of the Earth–Moon barycenter. The requirement is for a fully relativistic equation of motion, including the terms expressing the violations of general relativity with the post-Newtonian parameters, such as $\gamma, \beta, \zeta, J_{2\odot}, \eta, \alpha_1, \alpha_2$. Moreover, the mass of the Sun and possibly of some of the planets has to be improved. With 12 initial conditions for Mercury and the Earth–Moon barycenter, there are $\simeq 20$ parameters to be solved.

If it is possible to separate this portion of the problem from the determination of the Mercury-centric orbit, then it is a comparatively small computation, suitable for running a large number of tests exploring the combined effect of random and systematic errors (Milani *et al.* 2002). The observations can be introduced as two normal points, representing the range and range-rate data, for each pass of Mercury above the horizon. This simplified approach is useful in simulations, to identify the main problems and to assess the possible performances, see below; this does not imply that such a partitioning of the problem should be used in the operational processing of the real data. Five main problems have been identified with this approach.

(i)   The determination of the two planetary orbits and of the mass of the Sun results in an approximate rank deficiency of order 4.

(ii) There is a strong correlation, in the covariance matrix of the solution, between β and $J_{2\odot}$, degrading the marginal accuracy in the solution of both parameters.

(iii) The parameter γ appears also in the Shapiro effect, strongly dependent upon the minimum distance of the radio beam from the Sun. During a close superior conjunction, γ generates a Shapiro signal larger than that on the orbit of Mercury.

(iv) The parameter ζ describing $d(GM_\odot)/dt$ is very sensitive to the presence of a time-dependent systematic effect in the range measurement.

(v)   The orbit of the Earth–Moon barycenter has to be determined with such an accuracy that perturbations from asteroids are relevant.

The origin of these problems, and some possible solutions, are as follows.

(i) After eliminating the Sun's coordinates as dynamical variables by the use

of barycentric planetary positions, eq. (6.17), there are still three exact symmetries obtained by rotation of all the planets, as discussed in Section 6.6.
Even if the orbits of the other planets are not solved, but taken from the existing planetary ephemerides, because of the weak coupling with the orbits
of Mercury and the Earth there is still an approximate symmetry. Moreover,
a scale change is an approximate symmetry, provided the mass of the Sun is
simultaneously changed according to eq. (4.7). Thus it is necessary to add
four constraints, with the technique discussed in Section 6.1.

(ii) The main orbital effect of β is a precession of the argument of perihelion,
that is a displacement taking place in the plane of the orbit of Mercury;
$J_{2\odot}$ affects the precession of the longitude of the node, that is generates a
displacement in the plane of the solar equator. The angle between these
two planes is only $\epsilon = 3.3°$ and $\cos \epsilon = 0.998$, thus it is easy to understand
how the correlation between β and $J_{2\odot}$ can be 0.997. Short of using another
test body, with an orbit plane much more inclined than that of Mercury,
this correlation cannot be avoided. One possible way to mitigate this effect
is to use the equation derived by Nordtvedt (1970) for a generic theory of
gravitation under the assumption that it is a metric theory:

$$\eta = 4\beta - \gamma - 3 - \alpha_1 - \frac{2}{3}\alpha_2. \qquad (17.10)$$

When the values of γ, η and also of the preferred frame parameters (if included in the solution) are well determined, this equation acts essentially as
a strong constraint on the value of β, and as a result the variance of both β
and $J_{2\odot}$ is sharply reduced, see (Milani et al. 2002, Figure 6).

(iii) Because of the different time-scale, a superior conjunction experiment
with the specific purpose of strongly constraining the value of γ cannot be
simulated in the same way as the determination of the other post-Newtonian
parameters: a combined solution for the Mercury-centric orbit, for a local
correction to the orbit of Mercury and for γ has to be used. The quality of the
results, however, depends strongly on the assumptions. In every year there
are three superior conjunctions of Mercury as seen from the Earth, but each
individual superior conjunction produces a Shapiro S/N depending upon
the circumstances: in some cases Mercury is occulted by the Sun, in others
the radio waves are passing much farther from the Sun, thus the γ signal
contained in the Shapiro effect is much weaker. In (Milani et al. 2002) the
assumption used for the simulations was that for a comparatively short time
span (20 days) around the superior conjunction three ground stations were
available, distributed in longitude in such a way as to provide continuous
tracking; moreover, a conjunction in which Mercury was actually occulted by

the Sun was used. In the real mission, these assumptions may not be satisfied. A possible alternative is to perform a superior conjunction experiment during the interplanetary cruise phase, as in (Bertotti *et al.* 2003a). As discussed in Section 14.5, the handling of the non-gravitational perturbations is easier, the accelerometer is not really needed (the same experiment can be used as long-term calibration of the accelerometer). An experiment under conditions comparable to the Cassini one, but with the BepiColombo instrument (capable of very accurate range measurements, not just range-rate) could give results on γ more accurate by an order of magnitude. However, the conditions in the inner Solar System are more difficult for several reasons, including stronger radiation pressure and the need for a longer duration of the experiment.

(iv) The main effect of a change of either the universal gravitational constant G or the mass of the Sun $M_{\odot}$ by a fraction 10^{-13} in one year is a quadratic perturbation along-track, growing to $\simeq 15$ cm after one year for Mercury. If the range measurements contain a time-dependent bias with a quadratic signature, this results in a systematic error in the nominal solution for ζ. This argument was used to upgrade the requirements for the instrument to be used in the BepiColombo radioscience experiment, which now includes an internal calibration loop to measure the transponder delay. The mass of the Sun changes by a fraction $\simeq 7 \times 10^{-14}$ per year because of the mass shed as photons, and this contribution to ζ can be predicted with some accuracy. Less easy is to estimate accurately the mass shed by the Sun as charged particles, also because a large fraction of this phenomenon might occur near the poles of the Sun. Thus the determination of ζ from the orbit of Mercury is not a null experiment, but one in which there is a predicted value, although not a very accurate one. For estimated values of ζ of the order of a few parts in 10^{-13}, to discriminate between a change in G and the known change in $M_{\odot}$ might be difficult.

(v) The perturbations by the asteroid Ceres on the orbit of the Earth can be estimated from the Roy–Walker parameters of Table 4.1: the short periodic perturbations on the Earth's orbit should be $\simeq 2.2 \times 10^{-11}$ AU $\simeq 3$ m. Not to degrade the accuracy of the results obtained with ranging from Earth accurate to $\simeq 10$ cm, the mass of Ceres needs to be known with a relative error of the order of 0.01, and this appears to be feasible by using the deflections of the orbits of other asteroids passing near Ceres. The problem is that there are > 20 asteroids with mass at least $1/100$ of that of Ceres, $\simeq 150$ with mass at least $1/1000$, and for many of them the mass is largely unknown. The combined effect of these poorly known perturbations could degrade the orbital determination for the Earth; the effect on Mercury is one order of magnitude smaller. This problem is connected to that of observation weighting discussed in Section 5.8, because the accurate orbit determination of closely approaching asteroids is strongly affected by the astrometric data quality (Baer *et al.* 2008). A solution of the problem could come from future asteroid mass determinations, obtained with higher accuracy astrometric surveys either from the ground or from space.[8]

Table 17.2. *The standard deviations and full errors (including systematic effects) in two simulations of the Relativity Experiment with BepiColombo. Experiments A and B differ only in that B uses the Nordtvedt equation (17.10) as a constraint.*

[0.5ex]	Exp A (non-metric)		Exp B (metric)	
Parameter	RMS	True error	RMS	True error
$\beta - 1$	6.7×10^{-5}	2.2×10^{-4}	7.5×10^{-7}	2.0×10^{-6}
η	4.4×10^{-6}	1.5×10^{-5}	3.0×10^{-6}	7.9×10^{-6}
ζ	4.0×10^{-14}	5.2×10^{-13}	3.9×10^{-14}	5.3×10^{-13}
ΔJ_2	7.9×10^{-9}	2.8×10^{-8}	2.4×10^{-10}	2.1×10^{-9}
$\Delta M_\odot / M_\odot$	1.9×10^{-12}	5.9×10^{-12}	3.3×10^{-13}	1.0×10^{-12}

The results of two simulations performed by Milani *et al.* (2002) were as described in Table 17.2. The parameter γ was considered to be known at the 2×10^{-6} level as a result of a superior conjunction experiment. The systematic effects included in the error model of the data simulation included a time-dependent bias with a nonlinear growth to 50 cm in one year, and this affected all the parameters, as shown by the significant difference between the formal error, computed from the covariance matrix, and the true error, obtained as the difference with the values actually used in the simulation. For ζ the ratio true error/formal RMS is particularly large, as expected. For β, $J_{2\odot}$ the marginal uncertainty is degraded by a factor $\simeq 100$ in the non-metric case A, with respect to the metric case B, in which eq. (17.10) is used as constraint. This is made possible by the good determination of γ and η, and also of α_1, α_2 as shown by other simulations including also these parameters. In conclusion, the results of these simplified simulations are encouraging, which does not mean that all the problems have been solved.

17.6 Global data processing

We have described, in the previous three sections, three aspects of the Bepi-Colombo radioscience experiment: the gravimetry, rotation, and relativity

---

[8] The determination of many asteroid masses is one of the science goals of another ESA mission, the hyper-accurate astrometric survey **Gaia**, to be launched in 2012.

experiments. However, the main challenge of the BepiColombo orbit determination is to assemble all the observations and solve for all the relevant
parameters, in a complete and self-consistent way. This results in a comparatively large least squares fit, although not as large as some of the examples
in Chapter 16. With state-of-the-art computer hardware, neither the memory size nor the computational load are a problem. The difficult task is
to ensure that all the equations, representing physically heterogeneous phenomena, are accurate and consistent at the required level. In this last section
we discuss how this global data processing could be done.

The local–global decomposition

The range and range-rate observations are naturally decomposed into arcs,
one for each pass of Mercury above the horizon of the ground station(s). If
there is only one ground station, this means one arc per day. Additionally,
there are separate arcs of angular observations of the geodetic reference
points, taken with the high-resolution camera.

The Mercury-centric orbit cannot be propagated in an accurate way for
a very long time span, because of the poorly modeled non-gravitational
perturbations during the interval between two passes, and also because of the
possible attitude maneuvers. During a pass, the accelerometer calibration
parameters are obtained from the fit of the observations (mostly the range-
rate). A rough order of magnitude estimate could be as follows: over 1000 s
with a measurement of range-rate accurate to 3×10^{-4} cm/s, a constant
acceleration of $\simeq 3 \times 10^{-7}$ can be determined. With only one ground station,
in the interval of 14–16 hours between two passes the accelerometer will
record uncalibrated accelerations. In the simulations (Milani *et al.* 2003),
even assuming a priori calibration with 1% accuracy of the thermal signal
(the best case considered in Section 17.3), the mean error in the spacecraft
position propagated by one day was 3.8 m, a very substantial growth with
respect to the mean error in the initial conditions, which was < 10 cm.

That is, any attempt to propagate in a deterministic way the orbit while
it is not tracked results in an error which is far larger than the measurement
accuracy. Thus to the Mercury-centric orbit determination we can apply the
same argument used for the geocentric orbit of LAGEOS in Section 15.3, and
a good choice could be to use a multi-arc strategy, in which an independent
set of six initial conditions is solved for each arc.

Then we can find out which parameters are local to one arc, in the sense
of eq. (15.1), and which ones are global. The arc initial conditions and three
constant calibrations (one for each accelerometer sensitive axis) are local

parameters, for a total of $\simeq 9 \times 365$ for a nominal one-year mission. The geodetic coordinates of the reference points used to find the shift between images are local to an arc containing only the camera observations.

The harmonic coefficients ($31^2 - 4$ for degree and order up to 30), the planetary initial conditions with some masses and post-Newtonian parameters, the global range calibrations, the tidal coefficients, and Mercury's rotation parameters are global parameters.

Following the algorithm described in Section 15.1, the problem can be solved by steps. First the local normal matrix is inverted arc by arc, see eq. (15.2), then the global variables are corrected, eq. (15.4), last the local variables are corrected by eq. (15.5). The problem in this approach appears in the first step: the local-only normal matrix has an approximate rank deficiency of order 1.

Line of sight symmetry

The symmetry responsible for the weakness of the local only normal matrix is an approximate version of the exact symmetry found for extrasolar planets in Section 6.5. If the Mercury-centric orbit is rotated around an axis $\hat{\rho}$ in the direction from the Earth to the center of Mercury, then there would be an exact symmetry in the range and range-rate observations if $\hat{\rho}$ were constant. Given that $\hat{\rho}$ changes with time, the small parameter in the approximate symmetry is the angle by which $\hat{\rho}$ rotates (in an inertial reference system) during the arc time span (Bonanno and Milani 2002).

Different solutions can be adopted to stabilize the solution for the local parameters. A set of a priori observations weakly constraining the initial conditions (with an uncertainty of 3 m in position and 3 m/day in velocity) would be enough to stabilize the solution (Milani *et al.* 2001b). This is a simplified method used in a simulation; under operational conditions we would need to compute a lower accuracy long arc solution which does not contain the approximate symmetry (because the time span is comparable with the synodic period of Mercury and the Earth), and use it to weakly constrain the local initial conditions for a short arc. As an alternative approach, the initial conditions for two consecutive arcs could be weakly constrained together, by using a covariance matrix for the prediction at the next day enlarged[9] with respect to the deterministic covariance propagation of Section 5.5. A final choice among the different options has not yet been done, but will be dictated by the results of new rounds of full-scale simulations.

---

[9] This approach is closely related to the Kalman filter class of algorithms.

A complex experiment

We would like to conclude this discussion of the BepiColombo radioscience experiment with a few words on the context. Because of the expected extreme quality of the data, the BepiColombo tracking will contain an enormous amount of information, which could generate very interesting results on both the structure of Mercury and the theory of gravitation. However, we need to be well aware of what could go wrong, also because this is the only way to safeguard against this possibility.

As a matter of principle, unless there is available absolute a priori knowledge, with errors too small to significantly affect the measurements, all the parameters affecting the observations should be included in a global least squares fit. This avoids the risk of confusion between marginal and conditional uncertainty for correlated parameters of interest.

In practice some decomposition of the problem is unavoidable, in particular the one resulting from the fact that some data are acquired by other teams and transmitted with a not too complicated interface. For example, the ground station calibration parameters, the ground antenna motion, the spacecraft antenna position and motion, the tropospheric corrections, the spacecraft attitude, and the camera alignment will be measured by other teams. For most of these measurements the requirements for keeping up with the target accuracy of the BepiColombo data set are challenging.

In other words, the BepiColombo radioscience experiment is a system experiment, involving the performance of many spacecraft subsystems and of the ground station. The work of hundreds of specialists needs to be kept at a high level of quality. This chapter has given a short account of the work we have to do, but we need to be aware of the need to rely on the work of many others.

Conclusions

We conclude this book with some words of caution: as already anticipated in the preface, we never had the intention of writing a complete reference for all methods which could be used in orbit determination problems. The goal was to present to the readers something new which they may use, mostly from the research we have conducted ourselves with many coworkers and in many years. Please let us know if we have succeeded in this.

References

Aguirre-Martinez, M. and Sneeuw, N. (2003). Needs and tools for future gravity measuring missions, *Space Sci. Rev.* **108**, 409–416.

Aitken R. G. (1964). *The Binary Stars* (Dover Publication, New York).

Aksnes, K. (1976). Short-period and long-period perturbations of a spherical satellite due to direct solar radiation, *CMDA* **13**, 89–104.

Albertella, A. (1993). Calcoli geodetici sulla sfera con la serie di Fourier, Politecnico di Torino, D. Phil. thesis.

Albertella, A. and Migliaccio, F. (eds) (1998). *SAGE, Satellite Accelerometry for Gravity Field Exploration: Phase A Final Report* (International Geoid Service, Milano).

Albertella, A., Sansò, F. and Sneeuw, N. (1999). Band-limited functions on a bounded spherical domain: the Slepian problem on the sphere, *J. Geod.* **73**, 436–447.

Anderson, J.D., Colombo, G., Esposito, P.B., Lau, E.L. and Trager, G.B. (1987). The mass, gravity field, and ephemeris of Mercury, *Icarus* **124**, 337–349.

Anselmo, L., Bertotti, B., Farinella, P., Milani, A. and Nobili, A. M. (1983a). Orbital perturbations due to radiation pressure for a spacecraft of complex shape, *CMDA* **29**, 27–43.

Anselmo, L., Farinella, P., Milani, A. and Nobili, A. M. (1983b). Effects of the Earth-reflected sunlight on the orbit of the LAGEOS satellite, *Astron. Astrophys.* **117**, 3–8.

Arnold, V. (1976). *Mathematical Methods of Classical Mechanics* (Springer, Berlin).

Baer, J., Milani, A., Chesley, S.R. and Matson, R.D. (2008). An Observational Error Model, and Application to Asteroid Mass Determination, AAAS-DPS meeting 2008, abstract 52.09.

Balmino, G., Barriot, J. P. and Valés, N. (1990). Non-singular formulation of the gravity vector and gravity gradient tensor in spherical harmonics, *Manuscripta Geodetica* **15**, 11–16.

Bern, M. and Eppstein, D. (1992). Mesh generation and optimal triangulation. In *Computing in Euclidean Geometry*, eds. D.-Z. Du and F.K. Hwang (World Scientific), pp. 23–90.

Bertotti, B. and Iess, L. (1991). The rotation of LAGEOS, *J. Geophys. Res.* **96**, 2431–2440.

Bertotti, B., Iess, L. and Tortora, P. (2003a). A test of general relativity using radio links with the Cassini spacecraft, *Nature* **425**, 374–376.

Bertotti, B., Farinella, P. and Vokrouhlický, D. (2003b). *Physics of the Solar System* (Kluwer, Dordrecht).

Bini, D., Capovani, M. and Menchi, O. (1988). *Metodi numerici per l'algebra lineare* (Zanichelli, Bologna).

Bini, D. A. (1997). Numerical computation of polynomial zeros by means of Aberth method, *Numer. Algorithms* **13**, no. 3–4, 179–200.

Boattini, A., D'Abramo, G., Forti and G. Gal, R. (2001). The Arcetri NEO Precovery Program, *Astron. Astrophys.* **375**, 293–307.

Bonanno, C. (2000). An analytical approximation for the MOID and its consequences, *Astron. Astrophys.* **360**, 411–416.

Bonanno, C. and Milani, A. (2002). Symmetries and rank deficiency in the orbit determination around another planet, *CMDA* **83**, 17–33.

Bowell, E. and Muinonen, K. (1994). Earth-crossing asteroids and comets: ground-based search strategies, in *Hazards due to Comets and Asteroids*, ed. T. Gehrels (University of Arizona Press, Tucson), pp. 149–197.

Bowell, E., Hapke, B., Domingue, D., Lumme, K., Peltoniemi, J. and Harris, A.W. (1989). Application of photometric models to asteroids. In *Asteroids II*, eds. R. P. Binzel, T. Gehrels and M. S. Mathews (University of Arizona Press, Tucson), pp. 524–556.

Broucke, R. A. and Cefola, P. J. (1972). On the equinoctial orbit elements, *CMDA* **5**, 303–310.

Carpino, M., Milani, A. and Chesley, S. R. (2003). Error statistics of asteroid optical astrometric observations, *Icarus* **166**, 248–270.

Celletti, A. and Pinzari, G. (2005). Four classical methods for determining planetary elliptic elements: A comparison, *CMDA* **93**, 1–52.

Celletti, A. and Pinzari, G. (2006). Dependence on the observational time intervals and domain of convergence of orbital determination methods, *CMDA* **95**, 327–344.

Cesare, S. *et al.* (2005). *Laser Doppler Interferometry Mission: Final Report*, Alcatel Alenia Space Italia report No. SD-RP-AI-0445. 19 December 2005.

Cesare, S., Sechi, G., Bonino, L., Sabadini, R., Marotta, M., Migliaccio, F., Reguzzoni, M., Sansò, F., Milani, A. and Pisani, M. (2006). Satellite-to-satellite laser tracking mission for gravity field measurement. In *Gravity Field of The Earth*, Proceedings of the First International Symposium of the International Gravity Field Service, Istambul, 28 August – 1 September 2006.

Charlier, C. V. L. (1910). On multiple solutions in the determination of orbits from three observations, *MNRAS* **71**, 120–124.

Charlier, C. V. L. (1911). Second note on multiple solutions in the determination of orbits from three observations, *MNRAS* **71**, 454–459.

Chesley, S.R. (2005). Very short arc orbit determination: the case of asteroid 2004 FU$_{162}$. In *Dynamics of Populations of Planetary Systems*, eds. Z. Knežević, and A. Milani (Cambridge University Press), pp. 259–264.

Chesley, S.R. (2006). Potential impact detection for near-Earth asteroids: the case of 99942 Apophis, in *Asteroid, Comets, Meteors*, eds. Lazzaro, D. *et al.* (Cambridge University Press), pp. 215–228.

Chesley, S.R., Chodas, P.W., Milani, A., Valsecchi, G.B. and Yeomans, D.K. (2002). Quantifying the risk posed by potential Earth impacts, *Icarus*, **159**, 423–432.

Chesley, S. R., Ostro, S. J., Vokrouhlický, D., Čapek, D., Giorgini, J. D., Nolan, M. C., Margot, J.-L., Hine, A. A., Benner, L. A. M. and Chamberlin, A. B.

(2003). Direct detection of the Yarkovsky effect by radar ranging to Asteroid 6489 Golevka, *Science* **302**, 1739–1742.

Chesley, S.R., Vokrouhlický, D. and Matson, R.D. (2008). Orbital identification for asteroid 152563 (1992 BF) through the Yarkovsky effect, *Astron. J* **135**, 2336–2340.

Chodas, P.W. and Yeomans, D.K. (1996). The orbital motion and impact circumstances of Comet Shoemaker-Levy 9. In *The Collision of Comet Shoemaker-Levy 9 and Jupiter*, eds. K.S. Knoll *et al.* eds. (Kluwer, Dordrecht), pp. 1–30.

Cicalò, S. (2007). Determinazione dello stato di rotazione di Mercurio dallo studio del campo gravitazionale, University of Pisa, Master thesis.

Colombo, G. (1966). Cassini's second and third laws, *Astron. J* **71**, 891–896.

Colombo, O.L. (1989). Advanced techniques for high-resolution mapping of the gravitational field. In *Theory of Satellite Geodesy and Gravity Field Determination, Lecture Notes in Earth Sciences* **25**, eds. F. Sansò and R. Rummel (Springer, Berlin), pp. 335–369.

Colombo, G. and Shapiro, I.I. (1966). The rotation of the planet Mercury, *Astrophys. J* **145**, 296–307.

Conn, A.R., Gould, N.I.M. and Toint, Ph.L. (1992) *LANCELOT: a Fortran package for large-scale nonlinear optimization* (Springer, Berlin).

Coradini, A. *et al.* (1996). *MORO Moon ORbiting Observatory*, ESA SCI (96) 1, March 1996.

Cox, D. A., Little, J. B. and O'Shea, D. (1996). *Ideals, Varieties and Algorithms* (Springer, Berlin).

Crawford, R. T., Leuschner, A. O. and Merton, G. (1930). *Determination of Orbits of Comets and Asteroids* (McGraw Hill, New York).

Danby, J. M. A. (1988). *Fundamentals of Celestial Mechanics*, Second edition (Willmann Bell, Richmond VA).

Delaunay, B. (1934). Sur la sphere vide, *Izvestiya Akademii Nauk SSSR, Otdelenie Matematicheskii i Estestvennykh Nauk* **7**, 793–800.

de' Michieli Vitturi, M. (2004). Approximate gradient-based methods for optimum shape design in aerodynamic, University of Pisa, D. Phil. thesis.

Dufey, J., Lemaitre, A. and Rambaux, N. (2008) Planetary perturbations on Mercury's libration in longitude, *CMDA* **101**, 141–157.

Edmonds, A.R. (1957). *Angular Momentum in Quantum Mechanics* (Princeton University Press.)

European Space Agency (1999). Gravity field and steady-state ocean circulation mission (GOCE), ESA SP-1233(1), July 1999.

Evans, L. C. (1998). *Partial Differential Equations* (American Mathematical Society).

Everhart, E. and Pitkin, E. T. (1983). Universal variables in the two-body problem, *Am. J. Phys.* **51/8** 712–717.

Farnocchia, D. (2008). Orbite preliminari di asteroidi e satelliti artificiali, University of Pisa, Master thesis.

Ferraz-Mello, S. (1981). Estimation of periods from unequally spaced observations, *Astron. Astrophys.* **86**, 619–624.

Field, D.A. (1988). Laplacian smooting and Delaunay triangulations, *Commun. Appl. Math* **4**, 709–712.

Gauss, C. F. (1809). *Theoria motus corporum coelestium in sectionis conicis solem ambientum*, Hamburg; also in *Werke, siebenter band*, (1981, Olms Verlag, Hildesheim).

Granvik, K. and Muinonen, K. (2008). Asteroid identification over apparitions, *Icarus* **198**, 130–137.

Granvik, K., Muinonen, K., Virtanen, J., Delbó, M., Saba, L., De Sanctis, G., Morbidelli, R., Cellino, A. and Tedesco, E. (2005). Linking Very Large Telescope asteroid observations. In *Dynamics of Populations of Planetary Systems*, eds. Knežević, Z. and Milani, A. (Cambridge University Press), pp. 231–238.

Greenberg, R., Carusi, A. and Valsecchi, G.B. (1988), Outcomes of planetary close encounters – A systematic comparison of methodologies, *Icarus* **75**, 1–29.

Gronchi, G. F. (2002). On the stationary points of the squared distance between two ellipses with a common focus, *SIAM Journ. Sci. Comp.* **24/1**, 61–80.

Gronchi, G. F. (2005). An algebraic method to compute the critical points of the distance function between two Keplerian orbits, *CMDA* **93**, 297–332.

Gronchi, G. F. (2009). Multiple solutions in preliminary orbit determination from three observations, *CMDA* **103**, 301–326.

Gronchi, G. F. and Tommei, G. (2006). On the uncertainty of the minimal distance between two confocal Keplerian orbits, *DCDS-B* **7/4**, 755–778.

Gronchi, G. F., Tommei, G. and Milani, A. (2007). Mutual geometry of confocal Keplerian orbits: uncertainty of the MOID and search for Virtual PHAs. In *Near Earth Objects, our Celestial Neighbors: Opportunity and Risk*, eds. Milani, A. Valsecchi, G. B. and Vokrouhlický, D. (Cambridge University Press), pp. 3–14.

Gronchi, G. F., Dimare, L. and Milani, A. (2008). Orbit determination with the two-body integrals, submitted.

Hartmann, P. (1964). *Ordinary Differential Equations* (John Whiley, Hoboken, NJ).

Herrick, S. (1971). *Astrodynamics*, Vol. **1** (Van Nostrand Reinhold, London).

Hobson, E. W. (1931). *The Theory of Spherical and Ellipsoidal Harmonics* (Cambridge University Press).

Hoots, F. R. (1994). An analytical method to determine future close approaches between satellites, *CMDA* **33**, 143–158.

Iafolla, V. and Nozzoli, S. (2001). Italian spring accelerometer (ISA): a high sensitive accelerometer for BepiColombo ESA CORNERSTONE *Plan. Space Sci.* **49**, 1609–1617.

Iess, L. and Boscagli, G. (2001). Advanced radio science instrumentation for the mission bepiColombo to Mercury, *Plan. Space Sci.* **49**, 1597–1608.

Jazwinski, A. H. (1970). *Stochastic Processes and Filtering Theory* (Academic Press, New York).

Jeffreys, B. (1965). Transformations of tesseral harmonics under rotation, *Geophys. J.* **10**, 141–145.

Kaula, W. M. (1966). *Theory of Satellite Geodesy* (Blaisdell, Whaltham).

Kholshevnikov, K. V. and Vassiliev, N. (1999). On the distance function between two keplerian elliptic orbits, *CMDA* **75**, 75–83.

King-Hele, D. (1964). *Theory of Satellite Orbits in an Atmosphere* (Butterworths, London).

Kinoshita, H., Hori, G. and Nakai, H. (1974). Modified Jacobi polynomial and its applications to expansions of disturbing functions, *Ann. Tokyo Astron. Obs. (Sec. Ser.)* **14**, 14–35.

Knežević, Z. and Milani, A. (1998). Orbit maintenance of a lunar polar orbiter, *Planet. Space Sci.* **46**, 1605–1611.

Knuth D.E. (1998). *The Art of Computer Programming, Volume 3, Sorting and Searching* (Addison-Wesley, Reading, Massachussets).

Kristensen, L.K. (1995). Orbit determination by four observations, *Astron. Nachr.* **316/4**, 261–266.

Kubica, J., Denneau, L., Grav, T., Heasley, J., Jedicke, R., Masiero, J., Milani, A., Moore, A., Tholen and D., Wainscoat, R. J. (2007). Efficient intra- and inter-night linking of asteroid detections using kd-trees, *Icarus* **189**, 151–168.

Lemoine, F.G., Kenyon, S.C., Factor, J.K., Trimmer, R.G., Pavlis, N.K., Chinn, D.S., Cox, C.M., Klosko, S.M., Luthcke, S.B., Torrence, M.H., Wang, Y.M., Williamson, R.G., Pavlis, E.C., Rapp, R.H. and Olson, T.R. (1998). *The Development of the Joint NASA GSFC and NIMA Geopotential Model EGM96*, NASA/TP-1998-206861, (NASA Goddard Space Flight Center, Greenbelt, MD).

Leuschner, A. O. (1913a). A short method of determining orbits from 3 observations, *Publ. Lick Obs.* **7**, 3–20.

Leuschner, A. O. (1913b). Short methods of determining orbits, second paper, *Publ. Lick Obs.* **7**, 217–376.

Lucchesi, D. and Iafolla, V. (2006). The non-gravitational perturbations impact on the BepiColombo radio science experiment and the key rôle of the ISA accelerometer: direct solar radiation and albedo effects, *CMDA* **96**, 99–127.

Marchi, S., Momany, Y. and Bedin, L. R. (2004). Trails of solar system minor bodies on WFC/ACS images, *New Astron.* **9**, 679–685.

Mehrholz, D., Leushacke, L., Flury, W., Jehn, R., Klinkrad, H. and Landgraf, M. (2002). Detecting, tracking and imaging space debris, *ESA Bulletin* **109** 128–134.

Milani, A. (1999). The asteroid identification problem I: recovery of lost asteroids, *Icarus* **137**, 269–292.

Milani, A. (2002a). *Introduzione ai sistemi dinamici* (Editrice PLUS, Pisa).

Milani, A. (2002b). Celestial mechanics and the real Solar System: measurements, models and tests. In *Celestial Mechanics, St. Petersburg 2002*, IAU Transactions no. 8 (Institute of Applied Astronomy, St. Petersburg), pp. 133–136.

Milani, A. (2005). Virtual asteroids and virtual impactors. In *Dynamics of Populations of Planetary Systems*, eds. Z. Knežević, and A. Milani (Cambridge University Press), pp. 219–228.

Milani, A. and Knežević, Z. (1995). *Selenocentric Proper Elements, A Tool for Lunar Satellite Mission Analysis, version 2.0.* ESA, Final Report of Study 144506, G. Racca technical officer.

Milani, A. and Melchioni, E. (1989). Determination of a local geodetic network by multi-arc processing of satellite laser ranges. In *Theory of Satellite Geodesy and Gravity Field Determination*, Lecture Notes in Earth Sciences, **25**, eds. F. Sansò and R. Rummel (Springer-Verlag, Berlin), 417–445.

Milani, A. and Nobili, A. M. (1983a). On topological stability in the general 3–body problem, *CMDA* **31**, 213–240.

Milani, A. and Nobili, A. M. (1983b). On the stability of hierarchical 4–body systems, *CMDA* **31**, 241–291.

Milani, A. and Valsecchi, G.B. (1999). The asteroid identification problem II: Target plane confidence boundaries. *Icarus* **140**, 408–423.

Milani, A., Nobili, A. M. and Farinella, P. (1987). *Non Gravitational Perturbations and Satellite Geodesy* (Adam Hilger, Liverpool).

Milani, A., Carpino, M., Rossi, A., Catastini and G., Usai, S. (1995). Local geodesy by satellite laser ranging: a European solution, *Manuscripta Geodetica* **20**, 123–138.

Milani, A., Luise, M. and Scortecci, F. (1996) The lunar sub-satellite experiment of the ESA MORO mission, *Planet Space Sci.* **44**, 1065–1076.

Milani, A., Chesley, S.R., and Valsecchi, G.B. (1999). Close approaches of asteroid 1999 AN_{10}: Resonant and non-resonant returns. *Astron. Astrophys.* **346**, L65–L68.

Milani, A., La Spina, A., Sansaturio and M. E., Chesley, S. R. (2000a). The asteroid identification problem III. Proposing identifications, *Icarus* **144**, 39–53.

Milani, A., Chesley, S.R., and Valsecchi, G.B. (2000b). Asteroid close encounters with Earth: risk assessment. *Planet Space Sci.* **48**, 945–954.

Milani, A., Chesley, S.R., Boattini, A. and Valsecchi, G.B. (2000c). Virtual impactors: Search and destroy, *Icarus* **145**, 12–24.

Milani, A., Sansaturio, M. E. and Chesley, S. R. (2001a). The asteroid identification problem IV: Attributions, *Icarus* **151**, 150–159.

Milani, A., Rossi, A., Vockrouhlicky, D., Villani, D. and Bonanno, C. (2001b). Gravity field and rotation state of Mercury from the BepiColombo Radio Science Experiments, *Planet. Space Sci.* **49**, 1579–1596.

Milani, A, Vokrouhlický, D., Villani, D., Bonanno, C. and Rossi, A. (2002). Testing general relativity with the BepiColombo radio science experiment, *Phys. Rev. D* **66**, 082001.

Milani, A., Rossi, A. and Villani, D. (2003). *The BepiColombo radio science simulations*, Report to ESA, Version 2, 11 April 2003.

Milani, A., Gronchi, G. F., de' Michieli Vitturi, M. and Knežević, Z. (2004). Orbit determination with very short arcs. I admissible regions, *CMDA* **90**, 59–87.

Milani, A., Gronchi, G. F., Knežević, Z., Sansaturio, M. E. and Arratia, O. (2005a). Orbit determination with very short arcs. II identifications, *Icarus* **79**, 350–374.

Milani, A., Chesley, S. R., Sansaturio, M. E., Tommei, G. and Valsecchi, G. (2005b). Nonlinear impact monitoring: line of variation searches for impactors, *Icarus* **173**, 362–384.

Milani, A., Sansaturio, M. E., Tommei, G., Arratia, O. and Chesley, S. R. (2005c). Multiple solutions for asteroid orbits: computational procedure and applications, *Astron. Astrophys.* **431**, 729–746.

Milani, A., Rossi, A. and Villani D. (2005d). A timewise kinematic method for satellite gradiometry: GOCE simulations, *Earth Moon Planets* **97**, 37–68.

Milani, A., Gronchi, G. F., Knežević, Z., Sansaturio, M. E., Arratia, O., Denneau, L., Grav, T., Heasley, J., Jedicke, R. and Kubica, J. (2006). Unbiased orbit determination for the next generation asteroid/comet surveys. In *Asteroids Comets Meteors 2005*, eds. D. Lazzaro *et al.* (Cambridge University Press), pp. 367–380.

Milani, A., Gronchi, G. F. and Knežević, Z. (2007). New definition of discovery for Solar System objects, *Earth Moon Planets* **100**, 83–116.

Milani, A., Gronchi, G. F., Farnocchia, D., Knežević, Z., Jedicke, R., Dennau, L. and Pierfederici, F. (2008). Topocentric orbit determination: Algorithms for the next generation surveys, *Icarus* **195**, 474–492.

Mood, A.M., Graybill, F.A. and Boes, D.C. (1974). *Introduction to Statistics*, (McGraw-Hill, New York).

Mossotti, O.F. (1816–1818). Nuova analisi del problema di determinare le orbite dei corpi celesti, in *Mossotti, scritti* (Domus Galileiana, Pisa).

Moyer, T.D. (2003). *Formulation for Observed and Computed Values of Deep Space Network Data Types for Navigation* (Wiley-Interscience, Hoboken, NJ).

Mussio, L. (1984). Il metodo della collocazione minimi quadrati e le sue applicazioni per l'analisi statistica dci risultati delle compensazioni. *Ricerche di Geodesia, Topografia e Fotogrammetria* **4**, 305–338.

Nordtvedt, K. (1970). Post-Newtonian metric for a general class of scalar-tensor gravitational theories and observational consequences, *Astrophys. J.* **161**, 1059–1067.

Olsen, Ø. (2007). The constancy of the Pioneer anomalous acceleration, *Astron. Astrophys.* **463**, 393–397.

Pail, R., Plank, G. and Schuh, W. D. (2001). Spatially restricted data distributions on the sphere: the method of orthonormalized functions and applications, *J. Geod.* **75**, 44–56.

Peale, S.J. (1972). Determination of parameters related to the interior of Mercury, *Icarus* **17**, 168–173.

Peale, S.J. (1988). The rotation dynamics of Mercury and the state of its core, in *Mercury*, eds. Vilas, F., Chapman, C.R. and Matthews, M.S. (University of Arizona Press, Tucson), pp. 461–493.

Plummer, H. C. (1918). *An Introductory Treatise on Dynamical Astronomy* (Dover Publications, New York).

Poincaré, H. (1906). Sur la détermination des orbites par la méthode de Laplace, *Bulletin astronomique* **23**, 161–187.

Risler, J. J. (1991). *Méthodes mathematiques pour le CAO*, Collection Recherche en mathematiques appliquées, *RMA 18*, Masson.

Rossi, A. (2005) Population models of space debris. In *Dynamics of Populations of Planetary Systems*, eds. Z. Knežević, and A. Milani (Cambridge University Press), pp. 427–438.

Sansaturio, M. E., Milani, A. and Cattaneo, L. (1996). Nonlinear optimisation and the asteroid identification problem. In *Dynamics, Ephemerides and Astrometry of the Solar System*, eds. S. Ferraz Mello *et al.* (Kluwer, Dordrecht), pp. 193–198.

Simmons, J.F.L., McDonald, A.J.C. and Brown, J.C. (1985). The restricted 3-body problem with radiation pressure, *CMDA* **35**, 145–187.

Sitarski, G. (1968). Approaches of the parabolic comets to the outer planets, *Acta Astron.* **18/2**, 171–195.

Sneeuw (1991). Inclination functions: group theoretical background and a recursive algorithm. In *Reports of the Faculty of Geodetic Engineering*, **91.2** (Delft University of Technology, Delft).

Taff, L. G. and Hall, D. L. (1977). The use of angles and angular rates. I – Initial orbit determination, *CMDA* **16**, 481–488.

Tapley, B., Ries, J., Bettadpur, S., Chambers, D., Cheng, M., Condi, F., Gunter, B., Kang, Z., Nagel, P., Pastor, R., Pekker, T., Poole, S. and Wang, F. (2005). GGM02 – An improved Earth gravity field model from GRACE, *J. Geodesy* **79**, 467–478.

Tommei, G. (2005). Nonlinear impact monitoring: 2-dimensional sampling, in *Dynamics of Populations of Planetary Systems*, eds. Z. Knežević, and A. Milani, (Cambridge University Press), pp. 259–264.

Tommei, G. (2006a). Canonical elements for Öpik theory, *CMDA* **94**, 173–195.

Tommei, G. (2006b). Impact monitoring of near-Earth objects: theoretical and computational results, University of Pisa, D. Phil. thesis.

Tommei, G., Milani, A. and Rossi, A. (2007). Orbit determination of space debris: admissible regions, *CMDA* **97**, 289–304.

Valk, S. and Lemaitre, A. (2007). Semi-analytical investigations of high area-to-mass ratio geosynchronous space debris including Earth's shadowing effects, *Adv. Space Res.* **42**, 1429–1443.

Valk, S., Lemaitre, A. and Anselmo, L. (2007) Analytical and semi-analytical investigations of geosynchronous space debris with high area-to-mass ratios influenced by solar radiation pressure. *Adv. Space Res.* **41**, 1077–1090.

Valsecchi, G.B., Milani, A., Gronchi, G.F., and Chesley, S.R. (2003), Resonant returns to close approaches: analytical theory, *Astron. Astrophys.* **408**, 1179–1196.

Virtanen, J., Muinonen, K. and Bowell, E. (2001). Statistical ranging of asteroid orbits, *Icarus* **154**, 412–431.

Vokrouhlický, D and Milani, A. (2000). Direct radiation pressure on the orbits of small near-Earth asteroids: observable effects?, *Astron. Astrophys.* **362**, 746–755.

Vokrouhlický, D, Milani, A. and Chesley, S. R. (2000). Yarkovsky effect on small Near Earth asteroids: mathematical formulation and examples, *Icarus* **148**, 118–138.

Wagner, W. E. and Velez, C. E. (eds) (1972). *Goddard Trajectory Determination Subsystem Mathematical Specifications* (Goddard Space Flight Center, Greenbelt, MD).

Walker, I.W., Gordon Emslie, A. and Roy, A.E. (1980). Stability criteria in many-body systems I, *CMDA* **22**, 371–402.

Wallin, J.F., Dixon, D.S. and Page, G.L. (2007). Testing gravity in the outer solar system: results from transneptunian objects, *ApJ* **666**, 1296–1302.

Wetherill, G. W. (1967). Collisions in the asteroid belt, *J. Geophys. Res.* **72**, 2429–2444.

Whipple, A.L. (1995). Lyapunov times of the inner asteroids, *Icarus* **115**, 347–353.

Wigner, E.P. (1959). *Group Theory and its Applications to the Quantum Mechanics of Atomic Spectra* (Academic Press, New York).

Will, C.M. (1981). *Theory and Experiment in Gravitational Physics* (Cambridge University Press).

Winslow, A.M. (1964). An irregular triangle mesh generator, *Report UCXRL-7880*, National Technical Information Service, Springfield, VA.

Yeomans, D.K., Chodas, P.W., Keesey, M.S., Ostro, S.J., Chandler, J.F. and Shapiro, I.I. (1992). Asteroid and comets orbits using radar data, *Icarus* **103**, 303–317.

Index

Printed in the United States
by Baker & Taylor Publisher Services